GEOINFORMATICS
for Marine and
Coastal Management

GEOINFORMATICS
for Marine and
Coastal Management

Edited by
Darius Bartlett and Louis Celliers

CRC Press
Taylor & Francis Group
Boca Raton London New York

CRC Press is an imprint of the
Taylor & Francis Group, an **informa** business

CRC Press
Taylor & Francis Group
6000 Broken Sound Parkway NW, Suite 300
Boca Raton, FL 33487-2742

First issued in paperback 2019

© 2017 by Taylor & Francis Group, LLC
CRC Press is an imprint of Taylor & Francis Group, an Informa business

No claim to original U.S. Government works

ISBN-13: 978-1-4987-3154-6 (hbk)
ISBN-13: 978-0-367-87368-4 (pbk)

Library of Congress Cataloging-in-Publication Data

Names: Bartlett, Darius J., 1955- author. | Celliers, Louis, editor.
Title: Geoinformatics for marine and coastal management / Darius Bartlett and Louis Celliers.
Description: Boca Raton : CRC Press, 2017.
Identifiers: LCCN 2016026822| ISBN 9781498731546 (hardback : alk. paper) | ISBN 9781498731553 (ebook)
Subjects: LCSH: Oceanography—Remote sensing. | Oceanography—Geographic information systems. | Coastal zone management—Remote sensing. | Coastal zone management—Geographic information systems. | Coastal mapping. | Geoinformatics.
Classification: LCC GC10.4.R4 B37 2017 | DDC 331.91/640285--dc23
LC record available at https://lccn.loc.gov/2016026822

Visit the Taylor & Francis Web site at
http://www.taylorandfrancis.com

and the CRC Press Web site at
http://www.crcpress.com

Contents

Foreword

We humans are on an epic journey, to somewhere, where exactly no one knows for sure. The vehicle on which the trip is built is made from 1's and 0's. The transformations that will take place along the way will probably radically impact all facets of our world, including of course, our oceans.

Only a few years ago, I recall speaking to an information and communication technology (ICT) official about a marine mapping exercise which his organization had been conducting. He mentioned that they had a number of terabytes of data collected to date. *'That's a lot of data'* was the unsaid message. Terabyte storage media is now virtually two a penny. Coming down the track is maybe yottabytes (10^{24} bytes) of data storage, or more. Is this unlikely? Maybe, but probably not. Creating data storage, which is based on quantum phenomena holds the promise of data storage at density levels which are hundreds of times higher than the current highest density storage media. Furthermore, quantum computing provides a vista involving processing speeds that current supercomputer users can only dream about. As the saying goes … watch this space. These and other technical developments provide a basis on which currently almost unimaginable feats of information processing will become possible, and in due course routine in nature. In the not too distant future, maybe Big Data analytics could well be referred to as 'big Big Data analytics', until at some point in that future even the analytics undertaken in this 'big Big' Data environment become commonplace. What does all this mean for our marine environments, and where does this book fit in? I will get to that after addressing another issue.

Currently, there are over 7 billion people in the world, every one of whom, no matter where they live on Earth has multiple relationships with our oceans. The relationships can be obvious and direct such as those exhibited by professional fishermen, academics who research marine matters, or lifeguards on a local beach. The relationships, however, can be hidden. Pollution in tuna collected from the Gulf of Mexico has been found to limit the expression of a particular protein in cell membranes that identifies toxins and expels them. For example, a person sitting in the middle of continental North America, many hundreds of kilometers from the nearest ocean who eats some of this tuna from a can purchased in her local supermarket creates a specific ocean relationship between her cell membranes and the ocean. The number of individual/ocean relationships for each person if explored thoroughly would probably run into the many hundreds, if not many thousands or hundreds of thousands. Does this sound like an overstatement? Not if one thinks, for example, that very many items, or parts of items, consumed or used in everyday life have been transported over oceans on ships, each one of which leaves its footprint on the oceans through which they move.

Another example. I access the Internet, a routine occurrence, and in doing so access electromagnetic impulses which are travelling from wherever through cables lying on the seabed; each meter of cable has specific environmental impacts on its immediate surroundings. These impacts are in addition to the ocean impacts associated with making, laying and maintaining that meter of cable.

A picture that emerges is that of a vast, dense, entangled network of interrelationships between not just humans and their activities, but all living things and solid objects as well, and the world's oceans. My thoughts wander to sometime in the future when trillions of interlinked sensors, that 'thing' we used to call the Internet of Things back in the 2010s, contribute towards vast numbers of 1's and 0's (or qubits in a quantum environment), which describe the incredibly dense and real thicket of relationships between living things and objects with the oceans.

Humans will no longer have the faintest chance of seriously making sense of 'the thicket' and so onto center stage in the command performance role will be learning machines, machines using learning algorithms to help us humans begin to understand ocean-related interactions at both high, general and also subtle, nuance levels. Not only will machine learning systems help to untangle 'the thicket' but they will also provide a fast and sophisticated means of controlling many of the relationships. Analysis and control by machines. ... loud cries go out *'Big Brother; alienation, machine dictatorship; what about our privacy; STOP'*. But the march goes on, inexorably.

What has all this speculation and dreaming have to do with this book? Quite a lot actually.

Firstly, the focus is on spatial information related to our marine environments, a type of information which is of fundamental importance for untangling 'the thicket'. The mantra *'everything occurs somewhere'* that is so often repeated by geospatial professionals is correct, whether the 'everything' is an object or a process. Without location information, which is linked to the virtual blizzard of non-location data that is being generated in real time, people and learning machines would not have the faintest chance of gaining any really deep integrated understanding of 'the thicket' generally, or more specifically, related to our oceans. This book 'swims' in these vitally important spatial information waters.

Secondly, the book represents an important stepping stone in the journey into a better future outlined briefly above. Other than the many useful insights that humans will derive from the book, I have little doubt that at some point in the future, text analytics software will read and absorb every word in every chapter (it has my prior permission to skip this Foreword!) and incorporate the many issues, messages, facts, and observations to help not only build 'the thicket', but also into evolving ever more sophisticated analytical frameworks so as to probe and determine the interrelationships buried in 'the thicket'.

Thirdly, the chapters that follow highlight the fact that multiple stakeholders have marine and coastal interests. For some it is professional (scientists, managers and administrators, etc.); for some it is rooted in economic gain (the aforementioned fishermen, but also oil company executives, mariners, tourism executives, etc.); and for others it is less formal but no less important, such as those who live or holiday on the coast, or for whom the sea is an integral part of their culture or heritage.

In parallel with this stakeholder diversity, the chapters also illustrate the growing diversity of geoinformatics as a discipline, with for example, many types of technology now being used to collect and process data: GIS, remote sensing, sensors, above-water and in-water drones, laser scanners, humans with their mobile phones or other devices, and others. The book illustrates clearly how 'the thicket' becomes thicker all the time.

Fourthly, increasing diversity means increasing complexity, and along with this rise in complexity there is an equally important need for an integrated interdisciplinary approach to the management of our oceans and coasts, and the data and understanding on which evidence-based approaches to management should firmly rest. This book highlights this need very well.

The words *'integrated interdisciplinary approach'* are so easy to write, and sound so good and intuitively correct, but the reality of achieving this ideal is exactly the opposite, hard, difficult, often messy, full of complicated humans, and definitely time consuming. A key concept that can act as a beacon to guide us along the way is 'interoperability', interoperability of technologies (physical and software interoperability), interoperability of data (semantic and logical interoperability) or interoperability of the people involved (human interoperability and teamwork approaches).

Within the geospatial data arena, spatial data infrastructures (SDIs) provide a framework within which interoperability can take on a formal and organized form. Land-focused SDIs are increasingly being complemented by marine SDIs, a most welcome development. SDIs can provide a really useful basis for undertaking marine and coastal management initiatives. In addition to the places in this book where SDIs are dealt with directly, much of the other material could provide useful contextual material for developing and using SDIs within a management context.

Fifthly, on 1 January 2016, a really important global milestone was reached, namely, the coming into effect through the United Nations of what is commonly referred to as the Sustainable Development Goals 2030. Goal 14 reads as follows: *'Conserve and sustainably use the oceans, seas and marine resources'.* This goal has 10 associated targets that deal with a wide variety of matters, including pollution, ocean acidification, overexploitation of fish and other biological resources, declaration of conservation areas, aquaculture, tourism, the need to deal specifically with island nations and developing countries, as well as other issues. Target 14a states '*… increase scientific knowledge, develop research capacity and transfer marine technology …'.* This book is placed squarely in this Target 14a space. A key challenge for each reader is to work out how its insights and information can be leveraged to see how best to contribute towards achieving this globally accepted target.

Other than the direct reference to oceans in Goal 14, I would argue that many of the issues associated with our marine environment are linked, often quite directly but certainly indirectly, through 'the thicket' to all of the other goals, which stated succinctly, include Goal 1 *'No poverty'*, Goal 2 *'Zero hunger'*, Goal 3 *'Good health and wellbeing'*, Goal 4 *'Quality education'*, Goal 5 *'Gender equality'*, Goal 6 *'Clean water and sanitation'*, Goal 7 *'Affordable and clean energy'*, Goal 8 *'Decent work and economic growth'*, Goal 9 *'Industry, innovation and infrastructure'*, Goal 10 *'Reduced inequalities'*, Goal 11 *'Sustainable cities and communities'*, Goal 12 *'Responsible consumption and production'*, Goal 13 *'Climate action'*, Goal 15 *'Life on land'*, Goal 16 *'Peace, justice and strong institutions'*. Taking account of the issues of diversity, complexity and interoperability mentioned above, Goal 17 is of particular relevance; it reads *'Revitalise the partnership for sustainable development'*. Geospatial information, including of course marine and coastal information, is a critical part of the information infrastructure which is needed to begin to meet Goal 14, and also all of the other goals. This book can make a contribution in this regard.

Finally, on a personal level, far removed from the big picture 'thicket' issues, I have been deeply enriched by my interactions with the sea, particularly during my formative years. For at least four of my teenage years, I and a number of others from my neighborhood would run every day, school days and holidays, the 4.5 kilometers from home to South Africa's Durban Country Club beach. We would spend virtually no time on the beach but all the time out in the water, mainly body surfing, but also doing some board surfing. Rain, shine, north easterly or south westerly wind, summer or winter made no difference. I learned so many life-enhancing lessons about the sea, but also about myself and what it is to be human. Insights into currents, the incredible 'sardine run', the impact of wind on water, bluebottle (Portuguese Man o' War) invasions, temperature differences through the upper water column, and the weight and power of water are just some of the ocean insights gained during these years. On a human level, the meaning of a wide range of emotions became clear, panic (thinking that one is swimming right next to a large Zambezi shark, coming to the end of one's breath while still being churned around underwater and not knowing which way is up), fear (being on a ski boat being overturned by a large wave), delight and pride (careening across the face of a large wave with only one's body doing the surfing, putting all 10 toes over the leading edge of a surf board) are just a few. Understanding how time can apparently stand still, what it means to be very fit, how dreadful heat stroke can feel, who are one's real friends, are all valuable life lessons gained during this time. I was in love with the sea.

My final message, if at all possible, try to engage with the waters of our oceans in intimate, 'loving', exploratory, life-enhancing ways, and not just through the medium of 1's and 0's ... but along the way, don't forget to engage with this book, even if it is coming to you by means of 1's and 0's! It will be worth it.

Bruce McCormack
Bruce McCormack Consulting
Kilcullen, Ireland

Preface

This is a book about the oceans, seas and coasts of the world. The oceans are our final frontier on Earth. To our ancestors they were wild and unknown, unpredictable and dangerous, a place where monsters lurked in the depths. Yet, for coastal communities, the nature of the coastal fringe and the oceans, both rich in provision and powerful in destruction, has been part of the ancestral backdrop against which lives have been led, cultures have developed, and the fabric of society has been maintained. Historically, the high seas were part of the global commons, a shared heritage. Yet, like terrestrial commons before them, they have become progressively subject to containment, apportionment, claims and counter-claims, enclosures by virtual and legal boundaries, imposition of national and international regulations and, especially in recent times, exploitation by corporate interests and enterprises whose structures may frequently transcend national boundaries and jurisdictions.

Never before in human history has there been such a pressing demand for coastal and marine resources to fuel national and corporate growth and development. Furthermore, the alarming demand for resources from the sea is situated within the context of a rapidly changing global climate, increasingly complex institutional and governance arrangements and a growing awareness of the need for protection and sustainable use of resources. The task of rationally and reasonably managing multiple resource demands, and of reducing conflict and overlap, has become too complex and multidimensional for traditional approaches. A new 'box of tools' that can contribute to integrated solutions for social, economic, political and environmental objectives is needed.

This is also a book about computers and information technology or, to be more specific, about how an expanding range of spatially enabled information technologies including geographical information systems (GIS), remote sensing and sensors are contributing data and information needed to support humanity's complex and often conflicting relationships with the coasts and oceans. Despite millennia of engagement with the sea, humanity's knowledge of how the oceans work, the changes they might experience, or what happens below the waves, remains extremely limited. This, however, is gradually changing, and the traditional problems posed by a lack of ocean data are now being superseded by the opposite but equally challenging problem of processing, storing and analysing the growing quantities of complex, diverse and highly dynamic data, many of them inherently spatial, which are now becoming available.

Thirdly, this is a book about management, of the oceans and the coastal fringe, and also of data and information about them. We consider here 'management' not necessarily in the sense of top-down 'control', though that is also sometimes required, but also in the more bottom-up and community-focused sense of allowing all stakeholders and interest groups to have a say in the policies and decisions that might affect them. Increasingly, at sea as well as on land, the over-riding objective of management is to find frameworks that allow sustainable and integrated use of the oceans and the resources they provide so as to maximize opportunities while keeping risk and hazard to a minimum.

The book is loosely organized into three parts although, like most of the boundaries that exist within the ocean itself, the divisions between these are fluid and negotiable. We have deliberately not marked the intended transitions from one section to the next, preferring instead to let the readers decide for themselves how the chapters contribute to each part. However, from an editorial perspective, we begin with a focus on marine and coastal geospatial data and technologies, and how our understanding of the oceans may be acquired, organized and structured. Attention then turns in the second cluster of chapters to matters of marine and coastal governance, whether formal or informal, and the role of geoinformation technologies in supporting management and decision making. Then, finally, in the third section, we consider coastal and marine resources and the 'ocean/blue economy'.

We hope this book will have long-term value for practitioners and professionals, for researchers, as well as for students and others whose work and interests involve an engagement with the sea. It could not, of course, have come to completion without the assistance and support of many individuals and institutions. In particular, both editors would like to thank Irma Shaglia-Britton and her team at CRC Press/Taylor & Francis, for encouraging us to take on the project in the first place, and for continued guidance and support at all stages thereafter. It also goes without saying that we recognize and appreciate with gratitude the work of all the authors who made inputs to the chapters that follow. Without them, there would be no book.

At an individual level, Darius Bartlett thanks the College of Arts, Celtic Studies and Social Sciences at University College Cork, for granting him sabbatical research leave, during which the bulk of the editorial work was undertaken; and his colleagues within the Department of Geography, especially Fiona Cawkwell and Helen Bradley, who cheerfully adjusted their own teaching schedules and took on some of his duties in order to facilitate his absence. He also thanks colleagues at the Council for Scientific and Industrial Research (CSIR) in Durban and Stellenbosch for their welcome and hospitality, while a generous grant from the Marine Institute in Galway greatly assisted his travel to and stay in South Africa. Last, but definitely not least, many thanks to Mary-Anne for support and encouragement, not only during the months that this book was under development, but over the many years that preceded it.

Louis Celliers acknowledges the institutional support provided by the CSIR, as well as the Local Organizing Committee and the Scientific Committee of the 12th CoastGIS Symposium, held in Cape Town in April 2015. This instalment of the CoastGIS Series, a first for Africa, resulted in an opportunity for the content of the book to take shape, and my modest contribution to making it a reality.

Darius Bartlett
Louis Celliers

Editors

Darius Bartlett, an earth scientist and geomorphologist by training, has been teaching and researching at the interface between GIS and coastal zone science and management in the Department of Geography at the University College Cork in Ireland for over a quarter-century, and is a past director of the department's MSc in GIS and remote sensing and the MSc in applied coastal and marine management taught programmes. As a corresponding member of the Commission on Coastal Systems of the International Geographical Union, he helped establish the CoastGIS Biennial Symposium Series, in partnership with Ron Furness (then at the Australian Hydrographic Office and chair of the International Cartographic Association's Commission on Coastal Mapping), and hosted the first meeting as a collaboration between the two organisations in Cork in 1995 (the 12th meeting in the series was held in Cape Town in April 2015). He is the co-editor of *Marine and Coastal Geographical Information Systems* (with Dawn Wright, CRC Press/Taylor & Francis, 2000) and *GIS for Coastal Zone Management* (with Jennifer Smith, CRC Press/Taylor & Francis, 2005).

Louis Celliers is employed by the CSIR in a dual role as a principal researcher in, and heads the Coastal Systems Research Group of the Natural Resources and the Environment Business Unit. He completed his graduate (BSc Agric) and postgraduate studies (MSc Agric) at the University of Pretoria in 1998, followed by a PhD in marine ecology at the University of Natal, South Africa in 2001. The responsibilities of his portfolio include the integration of coastal and marine science with management of the coastal area by the three spheres of government in South Africa. His research interests and professional development strongly favour an approach that guides, integrates and translates good science to the benefit of society through the maintenance of natural systems. His main research interests, and focus of his published work, are in the area of integrated coastal management, coastal governance and institutions and science to policy processes for the improvement of coastal and marine management.

Contributors

Shankar Aswani is a professor at the Department of Anthropology and at the Department of Ichthyology and Fisheries Science (DIFS) at Rhodes University, Grahamstown, South Africa. His research focuses on a diversity of subjects, including property rights and common property resources, marine indigenous ecological knowledge/ethnobiology, vulnerability and resilience of coastal communities, human behavioral ecology of fishing, economic anthropology tourism, ethnohistory and applied anthropology. He is also involved in designing and establishing marine conservation programmes in Oceania. For this effort, he was awarded a prestigious Pew Fellowship in Marine Conservation in 2005, the first anthropologist to receive this award in the programme's history.

Tom Barry is the executive secretary for the Conservation of Arctic Flora and Fauna (CAFF), which is the biodiversity working group of the Arctic Council. The CAFF Secretariat is based in Akureyri, Iceland. Barry has a broad range of experience at the national and international levels dealing with strategic planning and organizational development, a primary focus of which has been Arctic issues, where he works with a diverse range of stakeholders throughout the Arctic. Barry was heavily involved in the Arctic Biodiversity Assessment, which creates a baseline for use in global and regional assessments of biodiversity and provides a basis to inform and guide future Arctic Council work. He is also closely involved in the implementation of the Circumpolar Biodiversity Monitoring Programme (CBMP), which is working to facilitate more rapid detection, communication and responses to the significant biodiversity-related trends and pressures affecting the circumpolar world.

Jarbas Bonetti is an associate professor at the Federal University of Santa Catarina, Brazil, where he is the head of the Coastal Oceanography Laboratory at the Department of Geosciences. He graduated in geography, completed an MSc in geological oceanography and obtained his PhD in physical geography, all from the University of São Paulo. He worked at the Oceanographic Institute of that university, was a visiting researcher at IFREMER (France) and a visiting fellow at the University of Wollongong (Australia). His main research interests are related to the use of spatial analysis techniques to understand the dynamics of coastal systems, with a focus on coastal vulnerability analysis and integrated seabed assessment of shallow marine environments.

Albert Caton has worked for more than 40 years as a fisheries biologist at the interface between fisheries science and fisheries management, at both the national and international levels. Since retiring, he has been a joint scientific editor of the four most-recent published proceedings of the International Fisheries Geographic Information Systems Symposia.

Tom Christensen is an advisor and conservation biologist, with 20 years of experience within research, consultancy and nature management. His key expertise is in the Arctic ecosystem and nature conservation, as well as on international cooperation on these topics. He has an in-depth knowledge about the ecological importance and sensitive areas in Greenland and has more than 15 years of experience with related international conventions and agreements, including the national implementation and follow-ups. He has, among others, been the project leader on projects about environmental consequences related to potential increased shipping in the Arctic and Greenland and possible ecosystem-based approaches to management in relation to this. He is currently the co-chair of the Arctic Council's (CAFF) Monitoring Programme Circumpolar Biodiversity Monitoring Programme (CBMP), and has worked on several other Arctic Council initiatives, projects and expert groups, related to the protection of the Arctic marine environment.

Mimi D'Iorio is the GIS Manager for NOAA's Marine Protected Areas Center. D'Iorio specializes in coastal GIS and remote sensing and holds undergraduate degrees in geography and environmental studies and a doctorate degree in earth sciences. She manages the MPA Inventory, a comprehensive spatial database of U.S. MPAs and coordinates participatory ocean use mapping efforts designed to collect spatial data on human uses of the ocean to inform coastal and marine resource management and decision making. D'Iorio has over 15 years of experience working with geospatial data, tools and related applications.

Thomas Furey is the Marine Institute Joint Programme manager for INFOMAR, Ireland's Seabed Mapping Programme. An NUI Galway graduate in earth science, and MSc graduate of Bangor University, he worked offshore in the oil and gas industry in 3D seismics, before returning to Ireland to take up a consultancy role in the Irish National Seabed Survey. He now manages the Advanced Mapping Services team at the Marine Institute and has a key role in supporting the technology, innovation and product development associated with INFOMAR and its vast data resources, as well as increasing the downstream use and application of the data.

Grace Goldberg manages implementation projects for the McClintock Lab, developers of SeaSketch (seasketch.org); she works with marine planning practitioners from around the world to design science-based ocean policy through a participatory process. These projects range from large-scale multisector marine spatial planning, to community marine reserve initiatives. She is trained as a scientist, interested in research questions that include human users in marine ecosystems, with relevance to spatial management and real conservation goals. Goldberg received an MS in marine systems and conservation from Stanford University's Earth Systems Programme.

Kate Gormley is a marine biologist with a BSc in applied marine biology and a PhD in marine spatial planning and conservation both from Heriot-Watt University (United Kingdom). Gormley has been working with the offshore energy industry since graduating with her BSc in 2004, with a focus on marine ecology, monitoring and impact assessment. Her PhD focused on mapping of priority marine habitats, impacts of climate change to their distribution and consequences for management. She has lectured at both undergraduate and graduate levels in GIS and is currently involved in a number of joint industry academic projects mapping industry environmental data. Her main research interest areas include offshore decommissioning, environmental legislation, marine spatial planning and knowledge exchange between industries and academia.

David R. Green is the director of the Aberdeen Institute for Coastal Science and Management, director of the MSc Degree Programme in geographical information systems, and the BSc Degree Programme in marine and coastal resource management at the Department of Geography and Environment, University of Aberdeen, Scotland, United Kingdom. He is past chairman of the Association for Geographic Information; past editor in chief of *The Cartographic Journal*; past president of the EUCC–The Coastal Union; past vice chair of the European Centre for Nature Conservation Scientific Committee; and currently a director and vice chair of EGCP Ltd. Green is also editor of the *Journal of Coastal Conservation, Planning and Management* (Springer). He is a specialist in the environmental applications of geospatial technologies with interests in geographical information systems, remote sensing (terrestrial and bathymetric), cartography/digital mapping, Internet and mobile GIS, coastal and marine resource management, hydrography, marine spatial planning (MSP), precision viticulture (PV) and unmanned aerial vehicle (UAV) technology.

Soffia Guðmundsdóttir is the executive secretary of the Protection of the Arctic Marine Environment (PAME), which is the ocean policy and non-emergency pollution prevention and control measures working group of the Arctic Council. She holds degrees in chemistry, environmental engineering and business and has served as a consultant to the Regional Seas Program of the United Nations: Environmental Program (UNEP). She also teaches a course on marine activities in the polar regions at the masters level in the Polar Law program at the University of Akureyri. Guðmundsdóttir has extensive experience and expertise in ocean-related issues, in particular on policy and strategic levels and was heavily involved in the development of the Arctic Marine Strategic Plan 2015–2025, which provides the framework to guide the actions of the Arctic Council and its subsidiary to protect Arctic marine and coastal ecosystems and to promote sustainable development. She has worked on a number of implementation plans that provide the framework to systematically follow-up strategic actions and recommendations in an effort to facilitate projects and activities through the applications of cooperative, coordinated and integrated approaches.

David Hardy holds both a BA and MSc in geology from Trinity College Dublin, and has been employed as a geologist with the Geological Survey of Ireland (GSI) in the Marine Geology and Geophysics Section since 2002. During this time, he has worked primarily with the Seabed Survey (INSS/INFOMAR) and has, consequentially, utilized a wide array of instrumentation and software pertaining to marine spatial data. Hardy has particular expertise in geodesy and surveying and has extensive knowledge of the methods and issues associated with the acquisition, processing and interpretation of marine geophysical and hydrographic datasets.

Kiyoshi Itoh is the president of the Environmental Simulation Laboratory, Inc. Kiyoshi and Nishida (see his biography) established the International Fishery GIS Society in 1997, and developed the menu-driven Marine GIS (Marine Explorer). His works focus on monitoring fisheries and marine environment using GPS data loggers and Marine GIS (Marine Explorer). The latest subject is the development of a real-time fisheries and marine monitoring system using acoustics under the water and the wireless communication network inland, incorporating Marine GIS (Marine Explorer).

Kári Fannar Lárusson is a programme manager at the Conservation of Arctic Flora and Fauna (CAFF) which is the biodiversity working group of the Arctic Council. He has a broad range of experience in Arctic issues and works with a diverse range of stakeholders across the Arctic. Some of the key activities that he is currently involved with include the Arctic Biodiversity Assessment (ABA) and implementation of the Circumpolar Biodiversity Monitoring Programme (CBMP), which facilitates more rapid detection, communication and responses to biodiversity-related trends affecting the circumpolar world. A key interest of Lárusson's is ecosystem-based management with a focus on the importance of coordinating and streamlining data flow to help shorten the time between when a change is observed and an appropriate response formulated.

Yassine Lassoued is currently a research software engineer at IBM Research Dublin. His primary focus is on information retrieval, semantics and natural language processing (NLP) problems, in the areas of healthcare, cognitive systems and the Internet of Things. Prior to joining IBM, Lassoued worked for 9 years for the Coastal and Marine Research Centre of University College

Cork, where his main research interests were in the areas of geographic information systems (GIS), coastal and marine data, geospatial standards, semantic interoperability, semantic web and ontologies. As part of this role, Lassoued led the design and development of several semantic interoperability systems, including the International Coastal Atlas Network Semantic Mediator, and the Marine Debris Data Portal Semantic Web Service.

Adam Leadbetter leads the data management team at the Marine Institute in Galway, Ireland. His research interests include making marine datasets from various organizations compatible through Semantic Web technology and in developing a 'Born Connected' approach in which data arrive from sensors in linked data formats. He is a co-chair of the Technical Committee of the International Coastal Atlas Network project of UNESCO's International Oceanographic Data and Information Exchange and is actively involved in the Vocabulary Services Interest Group of the Research Data Alliance.

Melanie Lück-Vogel is an interdisciplinary scientist with more than 15 years of experience in remote sensing analysis of land cover change and dynamics. She holds an MSc in biology (ecology and geobotany) from the University of Bonn, Germany, and a PhD in geography (remote sensing) from the University of Würzburg (2006), Germany, in collaboration with the German Space Agency, DLR. Since 2007, she has worked as a senior researcher at the CSIR in South Africa on projects targeting remote sensing–based assessments of ecosystem conditions and dynamics, since 2011 with a strong focus on coastal environments. She is also an extraordinary senior lecturer at the Department of Geography and Environmental Studies at Stellenbosch University, South Africa, where she supervises many postgraduate students. She also lectures on specialized remote sensing topics at several South African and international universities and institutions, for example, the International Oceans Institute, the United Nations University, the University of Cape Town and Stellenbosch University.

Brian Mathews is a Lieutenant Commander (Lt Cdr) and Operations Branch Officer in the Irish Naval Service where he has served for 19 years. He has held a number of appointments ashore and at sea, his latest at sea being the Commanding Officer of an Offshore Patrol Vessel L.E. *Aisling*. His academic qualifications include an honours degree in computer science with a postgraduate higher diploma in GIS. He is presently the Officer in Charge of the Naval Computer Centre at the Naval Base where he is responsible for all Naval ICT systems. He has delivered enhanced command and control information systems and maritime situational awareness capabilities to the Naval Service and is tasked with implementing the wider Defence Forces Joint Common Operating Picture. Lt Cdr Mathews represents Ireland in the European Maritime Surveillance community, MARSUR Networking, and is presently the MARSUR Management chairman. He lives near Crosshaven in

Co. Cork with his wife Deborah and two children, Ella and Mia, and in his spare time he enjoys sailing and running.

Will McClintock is a scientist at the University of California, Santa Barbara, Marine Science Institute. He was the director of the MarineMap Consortium and GIS lead for California's Marine Life Protection Act Initiative (2004–2010), and subsequently received a gift from Esri to assist with the development of SeaSketch (www.seaksetch.org). The McClintock Lab has supported MPA planning in British Columbia (with the Marine Planning Partnership of the North Pacific Coast), Barbuda, Curaçao and Montserrat (with the Blue Halo Initiative), Cook Islands (with the Cook Islands Government), the United Kingdom (with the Joint Nature Conservation Committee) and other geographies.

Bruce McCormack is a qualified town and regional planner who has worked in the public and private sectors in both South Africa and Ireland. He is currently based in Ireland, where he has been directly involved in spatial data matters. For 2 years he was a senior advisor for Spatial Data Infrastructures to the World Bank. For the 4-year period until May 2015, McCormack was the president of EUROGI, the European Umbrella Organisation for Geographic Information. On a number of occasions, he has been the president of the IRLOGI, the Irish Organisation for Geographic Information. Upon retiring in 2014, from the Irish Department of Environment, Community and Local Government, where he worked as a planner, McCormack established a consultancy, Bruce McCormack Consulting, which specializes in geospatial and planning matters. In his consultancy role, he has advised Irish geospatial-focussed businesses on new emerging business opportunities related to current and emerging technologies, including, for example, the Internet of Things and Big Data Analytics. Until recently, he has been the moderator for the Africa Community of Practice which is part of the UN Environment Programme initiative which has produced an Africa Environment Assessment for inclusion into the United Nations GEO-6 Global Environment Outlook.

Charise McKeon is a marine geologist with the Geological Survey of Ireland (GSI). She graduated from the University College Dublin (UCD) in 2002 with a BSc in environmental geochemistry and has since completed an MSc in geographical information systems from the University of Ulster (UU), Coleraine. McKeon works in the Marine and Geophysics Programme within GSI primarily on the Seabed Survey project (INSS/INFOMAR) and has substantial experience in geographic information systems, remote sensing and other geoinformatic technologies and software including visualization software. In addition to hydrographic data processing, she has also worked extensively on the creation of physical habitat maps of the seafloor and on the Shipwreck Inventory of the Irish National Seabed Survey and INFOMAR.

Lauren McWhinnie received a first-class degree in marine and freshwater biology from Edinburgh Napier University before going on to earn a PhD in marine spatial planning, management and aquaculture at Heriot-Watt University. McWhinnie's PhD work explored the application and development of spatial tools to aid marine spatial planning, adaptive management, decision making and aquaculture site selection methods. She has lectured at both undergraduate and graduate levels in a number of subjects including GIS and environmental management. Now at the University of Victoria, her current research focuses on quantifying and modelling current and future ship-source marine noise exposure to marine mammals and how we can use this knowledge to better inform policy makers, planners and stakeholders and ultimately aid in decision making.

Xavier Monteys is a researcher and marine geologist at the Geological Survey of Ireland where he joined the Seabed Mapping Programme in 2001. He graduated from the University Autònoma of Barcelona (UAB) in 1999, followed by postgraduate studies in GIS in Barcelona, 2000 (UPC), and an MSc in GIS and remote sensing (2001) at Maynooth University, Ireland. Since 2005, he has been working in the INFOMAR programme where he has been responsible for coordination of marine research within the program. He has 15 years experience of offshore research, participating in more than 20 deep-water expeditions. His main research interests, and focus of his published work, are in the area of marine acoustics, seabed characterization, gas in marine sediments and cold-water coral.

Anders Mosbech is head of research and advisory tasks at the Arctic Environment Group, Bioscience, Aarhus University. He has worked with biodiversity and environmental issues in Greenland for more than 25 years. His research interests include marine ecology, population development, oil spill sensitivity analysis, delineation and identification of key marine habitats including use of survey data, tracking data (satellite transmitters and data loggers) combined with oceanography and remote sense spatial data. He is the leader of the joint Baffin Bay Environmental Study Program, Aarhus University and Greenland Institute of Natural Resources. Currently, he is the Co-PI for the Arctic Council project Adaptation Actions for a Changing Arctic (Baffin Bay-Davis Strait Region) and Co-PI for the THE NOW PROJECT: Living Resources and Human Societies around the North Water in the Thule Area, NW Greenland, which is an interdisciplinary research project integrated by archaeologists, biologists and anthropologists (now.ku.dk).

Paul K. Murphy is a graduate from University College Cork (UCC), Ireland. Having grown up on the coast, sailing and exploring with his parents, he developed an innate fascination with the marine environment and the way in which it functions. This fascination was a key driver in his studies at university, where he completed both a BSc (Hons) in earth science and an MSc in applied coastal and marine management. His research interests arising from his studies include sedimentology, GIS, climate change and renewable energy. His contribution to this book is based in part on his master's degree, which he worked on with the guidance of his supervisor, Professor Andrew Wheeler. Since graduating in early 2016, he is pursuing a dream career in the renewable energy industry, undertaking a research master's at Gaelectric Developments Ltd., in collaboration with UCC and funded by the Geological Survey of Ireland, which is concerned with the identification of potential sites for renewable energy installations in the Irish Sea.

Tom Nishida is the president of the International Fishery GIS Society (http:// www.esl.co.jp/Sympo) and also an associate scientist at the National Research Institute of Far Seas Fisheries (Japan). His works focus on fisheries GIS and stock assessments (tuna and demersal fish). Nishida and Itoh (refer to his biography) established the International Fishery GIS Society in 1997 (Japan). The major objective of this society is to promote GIS and spatial analyses in fishery and aquatic sciences. Its primary activity is to organize the tri-annual 'International Symposium on GIS/Spatial Analyses in Fishery and Aquatic Sciences' and to publish its proceedings.

Noel E. O'Connor is currently a professor in the School of Electronic Engineering at Dublin City University (DCU) and a principal investigator in INSIGHT Centre for Data Analytics. O'Connor is also the academic director of DCU's Research and Enterprise Hub on Information Technology and the Digital Society. Since 1999, he has published over 400 peer-reviewed publications, made 11 standards submissions and filed 6 patents. He has graduated 23 PhD students and 5 master's students. He is an area editor for *Signal Processing: Image Communication* (Elsevier) and an associate editor for the *Journal of Image and Video Processing* (Springer). He has edited six journal special issues, including *Signal Processing: Image Communication, Multimedia Tools and Applications*, the *Journal of Web Semantics* and the *Journal of Embedded Systems*. He was awarded the DCU President's Research Award for Science and Engineering in 2010. Also in 2010, he was awarded Enterprise Ireland's National Commercialization Award for ICT. He is a member of the IEEE, Engineers Ireland and the IET.

Cathal Power is a Lieutenant Commander (Lt Cdr) and Operations Branch Officer in the Irish Naval Service where he has served for 23 years. He has held a number of appointments ashore and at sea, his latest at sea being the Commanding Officer of an offshore patrol vessel, L.E. *Ciara*. His academic qualifications include a master's of arts degree in leadership, management and defence studies, a postgraduate diploma in supply chain management (including a Lean Black Belt) and a postgraduate higher diploma in GIS. He is presently the Officer in Charge of the Naval Operations Centre at the Naval Base where he is responsible for the day-to-day tasking of the naval fleet of eight offshore patrol vessels, including ships deployed on overseas missions. His utilization of the Recognised Maritime Picture and its advanced capabilities has improved the situational awareness of naval operations, lending to a more efficient utilization of ships in the provision of maritime, defence and security operations at sea. Lt Cdr Power is the national representative on the MARSUR Operator Working Group (MOWG). He is married to Nichola and they are parents to two girls, Caoimhe and Cliona. In his spare time he enjoys triathlons, adventure races and (watching!) rugby.

Courtney Price is the communications manager at the Conservation of Arctic Flora and Fauna (CAFF), the biodiversity working group of the Arctic Council, a high-level intergovernmental forum to promote cooperation, coordination and interaction among the Arctic states, with the involvement of Arctic indigenous peoples. Price works with key partners to address international scientific and political audiences in formal and informal settings and was instrumental in organizing and hosting the highly successful Arctic Biodiversity Congress. She coordinates the Arctic Migratory Birds Initiative and is overseeing efforts to unite Arctic and non-Arctic countries in the conservation of declining populations of Arctic-breeding migratory birds. She joined CAFF to specialize in Arctic issues after 3 years as a science and technology liaison officer at Environment Canada, and 2 years in communications for the Canadian Wildlife Service, where she worked to promote the findings of the wildlife research community to departmental policy audiences. Previous to that she worked at the Centre for Canadian–Australian Studies at the University of Wollongong, Australia, as part of an international internship program of the Canadian government. Price has a degree in journalism, double major in mass communications, from Carleton University, Ottawa, Canada and is completing an MSc in global challenges through the University of Edinburgh, Scotland.

Fiona Regan is a professor in chemical science at Dublin City University and director of the DCU Water Institute. Regan studied environmental science and technology and later completed a PhD in analytical chemistry in 1994. Following postdoctoral research in optical sensing in DCU, in 1996, she took up a lecturing position at Limerick Institute of Technology. In 2002, she joined the School of Chemical Sciences as a lecturer in analytical chemistry; in 2008, she became senior lecturer and in 2009 became the Beaufort principal investigator in Marine and Environmental Sensing. Since 2010, Regan has led the establishment and development of SmartBay Ireland and coordinated an access fund to facilitate researchers in accessing the national test and demonstration infrastructure. She has published over 100 papers, graduated 14 PhD and 8 MSc students and currently manages a group of 8 people. Regan's research focuses on environmental monitoring and she has a special interest in priority and emerging contaminants as well as the establishment of decision support tools for environmental monitoring using novel technologies and data management tools. Her work includes the areas of separations and sensors (including microfluidics), materials for sensing and antifouling applications on aquatic (marine and freshwater) deployed systems.

Phillip Saunders is an associate professor, Schulich School of Law, Dalhousie University, and director of the Marine and Environmental Law Institute and Research Fellow at the Centre for Foreign Policy Studies. His teaching and research interests lie mainly in international marine and environmental law, maritime boundary delimitation, tort law and international fisheries law. Saunders was formerly with the International Centre for Ocean Development, as senior policy advisor and as a field representative, South Pacific, delivering ocean-related development programmes in the South Pacific, Caribbean and Indian Ocean. He acted as counsel for Nova Scotia in the arbitration of the boundary between the offshore areas of Nova Scotia and Newfoundland, and has advised the Canadian Department of Global Affairs on maritime boundary and related matters. He is an author and co-general editor of *International Law, Chiefly as Interpreted and Applied in Canada* (Toronto, Emond Montgomery, 8th ed., 2014).

Gill Scott is a marine geologist who has worked on a wide range of research projects for the INFOMAR Seabed Survey Programme within the Geological Survey of Ireland (GSI) since 2011. With a BSc in geological oceanography, a PhD in marine geology, both from Bangor University (United Kingdom) and an MSc in geographical information systems and remote sensing from Maynooth University (Ireland), she has also lectured at both the undergraduate and graduate levels in GIS, image analysis and oceanography. Her main research interests are marine geomorphology and the development of methods to improve seabed mapping from backscatter data.

Adam Weintrit is a professor, dean of the Faculty of Navigation (2008–2016) and the head of the Navigational Department (since 2003) of Gdynia Maritime University, Poland. He has over 25 years of experience in teaching and research; he works in the fields of maritime navigation, sea transport, telematics, geomantics, hydrography, cartography, geodesy and safety at sea. He has published more than 250 reports and papers in journals and conference proceedings. He is the author of 18 books on navigation, including 6 handbooks on GIS, ECDIS and electronic charts, and is a Fellow of the Royal Institute of Navigation (London), Fellow of the Nautical Institute (London), member of the Polish Cartographer Society and, since 2007, founder and editor-in-chief of *TransNav, The International Journal on Marine Navigation and Safety of Sea Transportation*.

Andrew J. Wheeler is a professor of geology at University College Cork and is currently the head of geology and vice head of the School of Biological, Earth and Environmental Sciences. Wheeler is fascinated by the geology of the deep ocean seabed and has published extensively over the past 20 years on marine sedimentation processes, marine sediment archives of environmental change, and especially the geology of cold-water coral reefs. As an ocean explorer and scientist, Wheeler has led numerous oceangoing expeditions. His discovery of a new hydrothermal black smoker system on the mid-Atlantic ridge at 3000 m water depth was featured on National Geographic Television. His research team's work on cold-water coral carbonates found the earliest records of glaciation in Ireland, 1 million years earlier than anticipated, and was featured in his TEDx Talk in 2014 – 'How a Grain of Sand Rewrote Our Ocean's History'. Wheeler is a member of the Royal Irish Academy Geosciences and Geographical Sciences Committee and a member of the Executive Management Committees of the Irish Centre for Research in Applied Geosciences (iCRAG) and the UCC Environmental Research Institute.

Colin D. Woodroffe is a coastal geomorphologist in the School of Earth and Environmental Sciences at the University of Wollongong. He has a PhD and ScD from the University of Cambridge, and was a lead author on the coastal chapter in the 2007 Intergovernmental Panel on Climate Change (IPCC) 4th Assessment Report. He has studied the stratigraphy and development of coasts in Australia and New Zealand, as well as on islands in the West Indies, and Indian and Pacific Oceans. He has written a comprehensive book on *Coasts, Form, Process and Evolution* (Cambridge University Press, 2003), co-authored a book on *The Coast of Australia* (Cambridge University Press, 2009), and is also co-author of the book *Quaternary Sea-Level Changes: A Global Perspective* (Cambridge University Press, 2014).

Dian Zhang graduated with a PhD from Dublin City University (DCU) in 2015 as a member of Professor Noel O'Connor and Professor Fiona Regan's research group where he developed a multimodal smart sensing framework for marine environmental monitoring. He is currently working as a postdoctoral researcher at Insight and Mestech Research Centre at DCU. In 2010, he won the Enterprise Ireland Commercialisation Award for ICT, which was presented by the Minister for Science, Technology and Innovation at The Big Ideas Showcase. In 2012, he was appointed as the lead developer of the IABA (Irish Amateur Boxing Association) Boxing Training Analysis Project, which won the 2012 DCU INVENT Commercialisation Awards, which was successfully licensed to Audio Visual Services Ltd. He is also the technical lead of the Environmental Protection Agency Ireland funded project Climate Change in Irish Media.

1

Geoinformatics for Applied Coastal and Marine Management

Darius Bartlett and Louis Celliers

CONTENTS

1.1 The Context: Coastal and Marine Spaces, and the Need to Manage Them

In the twenty-first century, the world is becoming increasingly dependent on the economic, social and environmental benefits derived from ocean and coastal services and resources. This is of course primarily true of those countries that have coastlines but, as Bruce McCormack has suggested in the preface to this book, even those communities that lie far inland from the sea will benefit as well. The 'ocean economy' has become a recognized and measured contributor to national gross domestic product (GDP) (Pauli, 2010; Economist Intelligence Unit, 2015). According to the national account of many countries, the ocean economy and marine-based industries may produce from 1–5% of their GDP (Park and Kildow, 2014). Not only has there been greater emphasis on implementing strategies to develop the economic sector, but there is also an increasing urgency for protection of marine resources. This is mirrored by Sustainable Development Goal 14, which is to conserve and sustainably use the oceans, seas and marine resources for sustainable development.

In 2012, for example, it was projected that the Irish ocean economy would deliver approximately 0.8% of GDP by 2014 (€1.4 billion) and employ 18,480 people (full-time equivalent [FTE]; Vega et al., 2012). For 2013–2014, the United States measured an ocean economy contribution of 2.2% to GDP (US$359 billion) and employed 137 million people (Kildow et al., 2016).

The ocean economy also offers opportunity for the promotion of economic growth, environmental sustainability, social inclusion and the strengthening of fisheries and aquaculture, renewable marine energy, marine bioprospecting, marine transport and marine and coastal tourism. All of these sectors offer growth and development opportunities for coastal states and, especially, for Small Island Developing States (SIDS; UNCTAD, 2014).

However, the growth and development opportunities of the ocean economy are tempered by increasing and complex challenges facing coasts and oceans (Kenneday et al., 2002; Hoegh-Guldberg et al., 2015). These include the unsustainable extraction of marine resources, marine pollution, alien invasive species, ocean acidification, climate change impacts and the physical alteration and destruction of coastal and marine habitats (UNDESA, 2014).

Human-induced increases in atmospheric concentrations of greenhouse gases are expected to cause much more rapid changes in the Earth's climate than have been experienced for millennia. These may have a significant effect on coastal ecosystems, especially estuaries and coral reefs, which are relatively shallow and already under stress because of human population growth and coastal developments. Climate change may decrease or increase precipitation, thereby altering coastal and estuarine ecosystems. Wind speed and direction influence production of fish and invertebrate species, such as in regions of upwelling along the U.S. West Coast. Increases in the severity of coastal storms and storm surges would have serious implications for the well-being of fishery and aquaculture industries. Sea-level rise may inundate or cause migration of important coastal ecosystems, such as mangroves and tidal mudflats, which may be important breeding habitats for fish and other deep-sea organisms in their larval stages. The immense area and the modest extent of our knowledge of the open ocean hamper predictions of how ocean systems will respond to climate change.

Further complexity of the ocean's economy arises from the multilayer regulatory framework under the United Nations Convention on the Law of the Sea (UNCLOS) and other national, regional and multilateral as well as sectoral governance regimes (UNCTAD, 2014). This requires the development of a more coherent, integrated and structured framework that takes account of the economic potential of all marine natural resources, which include seaways and energy sources located in the ocean space.

Within this context there is a growing need to know enough about natural systems, processes and rates of consumption in order to make wise decisions regarding the simultaneous protection and use of natural resources. Even though about 70% of the planet's surface is covered by ocean water, humans have only effectively explored less than 10% of this, and we have mapped even less than that in any detail. There is thus a growing urgency attached to the need for more comprehensive and reliable data related to the marine environment. This of course will require both political willingness, on an international level, and also corresponding economic investment but, when compared to other areas of major current outlay, the costs should not be seen as prohibitive: it was suggested recently that mapping the whole of the world's oceans would cost about US$3 billion (approximately the same as a single mission to Mars), and about 200 ship-years to complete (Holden, 2015).

1.2 The Role of Spatially Enabled Information and Communication Technologies (ICT)

In recognition of the many problems that have arisen in the past through inappropriate use and management of the coast, a new, more 'environmentally oriented' ethos of coastal management is emerging. Based on more holistic, trans-disciplinary and integrative principles (sometimes simply referred to as 'joined-up thinking'), this new approach is aimed at sustainable management of resources for the benefit of all stakeholders. A similar evolution may also be seen at work in the context of ocean management, with the prevailing sectorial approach being increasingly placed in the broader contexts of international law and agreements, and the emergence of marine spatial planning as a framework for addressing concerns and reconciling potentially conflicting interests.

For it to work, this new philosophy depends on a thorough understanding of the entities and relationships at work in marine and coastal systems. Governments and the scientific community alike have responded to this need, leading to a major paradigm shift in scientific research (Birkin, 2013), in which an earlier focus on experimentation and reasoning is being replaced by a preoccupation with analysis of the increasing volumes of data being made available at unprecedented spatial and temporal scales. For these and other reasons, spatially enabled information and communication technologies (ICT) are increasingly being used by coastal and marine scientists and administrators to assist them in their work.

Scientific applications have frequently been at the forefront of driving computer and information technology (Baru, 2011). In most domains, the pioneering applications of these tools were limited by the size and expense of computers, and by the limited storage, processing and graphic output capabilities of the machines available at the time (Biles et al., 1989; Marble, 2010). This was certainly the case for early applications of ICT to coastal and ocean science (Bartlett, 1999; Wright, 1999), where the challenges were further compounded by the complexity of marine and coastal environments, and the need to devise appropriate digital formats and structures to represent these particular geographies. But, as Marble and Peuquet observed, in most disciplines 'there are problems which prove intractable when first encountered, but which are reduced within a few years through the application of additional theoretical insights as well as significant amounts of hard work and luck' (Marble and Peuquet, 1983). As the chapters in this book will demonstrate, many of the early obstacles have now been addressed and, while a number of challenges remain, and much research and development still needs to be undertaken, the application of geospatial technologies to the demands of marine and coastal management is rapidly becoming routine.

Until the early 1980s, the main contribution of information technology to marine and coastal studies consisted primarily of stand-alone programmes, written in FORTRAN and other early languages, to address specific tasks in the fields of biological oceanography, chemical oceanography, geoscience, physical oceanography, navigation and charting, and the retrieval and editing of ocean data (for a comprehensive annotated listing of early computer programmes for oceanographic data management and analysis, including details of the programming languages used and the computers for which they were developed, see Dinger, 1970). By the early 1980s, however, collections of programmes for spatial data handling were being progressively brought together and packaged in the form of integrated, general-purpose geographical information system (GIS) toolboxes (Coppock and Rhind, 1991; Goodchild, 2000; Chrisman, 2005, 2006; Hoel, 2010). Although in their early days these systems were primarily aimed at terrestrial applications and their users, their potential to also address coastal and marine needs soon became recognized (Bartlett, 1993a,b, 2000; Wright, 2000). Since then, the range and scope of these applications have expanded rapidly as existing technologies have improved, new ones have emerged and extended methodological frameworks (Green and King, 2003; Bartlett and Smith, 2004; Wright et al., 2007; Green, 2010) have been developed.

Today, spatial information technologies are ubiquitous and no longer the preserve of the pioneer or the specifically trained specialist (see for example Carpenter and Snell, 2013). As well as their presence in the workplace, they are also pervasive and frequently to be found embedded in consumer products, including mobile phones, tablet computers, on desktop computers, in car dashboards, on the bridges of ships, on the wrists of athletes, incorporated into people's clothing and in a myriad of other locations that would have been barely imaginable just a few years ago. Furthermore, applications based on these technologies increasingly involve the convergence and integration of multiple elements drawn from an ever-widening range of possible ingredients, including geographical information systems (GIS), digital cartography, optical and microwave remote sensing, spatial database systems, Internet and mobile phone technologies, global satellite positioning systems, light detecting and ranging (LiDAR) and other laser-based survey techniques, gaming engines, digital photogrammetry, sensors and autonomous data collecting devices and others.

The term geoinformatics has been adopted independently by a number of geospatial and geoscience disciplines (Keller, 2011) to collectively describe these technologies, their applications and the scientific disciplines that underpin them, particularly since the closing years of the twentieth century (Gundersen, 2007). For Fotheringham and Wilson (2008:1), 'geoinformatics' is synonymous with the related concepts of geocomputation, geoprocessing and geographic information science (sometimes abbreviated to GISci, to distinguish it from geographical information systems, GIS – see, for example, Mark, 2002; Longley et al., 2011, 2015). It is also closely related to the word 'geomatics', defined by Gagnon and Coleman as '*a field of scientific and technical activities which, using a systemic approach, integrates all the means used to acquire and manage spatially referenced data as part of the process of producing and managing spatially based information*' (Gagnon and Coleman, 1990).

Technical and semantic interoperability between system components, along with greater accessibility of geoprocessing resources, '*improve the application of geospatial data in various domains and help to increase the geospatial knowledge available to society*' (Zhao et al., 2012). This is especially important for marine and coastal management, where multiple datasets, collected by different agencies through many different means, have to be integrated, compared and analysed together. These issues of integration and data compatibility are increasingly being addressed through the development of dedicated spatial data infrastructures (Bartlett et al., 2004; Longhorn, 2004; Strain et al., 2006; Wright, 2009), comprising standards for data, computer hardware and software; geoportals designed to make easier the task of data discovery and access and rules and regulations that define the legal, financial and institutional contexts within which geoinformatics methods and applications can and should operate.

1.3 Case Study: Cork Harbour

One of the largest natural harbours in the world, Cork Harbour on the south coast of Ireland (Figure 1.1) epitomizes both the great diversity of maritime activities prevalent in a modern economy, and also the range of applications of spatial information technologies to support these.

In physical terms, the harbour is a former river valley, flooded as world sea levels rose at the end of the last ice age, approximately 10,000 years ago. Its primary river, the River Lee, feeds into the harbour in the northwest, and Cork City, the second city of the Republic of Ireland, is built on former marshland at the river's lowest bridging point. Several other, smaller, towns and settlements are also to be found around the shores of the harbour.

Many locations within and around Cork Harbour are of major environmental and ecological significance. The waters of the harbour are home to grey seals (*Halichoerus grupus*), common seals (*Phoca vitulina*), and are occasionally visited by common dolphins (*Delphinus delphina*), bottlenose dolphins (*Tursops truncatus*) and other cetaceans. Closer to shore, and especially in the inner harbour area, the mudflats, salt marshes and shingle shorelines host large populations of feeding, breeding and, in some cases, over-wintering waders,

FIGURE 1.1
Cork Harbour, Ireland, looking south from the town of Cobh towards the harbour mouth and the open ocean. In the middle distance Haulbowline Island can be seen (home to the Irish Naval Service) and, behind it, wind turbines marking the position of the Irish Marine and Energy Research Cluster (IMERC). To the left of the photograph, the Whitegate oil terminal and refinery is in the distance, while on the right lies the industrial and port complex of Ringaskiddy. The Cobh cruise ship terminal is seen in the immediate foreground, on the seaward side of the railway tracks. (Courtesy of Darius Bartlett.)

wildfowl and other birds, as well as numerous otters (*Lutra lutra*). This has led to several sites being designated as Special Protection Areas, Ramsar sites and Special Areas of Conservation. While primary responsibility for designating these rests with the National Parks and Wildlife Service (a government agency), numerous voluntary bodies and non-governmental organizations (NGOs) also play important roles in monitoring them and ensuring their well-being. Several of these 'citizen science' initiatives use web-enabled GIS tools to facilitate the collection, analysis and reporting of observation data by members of the public: examples of this would include the annual Coastwatch survey (http://coastwatch.org/europe/survey/) and the BirdTrack online bird recording scheme, developed jointly by BirdWatch Ireland, the British Trust for Ornithology, the Royal Society for Protection of Birds, The Scottish Ornithologists' Club and the Welsh Ornithological Society (http://www.bto .org/volunteer-surveys/birdtrack/about). The role of volunteered geographical information in coastal and marine management is discussed by Goldberg, D'Iorio and McClintock, in an international context, in Chapter 7 of this book.

Historically, Cork Harbour has always been an important locus for international shipping and commerce and, by the end of the eighteenth century, the port of Cork was Ireland's leading transatlantic shipping port (Rynne, 2005). Initially, many of these port activities were located on the quaysides of Cork City itself but, by the early years of the nineteenth century, extensive silting up of the shipping channels led to the gradual migration of commercial and industrial activities down-river to the outer harbour area. Even today, periodic dredging of the main shipping channels is required, both in the inner and outer harbour areas, and several formerly thriving lesser fishing and trading ports around the harbour are now no longer viable. Regular surveying and monitoring of the bathymetry of the seabed and shipping channels, as described by Scott, Monteys, Hardy, McKeon and Furey, in Chapter 2, are essential to inform this task and are undertaken regularly (Figure 1.2); while

Cork Lower Harbour and Approaches

FIGURE 1.2
Cork Lower Harbour and Approaches, based on bathymetric data from the INFOMAR programme (see Chapter 2 of this volume) and satellite data from Esri. (Cartography and image compilation by Nicola O'Brien, Marine Institute, Galway, Ireland.)

a buoy containing sensors for a range of meteorological and marine environmental parameters is moored just off the harbour entrance. The increasingly important role of sensors for collecting data about the marine environment is discussed by Zhang, O'Connor and Regan, in Chapter 5.

Ireland's National Space Centre is located at Elfordstown, about 15 km inland to the north of the harbour (http://nationalspacecentre.eu/). Originally built in the early 1980s as a link in the Eutelsat network to facilitate Trans-Atlantic air traffic and telecommunications, the site was redeveloped and reopened in 2010 as Europe's most westerly teleport and data download station, where one of its primary areas of specialization is in the use of Earth observation for maritime surveillance. Lück-Vogel explores recent developments in satellite remote sensing and Earth observation as a further important source of geospatial data for marine and coastal management in Chapter 4; Saunders examines maritime surveillance from the perspective of international legal conventions and agreements in Chapter 9; and Mathews and Power explain how geoinformatics and telecommunications are assisting in the task of policing Irish territorial waters, as well as in search and rescue operations in the Mediterranean in Chapter 15.

About 10 million tonnes of freight is currently handled by the port of Cork annually. Ireland's only oil refinery, located on the eastern side of the harbour near its outlet to the sea, accounts for over 55% of this freight, and 28% of the ship tonnage (Port of Cork Company, 2015), while three container terminals in the harbour handle between them a considerable proportion of the rest. Across the water, on the western side of the harbour, lies the Ringaskiddy deepwater terminal, with container and bulk cargo handling facilities, as well as passenger and car ferry connections with France and, until recent years, the United Kingdom.

Ringaskiddy is also home to IMERC, the Irish Maritime and Energy Research Cluster, launched in 2010 as a joint initiative of University College Cork, Cork Institute of Technology and the Irish Naval Service (www.imerc.ie). This campus is intended to become a world-leading specialist centre of expertise for research and development in marine energy (renewables and offshore hydrocarbons); maritime information and communications technologies (ICT); shipping logistics and transport; maritime security and safety and yachting products and services. The campus includes the National Maritime College of Ireland, which specializes in addressing the training and research needs of the Irish and international maritime sector (http://www.nmci.ie/); the MaREI centre for marine and renewable energy (http://marei.ie/) and the LIR National Ocean Test Facility (http://www.lir-notf.com/). The role of geoinformatics for marine renewable energy is examined by Green in Chapter 12.

The town of Cobh, almost directly opposite the entrance to Cork Harbour, has important heritage and historical significance as the embarkation point for over 1 million emigrants, who set sail from Ireland to Britain, America, Canada, Australia and other places of refuge during, and especially after, the

Great Irish Famine of the 1840s and 1850s (Foster, 2005; Crowley et al., 2012); and also as the last port of call of the steamship *Titanic*, on its maiden voyage across the Atlantic. Today, Cobh has Ireland's only dedicated terminal for cruise liners (Figure 1.3), and is thus an important centre for the tourism and heritage sectors. In 2014, 58 cruise liners, with a gross tonnage of 3.5 million tonnes, brought over 120,000 passengers and crew to the region (Port of Cork Company, 2015), and a small but locally important trawler fishing fleet. In Chapter 14, Weintrit outlines the increasing role of electronic charting and automatic vessel identification systems in ensuring safe navigation of shipping; while in Chapter 13, Nishida, Itoh, Caton and Bartlett, discuss the specific role of geoinformatics in the context of the fishing industry.

Finally, in the centre of the harbour and directly across the channel from Cobh, lies the main base of the Irish Naval Service (INS) on Haulbowline Island. The INS has a number of defence and other tasks that bring an important international dimension to their work, including monitoring and protection of fisheries in one of the largest marine sectors within the European

FIGURE 1.3
Cruise ship docked at the Cobh cruise ship terminal, Cork Harbour. (Courtesy of Darius Bartlett.)

Union, policing operations in Irish territorial waters and, in recent years, actively supporting responses to the migrant crisis in the central and eastern Mediterranean (see Chapter 15 by Mathews and Power).

Although not as vulnerable to environmental hazards as several other coastal locations might be, the communities and settlements around Cork Harbour and area are subject to the impacts of storms, flooding and coastal erosion (Figure 1.4), all of which are projected to increase in frequency and severity under conditions of predicted climate change (Cummins and O'Donnell, 2005; Devoy, 2008; Kopke and O'Mahony, 2011). Cork City itself is particularly prone to flooding, due largely to its location at the head of the harbour and on the delta of the River Lee (Crowley et al., 2005). However the opening decade of the Twenty-first Century saw rapid expansion and development of many settlements around the harbour and, in some cases, houses appear to have been built in areas that longer-term community wisdom spanning multiple generations knew to be prone to flooding. Examples such as this illustrate the value of seeking out and incorporating traditional knowledge of place when drawing up plans for coastal development and management. In Chapter 10, Aswani outlines how approaches based on participatory GIS can assist in acquiring and using this traditional understanding of the sea and its characteristics; while the application of geoinformatics technologies for more formal assessment, reduction of and response to coastal hazard and vulnerability is discussed by Bonetti and Woodroffe in the final chapter of this book.

FIGURE 1.4
Floods in Cork City centre, November 2008. The flooding arose after a period of exceptionally heavy rain, which led to significantly increased discharge from the River Lee, which runs through the city, coupled with spring high tide, low barometric pressure and strong onshore winds. Floods around the entire harbour are increasing in frequency and impact, a trend that is likely linked to the effects of global climate change. (Courtesy of Darius Bartlett.)

The multiplicity of uses and economic activities in Cork Harbour and its hinterland, as outlined in this section, inevitably puts pressure on the resources and environments it contains. This in turn can (and occasionally does) lead to conflict and potential disputes between stakeholders. It also requires appropriate management frameworks to give coherence and longer-term vision to policies and decision making aimed at reducing conflict. For Cork this is particularly significant, since management of the harbour, its environment and its resources is devolved and shared across a number of statutory and other agencies, including the Marine Institute, the Environmental Protection Agency, the Office of Public Works, the County and (Cork) City councils and a number of other bodies. The concepts of integrated coastal zone management (ICZM) and marine spatial planning (MSP) offer frameworks through which integration and more sustainable management of coastal and ocean spaces might be achieved, and a number of studies (e.g. O'Mahony et al., 2009, 2014; Queffelec et al., 2009) have investigated how they might be applied within the harbour. The origins, applications and significance of ICZM and MSP, and the role of geoinformatics technologies in assisting these, is discussed in a wider context by McWhinnie and Gormley in Chapter 8, while Lassoued and Leadbetter (Chapter 6) explain how and why a consideration of ontologies is important in order to ensure compatibility and interoperability of databases, technologies and working practices when seeking to develop sustainable and inclusive approaches to coastal management.

In Section 1.1 of the present chapter, the point was made that large tracts of ocean floor have yet to be surveyed and mapped in any detail. The INFOMAR programme, outlined by Scott et al. in Chapter 2, is helping to address this need with regard to the Irish continental shelf while; in Chapter 3, Murphy and Wheeler discuss their analysis of the Gollum Channel complex on the Porcupine Bank, approximately 100 km from the Irish coast, and demonstrate how high-resolution bathymetric data from INFOMAR, along with advanced GIS tools, can support fundamental oceanographic science as well as economic purposes.

Looking even further afield, in Chapter 11, Barry, Christensen, Guðmundsdóttir, Lárusson, Price and Mosbech present the work of the Arctic Council and the role of geoinformatics technologies in supporting this. One of the main current concerns in the Arctic is the increasing extent of ice-melt in the summer months, widely accepted as an indicator and consequence of anthropogenic-induced climate change. One of the many implications of this is the prospect that shipping lanes through the Arctic Ocean, whether along the north coast of Russia or via the north of Canada, will remain ice-free and navigable for longer in the year, offering far shorter, cheaper and quicker trade routes between Europe and markets in Asia and the western seaboard of the United States. While this could clearly be of benefit for port authorities in Europe, expansion of these northern sea routes is likely to have adverse impacts on many ports in the southern hemisphere, including Durban, the principal commercial port in South Africa in terms of the number of ship calls and the number of containers handled (Jones, 2014), and currently a key node in shipping between the Atlantic and Indian Oceans. This example serves to further illustrate the increasingly important globalized nature of society's relationship with the sea, a relationship that spans and connects in so many ways the physical and natural environment with the socioeconomic and cultural aspects of twenty-first century living.

Inevitably, in selecting the chapters, examples and case studies for inclusion in this book, countless other applications of spatial technologies to marine and coastal management issues had to be omitted. Nonetheless, taken together, the chapters presented here do serve to illustrate the diversity of human uses of and demands on the world's oceans and coasts, the increasing importance of management strategies to enable the search for a more sustainable relationship between human society and the marine environment and, linking these, the growing role of geoinformatics technologies in facilitating this.

Acknowledgement

The authors warmly thank Nicola O'Brien of the Marine Institute in Galway, Ireland, for creating and making available the wonderful map of Cork Harbour reproduced as Figure 1.2 of this chapter.

References

Bartlett, D.J. 1993a. GIS and the coastal zone: An overview. In St. Martin, K. (Ed.), *Explorations in Geographic Information Systems Technology, Volume 3: Applications in Coastal Zone Research and Management*. Worcester, MA: Clark Labs and Geneva, Switzerland: United Nations Institute for Training and Research.

Bartlett, D.J. 1993b. *GIS and the Coastal Zone: An Annotated Bibliography*. National Center for Geographic Information and Analysis Report 93-9.

Bartlett, D.J. 1999. Working on the frontiers of science: Applying GIS to the Coastal Zone. Chapter 2 in Wright, D.J. and Bartlett, D.J. (Eds.), *Marine and Coastal Geographical Information Systems*. London: Taylor & Francis, pp. 11–24.

Bartlett, D., and Smith, J. 2004. *GIS for Coastal Zone Management*. Boca Raton, FL: CRC Press.

Bartlett, D.J., Longhorn, R., and Garriga, M. 2004. Marine and coastal data infrastructures: A missing piece in the SDI puzzle? 7th Global Spatial Data Infrastructure Conference, Bangalore, India.

Baru, C. 2011. Introduction to IT concepts and challenges. Chapter 2 in *Geoinformatics Cyberinfrastructure for the Solid Earth Sciences*, Kelleher, G.R. and Baru, C. (Eds.). Cambridge, England: Cambridge University Press.

Biles, G.E., Bolton, A.A., and DeRe, B.M. 1989. Herman Hollerith: Inventor, manager, entrepreneur – A centennial rememberance. *Journal of Management* 15(4), 603–615.

Birkin, M. 2013. Big data challenges for geoinformatics. *Geoinformatics & Geostatistics – An Overview* 1(1). http://dx.doi.org/10.4172/2327-4581.1000e101.

Carpenter, J., and Snell, J. 2013. Future trends in geospatial information management: The five to ten year vision. UN-GGIM. Retrieved from http://www.dbpia.co.kr/Journal/ArticleDetail/1563422.

Chrisman, N.R. 2005. Communities of scholars: Places of leverage in the history of automated cartography, *Cartography and Geographic Information Science* 32(4), 425–433. doi: 10.1559/152304005775194674.

Chrisman, N.T. 2006. *Charting the Unknown: How Computer Mapping at Harvard Became GIS*. Redlands, CA: Esri Press.

Coppock, J.T., and Rhind, D.W. 1991. The history of GIS. Chapter 2 in Maguire, D.J., Goodchild, M.F. and Rhind, D.W. (Eds.), *Geographical Information Systems – Principles and Applications*, Volume 1. London: Harlow Longman Scientific & Technical.

Crowley, J., Devoy, R.J., Linehan, D., and O'Flanagan, T.P. 2005. *Atlas of Cork City*. Cork, Ireland: Cork University Press.

Crowley, J., Smyth, W.J., and Murphy, M. 2012. *Atlas of the Great Irish Famine*. Cork, Ireland: Cork University Press.

Cummins, V., and O'Donnell, V. 2005. Cork Harbour and the Challenge of Change. In *Atlas of Cork City*, J. Crowley, R. Devoy, D. Linehan and P. O'Flanagan (Eds.). Cork, Ireland: Cork University Press, pp. 391–395.

Devoy, R.J.N. 2008. Coastal vulnerability and the implications of sea-level rise for Ireland. *Journal of Coastal Research* 24(2), 325–341. doi: http://dx.doi.org/10.2112/07A-0007.1.

Dinger, C. 1970. *Computer Programs in Oceanography*. National Oceanographic Data Centre Publication C-5 (Second Revision). Available online at https://ia600400.us.archive.org/25/items/computerprograms00ding/computerprograms00ding.pdf.

Economist Intelligence Unit. 2015. *The Blue Economy: Growth, Opportunity and a Sustainable Ocean Economy*. An Economist Intelligence Unit briefing paper for the World Ocean Summit 2015. Available online at http://www.economistinsights.com/sites/default/files/Blue%20Economy_briefing%20paper_WOS2015.pdf.

Foster, M. 2005. Queenstown/Cobh: An emigrant port. In *Atlas of Cork City*, Crowley, J. et al. (Eds.). Cork, Ireland: Cork University Press, pp. 246–250.

Fotheringham, A.S., and Wilson J.P. 2008. Geographic Information Science: An Introduction. Chapter 1 in Wilson, J.P. and Fotheringham, A.S. (Eds.). *The Handbook of Geographic Information Science*. Maldon, M.A.: Blackwell Publishing, pp. 1–8.

Gagnon, P., and Coleman, D.J. 1990. Geomatics. An integrated, systemic approach to meet the needs for spatial information. *CSIM Journal* 44(4), 377–382.

Goodchild, M.J. 2000. The current status of GIS and spatial analysis. *Journal of Geographic Systems* 2, 5–10.

Green, D.R. (Ed.). 2010. *Coastal and Marine Geospatial Technologies*. Dordrecht, the Netherlands: Springer Science and Business Media.

Green, D.R., and King, S.D. (Eds.). 2003. *Coastal and Marine Geo-Information Systems*. Dordrecht, the Netherlands: Springer Science and Business Media.

Gundarsen, L.C. 2007. Envisioning a geoinformatics infrastructure for the earth sciences: Technological and cultural challenges. In Brady, S.R., Sinha, A.K. and Gundersen, L.C. (Eds.), *Geoinformatics 2007 – Data to Knowledge,* Proceedings U.S. Geological Survey Scientific Investigations Report 2007-5199, p. 1.

Hoegh-Guldberg, O. et al. 2015. Reviving the ocean economy: The case for action – 2015. WWF International, Gland, Switzerland. Geneva.

Hoel, E., Gillgrass, C., and McGrath, M. 2010. History of GIS. The Commercial Era: 1980 to 2010 (Version 19). Available at http://www.docfoc.com/1-history-of -gis-the-commercial-era-1980-to-2010-erik-hoel-craig-gillgrass.

Holden, J. 2015. Space a distraction from the discoveries awaiting us in Earth's unexplored oceans. *Irish Times*, October 5, 2015. Available online at http://www .irishtimes.com/business/space-a-distraction-from-the-discoveries-awaiting -us-in-earth-s-unexplored-oceans-1.2376395.

Jones, T. 2014. Maritime transport and harbours. Chapter 6.3 in *Ugu Lwethu – Our Coast. A Profile of Coastal KwaZulu Natal*, Goble, B.J., van der Elst, B.P. and Oellermann, L.K. (Eds.). Cedara: KwaZulu Natal Department of Agricultural and Environmental Affairs and the Oceanogaphic Research Institute, pp. 98–101.

Keller, G.R. 2011. Science Needs and Challenges for Geoinformatics. Chapter 1 in Keller, G.R. and Baru, C. (Eds.). *Geoinformatics Cyberinfrastructure for the Solid Earth Sciences*. Cambridge, U.K.: Cambridge University Press, pp. 3–9.

Kenneday, V.S., Twilley, R.R., Kleypas, J.A., Cowan, Jr., J.H., and Hare, S.R. 2002. Coastal and marine ecosystems and global climate change potential effects on U.S. resources. *Pew Center on Global Climate Change*.

Kildow, J.T., Colgan, C.S., Johnston, P., Scorse, J.D., and Farnum, M.G. 2016. State of the U.S. ocean and coastal economies 2016 update. National Ocean Economics Program (NOEP) of the Center for the Blue Economy at the Monterey Institute of International Studies.

Kopke, K., and O'Mahony, C. 2011. Preparedness of key coastal and marine sectors in Ireland to adapt to climate change. *Marine Policy* 35(6), 800–809. http://dx.doi .org/10.1016/j.marpol.2011.01.008.

Longhorn, R.A. 2004. Coastal spatial data infrastructure. Chapter 1 in *GIS for Coastal Zone Management*, Bartlett, D. and Smith, J. (Eds.). Boca Raton, FL: CRC Press, pp. 1–16.

Longley, P.A., Goodchild, M.F., Maguire, D.J., and Rhind, D.W. 2011. *Geographic Information Systems and Science*, 3rd ed. Hoboken, NJ: John Wiley & Son.

Longley, P.A., Goodchild, M.F., Maguire, D.J., and Rhind, D.W. 2015. *Geographic Information Science and Systems*, 4th ed. Hoboken, NJ: Wiley.

Marble, D.F. 2010. An early excursion into computational geography. *Information Systems*, 1–25.

Marble, D.F., and Peuquet, D.J. 1983. The computer and geography: Some methodological comments. *The Professional Geographer* 35(3), 343–344.

Mark, D.M. 2002. Geographic information science: Defining the field. In *Foundations of Geographic Information Science*, Duckham, M., Goodchild, M.F. and Worboys, M.F. (Eds.). London: Taylor & Francis, pp. 3–18.

O'Mahony, C., Gault, J., Cummins, V., Köpke, K., and O'Suilleabhain, D. 2009. Assessment of recreation activity and its application to integrated management and spatial planning for Cork Harbour, Ireland. *Marine Policy* 33(6), 930–937. http://dx.doi.org/10.1016/j.marpol.2009.04.010.

O'Mahony, C., Köpke, K., Twomey, S., O'Hagan, A.M., Farrell, E., and Gault, J. 2014. Integrated coastal zone management in Ireland – Meeting water framework directive and marine strategy framework directive targets for Ireland's transitional and coastal waters through implementation of integrated coastal zone management. Report prepared under contract for Sustainable Water Network (SWAN).

Park, K.S., and Kildow, J.T. 2014. Rebuilding the classification system of the ocean economy. *Journal of Ocean and Coastal Economics* 2014(4). doi: http://dx.doi.org/10.15351/2373-8456.1001.

Pauli, G. 2010. The Blue Economy 10 Years, 100 Innovations, 100 Million Jobs Report to the Club of Rome. Taos, NM: Paradigm Publications.

Port of Cork Company. 2015. *Port of Cork Annual Report, 2014.* Available online at http://www.portofcork.ie/index.cfm/page/annualreports1?twfId=688&download=true.

Queffelec, B., Cummins, V., and Bailly, D. 2009. Integrated management of marine biodiversity in Europe: Perspectives from ICZM and the evolving EU maritime policy framework. *Marine Policy* 33(6), 871–877. http://dx.doi.org/10.1016/j.marpol.2009.04.016.

Rynne, C. 2005. Industry 1750–1930. Chapter 16 in *Atlas of Cork City*, Crowley, J. et al. (Eds.). Cork, Ireland: Cork University Press, pp. 183–189.

Strain, L., Rajabifard, A., and Williamson, I. 2006. Marine administration and spatial data infrastructure. *Marine Policy* 30(4), 431–441. doi:10.1016/j.marpol.2005.03.005.

UNCTAD. 2014. The oceans economy: Opportunities and challenges for small island developing states. United Nations Conference on Trade and Development, New York.

UNDESA, UN-DOALOS/OLA, IAEA, IMO, IOC-UNESCO, UNDP, UNEP, UNWTO. 2014. How oceans- and seas-related measures contribute to the economic, social and environmental dimensions of sustainable development: Local and regional experiences. Online publication. https://sustainabledevelopment.un.org/content/documents/1339Non_recurrent_e_publication_Oceans_final%20version.pdf.

Vega, A., Hynes, S., and O'Toole, E. 2012. Ireland's ocean economy reference year 2012. Socioeconomic Marine Research Unit (SEMRU), NUI Galway.

Wright, D., and Bartlett, D.J. (Eds.). 1999. *Marine and Coastal Geographical Information Systems.* London: Taylor & Francis, p. 320.

Wright, D.J., Blongewicz, M.J., Halpin, P.N., and Breman, J. 2007. *ArcMarine: GIS for a Blue Planet.* Redlands, CA: ESRI Press, p. 202.

Wright, D.J. 2009. Spatial data infrastructures for coastal environments. Chapter 5 in *Remote Sensing and Geospatial Technologies for Coastal Ecosystem Assessment and Management*, Yang, X. (Ed.), Lecture Notes in Geoinformation and Cartography. doi:10.1007/978-3-540-88183-4 5. Berlin: Springer-Verlag.

Zhao, P., Foerster, T., and Yue, P. 2012. The geoprocessing web. *Computers & Geosciences* 47, 3–12. http://dx.doi.org/10.1016/j.cageo.2012.04.021.

2

Mapping the Seabed

Gill Scott, Xavier Monteys, David Hardy,
Charise McKeon and Thomas Furey

CONTENTS

2.1 Introduction

Throughout human history, most official terrestrial cartography was militarily, commercially and politically motivated and the purview of the world's most powerful nations. Indeed when, in 1846, Ireland became the first country in the world to be mapped at a resolution of 6 inches to 1 mile (1:10,560) by the British Army it was for the purposes of land taxation (Andrews, 1975; Cadhla and Cuív, 2007). Similarly, most seabed maps were, initially, produced by the world's maritime superpowers for sovereign purposes (Reidy and Rozwadowski, 2014). Nonetheless, some mapping was driven by scientific curiosity, notably the first lead and line-based global seabed map produced following the famous H.M.S. Challenger expedition (1872–76; Murray, 1895). Despite the remarkably low number of measurements (less than 400) a map of the global ocean seafloor was attempted and expedition scientists were able to recognize that 'The bottom of the ocean it appears is as varied as the land for there are valleys and mountains, hills and plains all across the Atlantic' (Matkin and Rehbock, 1992). Despite the rapid scientific advances made in the years following the Challenger expedition, lead and line surveys remained the main technique for measuring water depth until the 1920s (Höhler, 2002).

Echosounders determine the depth of the seabed by measuring the time taken for transmitted sound (sonar) to be returned to the receiver (Bloomer et al., 2012) and, with their inception in the years following the first World War, it became possible to make bathymetric measurements remotely and speedily, initiating a new era of marine map-making (Höhler, 2002). Technological developments and data collection remained largely the preserve of the military but the results were sometimes made available to the scientific community, as was the case for the data used to create the first detailed map of the Atlantic Ocean by Marie Tharp and Bruce Heezen (Elmendorf and Heezen, 1957; Heezen et al., 1959) whose achievements are all the more remarkable given the sparsity of data available to them, leading some to describe the work as part physical science, part geological intuition (Doel et al., 2006).

During the 1960s, the U.S. Navy began using arrays of sonars, 'multibeam', which produced soundings for swathes of the seabed giving improved and continuous seabed coverage (Vilming, 1998). However, it was only with declassification of the technology in the 1970s that commercial multibeam systems were developed. Their evolution continued through the 1980s and 1990s resulting in ever higher-resolution mapping, and systems began to include acoustic backscatter recording capabilities, thus giving information on seabed type as well as its topography.

During the same period, advances in space-technology also began to impact on seabed mapping. Initially, this was also driven and controlled by the military but with declassification of the American global positioning system (GPS) in the 1980s and the subsequent development of differential GPS, sub-metre positional accuracies became possible and seabed mapping began to approach and even exceed the level of detail possible on land. The full declassification of GEOSAT altimetry in 1995 heralded civilian mapping of the seafloor from space (Smith and Sandwell, 1997).

Safe navigation remains one of the primary drivers for seabed mapping, particularly in the nearshore areas that have traditionally been of greatest interest to mariners, and hence to governments and agencies tasked with maritime matters. But, as the ability to commercially exploit deeper waters has grown, governments have increasingly sought to explore, understand, claim jurisdiction over and protect their off-shore resources as well. The extent of a nation's sovereignty became subject to international law under the terms of the United Nations Convention on the Law of the Sea treaty (UNCLOS; UN

General Assembly, Convention on the Law of the Sea, 1982), which was agreed in 1994 and ratified by Ireland in 1996 (UN Division for Ocean Affairs and the Law of the Sea, 2015). The Convention defines a number of divisions of maritime territory that might fall within a nation's jurisdiction, including the Exclusive Economic Zone (EEZ) within which a nation has exclusive exploitation rights to any resources extending from the sea surface, through the water column, to the seabed and its underlying geological formations. This zone extends up to 200 nautical miles from the coast or to the limits of the continental shelf, whichever is the greatest. It is clear from this that the determination of the continental shelf and its extent requires accurate and verifiable mapping, and so UNCLOS became a spur to national survey programmes.

There is also an onus on nations to protect their marine waters from the negative impacts of human activities including climate change, pollution releases and over-exploitation, while they also have responsibilities to ensure effective defence and law enforcement, including fisheries management and actions against piracy, smuggling and other illegal activities that might take place within their jurisdictions. These tasks and responsibilities are often driven by legislation. Seabed mapping is the foundation upon which effective management of the marine environment is realized, and is central to the identification and resolution of conflict between human uses and the environment (Ehler and Douvere, 2009).

2.1.1 The Irish Case: From Deep Waters to Coastal Areas

Prior to 1991, very little was known about Ireland's off-shore environments and resources, and most of the seabed had never been comprehensively surveyed or mapped. Following Irish independence from the United Kingdom in 1922, the UK Hydrographic Office (UKHO) continued to produce charting for Irish waters and the data collected during this period was compiled in the Admiralty Charts with source data accreditation shown on each chart. One such source was the Merchant Marine who updated charts via the UKHO Hydrographic note (H-note) system which enabled, and still enables, chart errors and dangers to navigation to be reported by seafarers. The Royal Navy continued to survey selected Irish sea areas with Ireland's agreement, but this was *ad hoc* and localized; by 2003 bathymetric data for Dublin Bay was still accredited to Captain William Bligh, of Mutiny on the Bounty fame, and dated from the nineteenth century. The UKHO continues to produce Irish navigation charts to this day though it no longer undertakes surveys directly.

In 1996, the Irish government contracted a commercial company, Fugro, to survey the continental margins to the west of Ireland as part of a submission to the United Nations Convention on the Law of the Sea (Government of Ireland, 2005, 2009; Joint Submission, 2006). This revealed that Ireland's continental shelf extends more than 1000 km from land and that, potentially, over 800,000 km^2 of seafloor might fall within an Irish designation, an area some 10 times greater than Ireland's land mass (Figure 2.1).

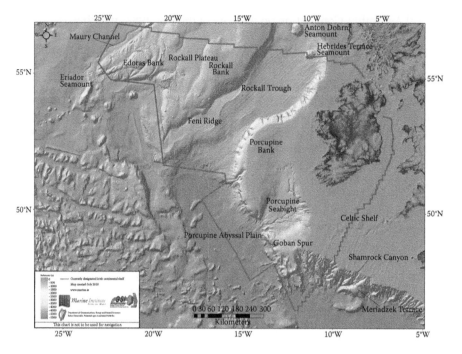

FIGURE 2.1
The 'real map of Ireland': The real map of Ireland shows Ireland's marine territory of over 220 million acres which is 10 times the size of the island of Ireland. The designated Irish Continental Shelf, delineated on the map, shows Ireland's current territorial waters, which extend out across the Atlantic Ocean and include parts of the Irish and Celtic Seas. This area also includes one of the largest marine exclusive economic zones in the European Union.

As a direct consequence of the submission to UNCLOS, and to support the application, the Irish National Seabed Survey (INSS) was established in 1999 with the objective of mapping all of Ireland's EEZ deeper than 200 m (excluding the already mapped margins) to International Hydrographic Standards (Figure 2.2). This programme was the largest scientific project ever funded in the history of the Irish State. Advice on survey strategy and data management was sought from Larry Mayer and John Hughes Clarke at the Ocean Mapping Group, University of New Brunswick in Canada while advice on programme strategy was provided by an Irish-based consultancy firm, the CSA group (now SLR Consulting). Initial surveys were carried out by a private contractor, Global Ocean Technologies Ltd. (Gotech; now ceased operating), but from 2003 the majority of survey work was carried out by INSS surveyors from aboard the Marine Institute vessel, the R.V. *Celtic Explorer* (Figure 2.3). The programme was managed by the Geological Survey of Ireland (GSI) in conjunction with the Marine Institute, and operated from 1999–2005 (The CSA Group, 2005). In addition to carrying out the largest mapping programme of an EEZ at that time and collecting 5.5 terabytes of data, the programme revealed much about Irish

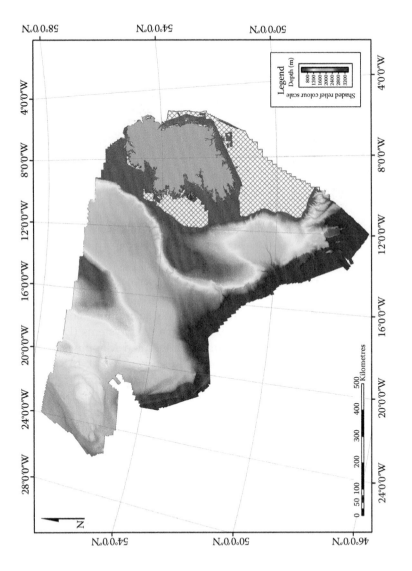

FIGURE 2.2
Coverage image, showing the extent of the Irish EEZ area which has been bathymetrically surveyed by the INSS and INFOMAR programmes (phase I) to the end of 2014. The second phase of the INFOMAR programme will see mapping of the remaining areas of the southern and western shelf and nearshore areas particularly on the east coast.

FIGURE 2.3
The largest Irish survey vessel, the R.V. *Celtic Explorer* (65.5 m) which came into service in 2003. This vessel can accommodate up to 35 personnel and endure 35 days at sea. (Courtesy of INFOMAR.)

deepwater areas including detailed information on seamounts, cold-water coral mounds, iceberg scour marks and extensive canyon systems (Dorschel et al., 2010). In 2006, the successor programme, Integrated Mapping for the Sustainable Development of Ireland's Marine Resource (INFOMAR), was launched to map inshore waters. The technologies and operational requirements encountered for INFOMAR differed from those used in the INSS in a number of aspects; in particular, shallow water systems as used in INFOMAR are capable of much higher resolution than the deepwater systems used for INSS, but there is significant increase in the time needed to achieve the same coverage due to reduced beam footprint (see Dorschel et al., 2010 for a more detailed explanation). Surveying very shallow areas is also challenging given the shoaling risks to boats and equipment, particularly when working close to shore, and the risk of disruption by other shipping. Phase I of INFOMAR ended in 2015 and saw the high spatial resolution (between 1 m and 10 m) mapping of 26 bays and 3 sea areas designated as priority. Phase II will run from 2016–2025 and proposes to survey the remaining Irish seabed areas.

A number of vessels have been deployed by the seabed mapping programme over the years with six currently in operation. These range from the largest, the R.V. *Celtic Explorer* (65.5 m) (Figure 2.3) to the mid-sized R.V. *Keary* (15.5 m) (Figure 2.4) through to the smallest vessel the R.V. *Geo* (7.4 m; http://www.infomar.ie/surveying/Vessels.php) (Figure 2.5). While larger

FIGURE 2.4
The GSI survey vessel the R.V. *Keary* (15.5 m) used in the INFOMAR programme, which can accommodate four crew members. Inset: The sonar head (Kongsberg EM2040), which is fixed below the boat to the hull. (Courtesy of INFOMAR.)

vessels have greater weather and sea state tolerance, their large drafts and turning circles make them unsuitable for shallow waters. Further discussion of choice of vessel size can be found in the next section.

2.2 Ship Borne Seafloor Mapping

2.2.1 Planning

Practical considerations for ship-based data acquisition begin with the identification of critical vessel and survey system characteristics required. Aside from scientific requisites and budgetary constraints, the ideal vessel for a given project is dependent on the anticipated water depth, weather conditions, and the endurance and self-sufficiency needed. The advantages of larger vessels are that scientific instrumentation is often permanently fixed to the vessel with systems configuration and interfacing supported by vendors (see Figure 2.4). Where this is the case, procedures for operation are often well-documented and routine (but are often bespoke to the vessel and systems installed). Smaller vessels commonly utilize less permanent fixtures

FIGURE 2.5
The GSI rigid-hulled inflatable boat (RHIB) the R.V. *Geo* (7.4 m) used in the INFOMAR programme, which boasts no accommodations. The sonar head (Reson Seabat T20-P) is attached to the boat on two arms which can be raised and lowered into the water so that it can easily be detached. (Courtesy of INFOMAR.)

of equipment (see Figure 2.5) both to facilitate regular maintenance and to enable changing configurations based on specific project requirements. Procedures on smaller ships are often less established, and operators may require greater experience/training to assess and troubleshoot performance. Nonetheless, the disadvantages of using a smaller vessel may be offset by much lower daily operating costs, greater efficiency in line turns, and greater suitability for shallow water operations. Regardless of the vessel chosen, adequate time and effort need to be allowed for the configuration and calibration of survey systems and for rigorous QA/QC of results.

During the planning stage, consideration should also be given to identifying and adopting appropriate standards. For bathymetric data, the International Hydrographic Office (IHO) provides standards applicable for all water depths (International Hydrographic Organisation, 2005, 2008). Data acquired to these standards are readily useable for traditional hydrographic charts and related products (e.g. see Chapters 3 and 15). There are no IHO standards as yet for acquisition of data to be used for marine geology and biological habitat mapping but guidelines have been produced by a number of international forums, for example, GeoHab (Lurton and Lamarche, 2015).

2.2.2 Navigation and Positioning

The problem of accurately locating and navigating ships, and especially submarines, at sea was one of the original catalysts that led to the development of the global positioning system in the United States (see Guier and Weiffenbach, 1998). For ship-based mapping, Global Navigation Satellite Systems (GNSS) positioning technologies have effectively superseded all other positioning techniques. Over most of coastal Europe and North America, freely available satellite-based augmentation systems (e.g. EGNOS, WAAS) improve the performance of GPS systems to the sub-metre level; elsewhere, nationally maintained or commercial differential GPS services provide the same precision, sufficient for real-time positioning of survey craft. Commercial GPS corrections provide even greater accuracy, with real-time navigation solutions approaching decimetre accuracies and providing sufficient vertical accuracy to reliably remove variations due to tidal state from data from waters deeper than 20 m. For shallow waters, where greater accuracy is required, high precision corrections such as real-time kinematics (RTK) GPS are used.

Software packages, such as Hypack, Reson PDS200 and QPS Qinsy (as used on Irish survey vessels) serve to translate coordinates from the GPS receiver to various points/instruments distributed throughout the ship, taking into account the vessel's heading and motion. This software also allows for the planning and guidance of survey lines/areas, fusion of the data feeds from survey instrumentation and online visualization of results. Geolocation of the survey data to the correct position on the seafloor is influenced by a number of factors. Good explanations of how this is done can be found in the IHO Manual on Hydrography (International Hydrographic Organisation, 2005).

The location of towed sensors relative to the vessel can be readily estimated using layback (cable-out) calculations or more rigorously computed by employment of ultra short baseline (USBL) instruments (Blondel and Murton, 1997, pp. 18–25). USBLs acoustically measure the range and bearing of towed instruments, for example, sidescan sonars or seabed samplers, to which a transponder has been attached. However, they require careful setup and calibration.

2.2.3 Sensors

Several sensor types are in routine use for seabed mapping. The most prominent of these are multibeam echosounders (MBES), which commonly achieve a swath of coverage of between 3 and 6 times the water depth. While MBES provide depth information primarily, the intensity of the returning signal (termed backscatter) also provides an indication of sediment composition and character. Details on their operation have been covered extensively in the literature (e.g. Clarke et al., 1996, Clarke, 2012) and several vendors provide a range of systems specialized from shallow waters to full ocean depths to all water depths. The Irish surveys have been undertaken with Kongsberg instrumentation for the most part but also R2Sonic and Reson multibeam systems. Available survey systems are continuously improving in terms of the resolution and data density they provide (Clarke, 2012). For example, at the start of the INSS programme, mapping for 200 m water depth would have been produced at a spatial resolution of circa 20 m (e.g. Thébaudeau et al., 2015). Now the resolution is circa 2 m (Blondel et al., 2015). Recent key advances include the capability to record reflectivity within the water column as well as/instead of at the seabed (allowing for the 3D mapping of fish shoals, hydrocarbon plumes, etc.) and improvements in the repeatability of backscatter readings.

While MBES systems can provide a wealth of high quality data, they also place the greatest demands on ancillary sensors in terms of the precision and accuracy required of positioning systems, motion sensors, sound-velocity profiles (SVPs) and tidal measurements, the processing capabilities required of computer resources and in terms of the increased complexity of crew training and survey procedures.

2.2.4 Processing

The processing required for a particular dataset is highly dependent on the systems used, water depths mapped, planned usage of data and on any limitations or issues encountered during acquisition. However, processing flows will, in general, comprise the following steps:

- Removal of the impacts of vessel motion by merging MBES readings with those derived from motion sensors and navigation systems, based on common timestamps.

- Conversion of raw MBES readings (commonly presented as acoustic travel times and angles) to depth values. This is achieved with reference to measured sound velocity profiles, to account for variability in the speed of sound in water resulting from thermohaline conditions (International Hydrographic Organisation, 2005).

- Application of vertical corrections resulting from tide variations and changes in vessel dynamics (such as changes in draft due to fuel loading or speed).

- Reduction and presentation of depths relative to a specified vertical datum (e.g. mean sea level). Data cleaning, to remove outliers and bad data points, is often achieved using a combination of manual and statistical methods (Calder and Mayer, 2003) and a range of visualization techniques.

While survey standards, including processing, are detailed in the IHO Standards for Hydrographic Surveys (International Hydrographic

Organisation, 2008) additional guidance, in a convenient modular format, can be found in the European Union INIS-Hydro project technical specification document (INIS Hydro, 2013) to which INFOMAR contributed (see also Hare, 2015).

Modern processing software, such as CARIS HIPS and SIPS as used by INFOMAR, in conjunction with current computer hardware, greatly facilitates the handling and processing of previously unimaginable data volumes; datasets derived from a single sortie can exceed 500 Gb in size while those from a complete survey might amount to many terabytes of data to be processed and integrated. Because of this, processing of MBES bathymetric data is commonly the most challenging and time-consuming step and one which, in the experience of the INFOMAR programme, varies considerably between deep and shallow waters.

Processing of data from deeper waters (>200 m) is relatively straightforward. Errors due to tidal variation are negligible relative to the water depths being mapped and can be readily addressed using modelled tides (e.g. POLPRED for Ireland and the UK). Variations in the thermohaline properties of the water column, captured by the SVPs, are limited and broad scale so that only sparse SVP profiles are needed. Data volumes are relatively low, enabling fast processing turnaround. For data from the shelf (50–200 m water depth) handling of tidal and thermohaline variations when processing becomes more complex. As surveying moves to shallower waters, the data will need to be supplemented with and cross-referenced to data from offshore seabed tide recorders and onshore tide gauges, or GPS systems with sufficient vertical precision, to successfully remove the impact of tide variation. In these waters, the SVP regime becomes more complex and will require sampling on a daily basis, or even more frequently dependent on observable impacts on data quality, to correct for errors and uncertainty. Data volumes increase dramatically as water depths decrease but remain manageable. For shallow waters (<50 m) however, processing of MBES data becomes much more complex and time consuming. Data volumes increase substantially and removal of 'bad' data must be performed more cautiously given the importance of correct bathymetric information for shipping. Considerably more attention is needed when applying tidal corrections and in the handling of sound velocity variations, if high quality outputs are to be achieved. Insufficient sampling of the SVP regime is often the most significant source of error in data acquired in shallow waters so it is critical that survey crews are aware of the importance of such readings and regularly assess if they are collecting sufficient readings to capture the temporal and spatial variability of the SVP regime; SVP readings may be required on an hourly or sub-hourly basis and planned lines may need to be adjusted (in terms of orientation, sequence and geographical extent) to minimize the impact of highly variable SVP regimes.

'Bad' data can arise due to MBES system characteristics, acoustic interference, aeration and biota (both in the water column and near the seafloor) but can be addressed either by manual identification and removal or using statistical methods, such as the CUBE algorithm (Calder, 2003). In most cases, a combination of approaches is needed and facilitated by available software. The effort expended in this step is partially dependent on subsequent data usage; data for hydrographic purposes in shallow water requires far greater cleaning effort than, for example, that acquired for habitat mapping in deep water. The Irish survey experience is that data collected for hydrographic purposes has approximately a 1:1 ratio of collection time against processing time for deep waters compared to 1:3 for shallow waters while data collected for biological purposes requires approximately half the processing time of that used for hydrographic mapping.

Processing also provides an opportunity to resolve issues encountered during data acquisition. These might include selecting the optimal SVP to be applied after rather than during acquisition; resolving temporary issues with tide measurements by comparison with overlapping data and crosslines; or improving the performance of motion correction by correcting the relative offsets and orientations of sensors.

2.3 Satellite Derived Bathymetry

The role of satellite remote sensing in coastal and marine management more generally is discussed in Chapter 4 of the present volume. The focus here is specifically on the use of data obtained from orbiting platforms for marine survey and mapping the seabed.

Satellite derived bathymetry (SDB) is based on two different types of measurements: radar altimetry for deep ocean waters and multispectral optical reflectances for shallow coastal areas, generally up to a maximum of 30 m water depth (Gao, 2009).

Radar altimetry provides very accurate centimetre-scale topographic measurements of the sea surface. Pulses of microwave energy (of the order of 3–38 GHz) are emitted from the orbiting platform, and are reflected off the sea surface. By measuring the return time for the signal, and comparing this to the known position of the satellite, the distance and hence the elevation of the sea surface may be computed. Ignoring meteorological and oceanographic processes, ocean surface height co-varies with the seafloor below due to local and regional variations in the Earth's gravity (see Sandwell and Smith, 1999); large positive features on the seabed, for example, an underwater mountain chain, will be denser than the surrounding water, and will result in a locally enhanced gravitational attraction. This will be seen as a bulge in the sea surface when measured from above. In contrast, large negative features such as submarine canyons will have locally reduced gravity values, and will be seen as dips in the sea surface.

Data from the first orbiting satellite systems, carried on platforms such as Seasat, GeoSat, ERS-1, and ERS-2 made it possible to identify ocean floor structures larger than 10–15 km for many remote sea areas (e.g. Smith and Sandwell, 1997). Later systems (TOPEX/Poseidon, Jason-1 and 2), gave much improved resolutions and recently Sandwell et al. (2014) were able to identify seamounts just 1–2 km wide. While satellite altimetry is unlikely to ever achieve the resolution of ship-borne systems, it offers comprehensive coverage for remoter areas and can provide an invaluable survey planning tool (Pe'Eri et al., 2014). For example, the General Bathymetric Chart of the Oceans (GEBCO), which includes satellite altimetry information in its global bathymetric dataset, is widely used by surveys, including the Irish programme, for planning.

For shallow waters the extraction of bathymetry from multispectral satellite imagery applies the principle that as light passes through water it becomes attenuated to differing degrees, depending on wavelength of the light, so that shallower areas appear brighter and deeper areas darker. As a consequence, algorithms have been developed (Lyzenga et al., 2006) to extract bathymetry based on the optical part of the light spectrum. However, water turbidity can interfere with light penetration and so ground truthing in the form of turbidity measurements is a requirement for this method and sun glint and atmospheric corrections also need to be applied. Data from various space-borne platforms (e.g. Landsat, QuickBird, SPOT and WorldView-2) have been used to derive shallow bathymetry, and the algorithms used and corrections applied have evolved.

Optical satellite imagery has proved a useful tool for predicting water depth in Irish coastal waters: Monteys et al. (2015) applied spatial statistics to satellite imagery from Dublin Bay to produce water depth predictions. While the prediction accuracies mean this method is unsuitable for navigation charts (5% of predictions differ from actual depth by +/–1 m) results obtained from these methods are useful for applications such as environmental monitoring, seabed mapping and coastal zone management. Work is ongoing to develop better models and improve predictions but these are unlikely to exceed sub-metre accuracies. In general, the advantage of satellite bathymetry is that it is timely, cost-effective, and can provide quality-controlled information for mapping, monitoring and managing coastal environments.

2.4 Recent Technological Advances

2.4.1 Unmanned Aerial Vehicles

Unmanned aerial vehicles (UAVs), also known as drones, are aircraft whose flight is controlled either remotely or autonomously by on-board computers. As a seabed survey platform they offer particular promise for the intertidal zone, sometimes called the 'white ribbon' in reference to the lack of data relative to the land and subtidal areas which bound it. Recent advances in UAV technology, specifically the emergence of survey-grade mapping drones, mean that a novel approach for addressing coastal and intertidal zone mapping may be feasible. While UAVs cannot survey submerged areas without dedicated high-cost spectral imaging systems, INFOMAR is currently exploring the possibility that, by careful planning of survey operations to coincide with low spring tides, crucial coastal zone areas can be topographically mapped to absolute accuracies of a few centimetres using UAV photogrammetry and standard RGB SLR cameras.

In September 2015, a proof-of-concept study was carried out by Greenlight Surveys (www.greenlightsurveys.com) in conjunction with the INFOMAR programme in Youghal, Co. Cork, in order to undertake an effective comparison between data generated using UAV photogrammetry and vessel-mounted sonar bathymetry. Results showed agreement with a static (fixed) difference of 0.25 m between both datasets. On the basis of this study, INFOMAR has committed to building capacity in this area through capital expenditure and a research initiative which will determine the effectiveness of this technique for large-scale regional mapping in the coastal and intertidal zone.

2.4.2 Synthetic Aperture Sonar

Synthetic aperture sonar (SAS) is an enhanced form of sideways looking sonar which represents a significant improvement in terms of resolution over conventional sonar systems. In essence, SAS works by combining successive pings coherently along a known track in order to increase the azimuth (along-track) resolution (Hayes, 2009), producing imagery of almost photographic quality. This, in turn, offers huge potential for improvements in the automatic detection and classification of seabed objects and of seabed type.

SAS systems are appropriate for targeted, high-resolution work and have been used by the military of NATO countries among others, for mine-detection (Pinto and Bellettini, 2007). They are not suitable for mapping of large areas and are, currently, very expensive compared to conventional side-scan systems.

2.4.3 Autonomous Surface Vehicles

Autonomous surface vehicles (ASVs) are surface-operating unmanned vehicles with independent propulsion, and integrated navigation systems, which can be used as platforms for acquiring data. While ASVs have been in use by the military since World War II, they have only recently become widely available commercially.

Boosted by advances in technology, particularly high-accuracy navigation systems, high precision GPS solutions (e.g. GPS-RTK) and energy systems, ASVs are ideal for carrying instrumentation that currently relies on ship-based platforms; mapping sensors such as sonar heads can easily be mounted to acquire bathymetric data. ASVs can operate independently for up to several months at a time and it may, in the future, be possible for them to operate almost indefinitely (Townsend and Shenoi, 2015). However, in practice, ASVs must be monitored because they may interfere with other shipping (Benjamin et al., 2006). They show particular promise for nearshore coastal environments and very shallow water depths as they are more efficient and carry less operational risks than conventional ship-based operations.

2.5 Seabed Mapping and Management of Ireland's Maritime Areas: Highlights and Future Challenges

The work of the Irish seabed survey has many ancillary economic and social benefits. In an external evaluation of the INFOMAR programme (Phase I) by Price Waterhouse Coopers it was estimated that, by 2029, it will have provided a return of between 4 to 6 times the programme cost (Price Waterhouse Coopers, 2008, 2013). This calculation was based on the programme's anticipated value to the commercial sector (e.g. fishing, aquaculture, biodiversity, renewable energy, energy exploration and aggregates) and the knowledge economy, in addition to the avoidance of EU fines for legal non-compliance. Less measurable, but no less real, are the programme's societal and scientific impacts. Here we have selected a couple of programme highlights to illustrate these contributions.

2.5.1 Shoals

Bathymetric shoals pose a significant hazard to shipping, particularly in harbours and their approaches. The INFOMAR programme, as part of its high resolution multibeam bathymetric surveys, makes every effort to map these areas and report uncharted shoals to UKHO for inclusion in future chart editions. This is generally done through the issue of H-notes, the system which facilitates the reporting of navigational hazards to shipping. Any mariner on discovering a danger to shipping, whether new or known but incorrectly charted, can submit a report using the H-note form which is available from the UKHO website or via the H-note app. New charts are amended to include the reported hazards once they have been evaluated by the UKHO, but older charts need to be annotated manually by their owner so the details are also included in the Notice to Mariners communications issued regularly by the UKHO to subscribers of this service. To date at least 100 H-notes have been submitted by the Irish seabed survey.

In 2013, INFOMAR surveyed Dingle Harbour and approaches in the west of Ireland. At Reenbeg Point outside the harbour entrance, a shoreward extension of the rock outcrops Crow and Colleen-oge Rocks was discovered by the survey to shoal to 0.8 m LAT (lowest astronomical tide). This differed considerably from the depth of 11.9 m at which it was charted (see Figure 2.6) thus posing a very real danger to shipping in the area. Such was the perceived threat that the UKHO issued a Radio Navigational Warning. For Ireland, these warnings are coordinated by the Marine Rescue Coordination Centre and are broadcast every three hours on Marine VHF following the Sea Area Forecast. Only the most significant and/or immediate threats are broadcast in this way but the Reenbeg shoal H-note continued to be broadcast for several weeks until it was deemed that all responsible mariners should have received and incorporated their H-note update onto their chart.

2.5.2 Shipwrecks

As an island nation, the number of shipwrecks in Irish waters is high – in the region of 13,000 according to the Shipwreck Inventory of Ireland compiled by the Underwater Archaeology Unit (UAU). INFOMAR, in cooperation with UAU, maintains a separate, more detailed geodatabase for those wrecks it has surveyed, to date over 300, some of which were known, while some have never previously been charted. The database comprises accurate coordinate information and a comprehensive description of wreck condition, extent, dimensions and water depth. This level of detail has been made possible by the migration over the survey programme from deep water systems to higher resolution, shallower water surveying and the coeval advancements in survey technologies.

During routine seafloor mapping, objects that might be wrecks are identified and checked against the UKHO H-note listing. If this confirms the likely presence of a wreck, the site is resurveyed in accordance with UKHO guidelines. These direct that the object be 'boxed in' by running four additional survey lines over the area while simultaneously logging water column data to check for protuberances such as masts which could be a shipping hazard. Onboard ship, the data are carefully evaluated by the data processor to ensure the object is a shipwreck rather than another type of seabed feature, for example, a rock outcrop.

Processing is carried out by INFOMAR using CARIS HIPS and SIPS hydrographic software and a bathymetric surface is created to an appropriate, system-dependent, vertical resolution that can be as high as 10 cm. XYZ data points are exported to Fledermaus visualization software and used to create 3D datasets giving a greater insight to the wreck, its condition and the surrounding site.

In collaboration with the UAU, INFOMAR has surveyed many significant shipwrecks, for example the RMS *Lusitania* (Figure 2.7), greatly contributing to the sum of knowledge about these wrecks and highlighting the need for protection of underwater archaeology sites (Brady et al., 2012). The majority of

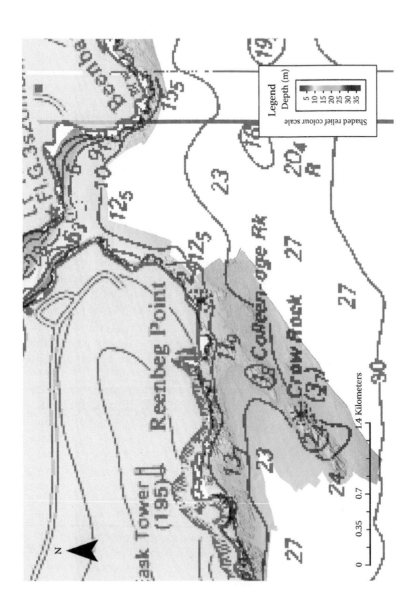

FIGURE 2.6

The results of the bathymetric survey of Dingle by INFOMAR in 2013 which identified a dangerous rock outcrop lying north east of Colleen-Oge Rock and shoaling to 0.8 m LAT overlain with the UKHO navigation chart 2789-0 on which the shoal was charted at 11.9 m depth. (Courtesy of INFOMAR with Admiralty Chart reproduced by permission of the Controller of Her Majesty's Stationery Office and the UK Hydrographic Office.)

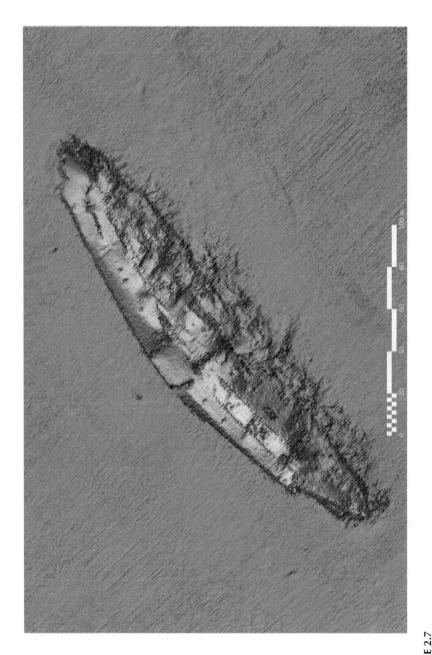

FIGURE 2.7
Image of the RMS *Lusitania* created from multibeam sonar measurements using the latest INFOMAR shallow water system, EM2040, on the R.V. *Keary* (25 cm resolution). (Courtesy of INFOMAR.)

wrecks surveyed are known but, for those that were unknown or unnamed, valuable verification information has been provided to the UAU database. This work is ongoing and it is envisaged that the shipwreck inventory will continue to benefit not only maritime heritage and shipping but other sectors such as marine tourism.

2.5.3 Habitats and Bedforms

As participants in the European project MESH (Mapping European Seabed Habitats, 2004–2008), a number of previously identified, ecologically important sites were mapped to better resolutions by INFOMAR including the Hempton Turbot Bank (Figure 2.8) located approximately 15 km off the coast of Donegal, northwest Ireland. As a 'submerged sandbank' it is designated a Special Area of Conservation under the European Union's Habitats Directive (Council of the European Communities, 1992) and has been shown to support a thriving benthic community (NPWS, 2015).

High-resolution mapping has revealed the bank to consist of mega-scale sand waves. It is permanently submerged but shoals to 15 m below the surface in water depths of 30–40 m. Sand waves can develop where there is a combination of high sediment supply and sufficiently strong currents. However, it is often the case that such bedforms have formed partly as a consequence of underlying glacial features and this is true of the Hempton Turbot Bank (Evans et al., 2015).

In addition to their ecological significance, the sand waves are notable for their extraordinary size, reaching maximum amplitudes of 20 m, wavelengths of 326 m and comprising sinuous wave crests which can extend up to 5 km (Evans et al., 2015). This is comparable, for example, to the Banner Bank formed in the exceptional tidal streams of the Bay of Fundy, Canada, which measure up to 19 m in amplitude, though the Banner Bank is associated with a headland (Li et al., 2014) and double the height of sand waves found in the mouth of San Francisco Bay which have been described as being among the largest in the world (Barnard et al., 2006).

The asymmetry displayed by the Hempton sand waves suggests migration in a southwest direction and recent repeat surveying of the area has confirmed that the sand waves are moving at rates of up to 7 m per year, implying very high sediment transport rates and very strong currents, in agreement with the modelled hydrodynamics for the area (Evans, pers. comm.).

2.5.4 Future Challenges

The Irish seabed survey has, from its inception, served the needs and interests of the Irish State and the programme is regularly reviewed for its wider impact and public engagement. As a consequence, the future challenges identified by the programme are intrinsically linked to societal challenges.

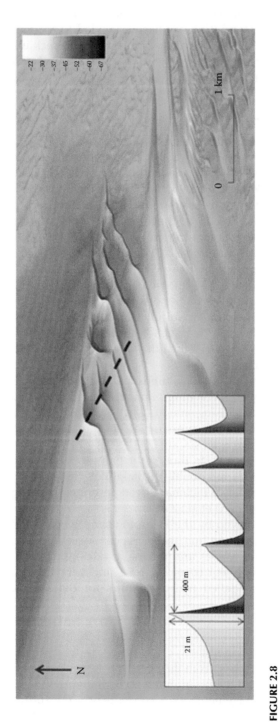

FIGURE 2.8
The Hempton Turbot Bank sand wave, shown at 3 m resolution. Inset: Profiles transect across the waves showing both their exceptionally large height and wavelengths and their asymmetry. Axes are in metres and the vertical is greatly exaggerated. (Courtesy of INFOMAR.)

Ireland has recognized that her ocean areas are an asset with vast, untapped economic potential offering significant possibilities for marine enterprises and sectors (Marine Coordination Group, 2012). The term 'Blue Growth' has been coined in Europe for this development and in 2011 the European Commission adopted a Communication on Blue Growth (European Commission, 2014) outlining the details of this potential while related reports identify that innovation in the 'blue economy' has been held back by 'a lack of information about the sea, the seabed and the life it supports' (Ecorys et al., 2012).

It is also now widely acknowledged that there are essential ecosystem services provided by the oceans, for example, food production, flood defence and absorption of CO_2 (Costanza et al., 1997; UNEP, 2006). These services are under severe threat. Globally, the pressures exerted by anthropogenic activities on the oceans, both direct and indirect, have never been greater: to date 41% of the world's oceans have been highly impacted by human activities (Halpern et al., 2008); climate change has already decreased ocean productivity, altered species distributions and increased vulnerability to disease (Hoegh-Guldberg and Bruno, 2010); and climate change-induced sea level rise threatens coastal regions, often the most populated areas.

Clearly, there is a need to underpin Blue Growth with good information and coordination whilst simultaneously protecting those ecosystem services from which we already benefit. To do so, Ireland, like many countries, is adopting an ecosystem-based management approach to marine spatial planning (MSP) defined by UNESCO as '... a public process of analysing and allocating the spatial and temporal distribution of human activities in marine areas to achieve ecological, economic and social objectives that have been specified through a political process.' Internationally, the main driver for protection, monitoring and MSP is legislation, for example the Oceans Act of Canada (Government of Canada, 2016). For northwest Europe, key legislation includes the OSPAR convention, the Habitats Directive and the Marine Strategy Framework Directive (MSFD) in particular (OSPAR Commission, 1992; Council of the European Communities, 1992, 2008).

Under the EU MSFD, each member state is required to develop a national marine strategy which includes the establishment of a monitoring programme. Underpinning this is the need for good and repeated marine mapping at a range of levels (species, habitat and ecosystem). Recognising this, Ireland has devised an Integrated Marine Plan (Marine Coordination Group, 2012) and work is currently under way to develop an Atlas of the Irish Marine Environment as a tool on which policy and protective actions can be based. The Atlas will include species and biodiversity mapping and habitat classification for which good seabed data, as derived from INSS and INFOMAR, will be essential.

Irish seabed mapping is already delivered to a wide range of marine stakeholders and feeds into existing pan-European geodatabases (e.g. EMODnet, IMAGIN, MESH, GeoHab) but delivering good bathymetric mapping is only part of the challenge for INFOMAR phase II. Realising good MSP and producing relevant seabed maps requires proper knowledge of the seafloor and thorough consideration of end-user needs. The production of 'value-added' maps such as habitat or seabed sediment is still an active area of scientific research, with new and improved methods in constant development (e.g. see Chapter 3 in the present volume). There are issues associated with resolution which need consideration; the mapping produced for a particular area may be of limited value to a particular end-user if the resolution is too low or may contain too many artefacts for a particular exercise if the resolution is too high. In addition, seabed mapping should be more properly understood as seabed prediction based on proxy measurements. This means that the prediction uncertainties for some mapping may be too great for the intended use. To address these issues, INFOMAR is engaging with stakeholders and constantly working to produce better and more relevant mapping products.

2.6 Summary

Despite being an island nation, the neglect by Ireland of her maritime assets and heritage has often been remarked upon (see, e.g. the National Maritime Museum of Ireland website http://www.mariner.ie/museum-3-2/nmmi/). That in 2003 the best available charts for Dublin Bay relied on data collected by Captain William Bligh in the very early 1800s gives some indication of the scale of the achievements of the Irish survey since the INSS programme began in 1999. Today, Ireland's extensive seabed areas are among the best surveyed in the world. Several factors have made this possible including the coincidence of a buoyant and confident Irish boom time economy with the impetus provided by UNCLOS to identify and claim the national EEZ. This has now been succeeded by the drive to meet societal challenges associated with the marine and coastal environments in particular. Underpinning this are the huge advances in geoinformatics during the last century, development which continues apace, and, crucially, the declassification and commercialisation of incipient geoinformatics technologies, opening the way for a small nation like Ireland to become a world leader in national seabed mapping.

References

Andrews, J.H., 2002. *A Paper Landscape: The Ordnance Survey in Nineteenth-Century Ireland*. Dublin, Ireland: Four Courts Press Ltd.

Barnard, P.L., Hanes, D.M., Rubin, D.M. and Kvitek, R.G., 2006. Giant sand waves at the mouth of San Francisco Bay. *Eos, Transactions American Geophysical Union*, 87(29), 285–289.

Benjamin, M., Curcio, J., Leonard, J. and Newman, P., 2006. Protocol-based COLREGS collision avoidance navigation between unmanned marine surface craft. *Journal of Field Robotics*, 23(5).

Blondel, P. and Murton, B.J., 1997. *Handbook of Seafloor Sonar Imagery*. Chichester, UK: Wiley.

Blondel, P., Prampolini, M. and Foglini, F., 2015. Acoustic textures and multibeam mapping of shallow marine habitats – Examples from Eastern Malta. In: *Proceedings of the Institute of Acoustics*, Vol. 37. Institute of Acoustics. Available online at http://opus.bath.ac.uk/47106/1/Blondel_etal_IOAProc_2015.pdf.

Bloomer, S., Monteys, X. and Chapman, R., 2010. Multifrequency classification and characterization of single beam echosounder data offshore Ireland. *Canadian Acoustics*, 38(3), 46–47.

Brady, K., McKeon, C., Lyttleton, J. and Lawler, I., 2012. *Warships, U-boats and Liners – A Guide to Shipwrecks Mapped in Irish Waters*. Department of the Arts, Heritage and the Gaeltacht, Geological Survey of Ireland and Stationery Office Dublin Ireland, Dublin.

Cadhla, S.Ó. and Cuív, É.Ó., 2007. *Civilizing Ireland: Ordnance Survey 1824– 1842: Ethnography, Cartography, Translation*. Kildare, Ireland: Irish Academic Press.

Calder, B., 2003. Automatic statistical processing of multibeam echosounder data. *Center for Coastal and Ocean Mapping*. Paper 980. Available online at http://scholars .unh.edu/ccom/980.

Calder, B.R. and Mayer, L.A., 2003. Automatic processing of high-rate, high-density multibeam echosounder data. *Geochemistry, Geophysics, Geosystems*, 4(6).

Clarke, J.E.H., 2012. Optimal use of multibeam technology in the study of shelf morphodynamics. Sediments, morphology and sedimentary processes on continental shelves: Advances in technologies, research and applications. *International Association of Sedimentologists Special Publication*, 44, 3–28.

Clarke, J.E.H., Mayer, L.A. and Wells, D.E., 1996. Shallow-water imaging multibeam sonars: A new tool for investigating seafloor processes in the coastal zone and on the continental shelf. *Marine Geophysical Researches*, 18(6), 607–629.

Costanza, R., d'Arge, R., de Groot, R., Farber, S., Grasso, M., Hannon, B., Limburg, K. et al., 1997. Ecosystem services. The value of the world's ecosystem services and natural capital. *Nature*, 387, 253–260.

Council of the European Communities, 1992. Council Directive 92/43/EEC of 21 May 1992 on the conservation of natural habitats and of wild fauna and flora. *Official Journal of the European Communities, series L*, 206(1992), 7–50.

Council of the European Communities, 2008. Directive 2008/56/EC of the European Parliament and of the Council of 17 June 2008 establishing a framework for community action in the field of marine environmental policy (Marine Strategy Framework Directive). *Official Journal of the European Union, Series L*, 164/19 25.6.2008.

Doel, R.E, Levin, T.J. and Marker, M.K., 2006. Extending modern cartography to the ocean depths: Military patronage, Cold War priorities, and the Heezen-Tharp mapping project, 1952–1959. *Journal of Historical Geography* 32, 605–626.

Dorschel, B., Wheeler, A.J., Monteys, X. and Verbruggen, K., 2010. *Atlas of the Deep-Water Seabed: Ireland*. New York: Springer Science and Business Media.

Ehler, C. and Douvere, F., 2009. *Marine Spatial Planning: A Step-by-Step Approach Toward Ecosystem-Based Management*. Intergovernmental Oceanographic Commission and Man and the Biosphere Programme. IOC Manual and Guides No. 53, ICAM Dossier No. 6. Paris: UNESCO.

Elmendorf, C.H. and Heezen, B.C., 1957. Oceanographic information for engineering submarine cable systems. *Bell System Technical Journal*, 36(5), 1047–1093.

European Commission, 2014. Communication from the Commission to the European Parliament, The Council, The European Economic and Social Committee and the Committee of the Regions, Innovation in the Blue Economy: Realising the potential of our seas and oceans for jobs and growth. *COM(2104) 254 final/2*. Available online at http://eur-lex.europa.eu/legal-content/EN/TXT/PDF/?uri =COM:2014:254:REV1andfrom=EN.

Ecorys, Deltares and Oceanic Developpement, 2012. Blue growth: Scenarios and drivers for sustainable growth from the oceans, seas and coasts. Available online at http://ec.europa.eu/maritimeaffairs/documentation/studies/documents /blue_growth_third_interim_report_en.pdf.

Evans, W., Benetti, S., Sacchetti, F., Jackson, D.W., Dunlop, P. and Monteys, X., 2015. Bedforms on the northwest Irish Shelf: Indication of modern active sediment transport and over printing of paleo-glacial sedimentary deposits. *Journal of Maps*, 11(4), 561–574.

Gao, J., 2009. Bathymetric mapping by means of remote sensing: Methods, accuracy and limitations. *Progress in Physical Geography*, 33(1), 103–116.

Government of Canada, 2016. Oceans Act (S.C. 1996, c.31). Published online at http://laws-lois-justice.gc.ca/eng/acts/o-2.4/page-1.html#h-1.

Government of Ireland, 2005. Ireland Submission to the Commission on the Limits of the Continental Shelf pursuant to Article 76, paragraph 8 of the United Nations Convention on the Law of the Sea 1982 in respect of the area abutting the Porcupine Abyssal Plain. Available online at http://www.un.org/depts /los/clcs_new/submissions_files/irl05/irl_exec_sum.pdf.

Government of Ireland, 2009. Ireland Submission to the Commission on the Limits of the Continental Shelf pursuant to Article 76, paragraph 8 of the United Nations Convention on the Law of the Sea 1982 in respect of the area abutting the Hatton-Rockall Area. Available online at http://www.un.org/depts/los /clcs_new/submissions_files/irl09/irl09_exsum.pdf.

Guier, W.H. and Weiffenbach, G.C., 1998. Genesis of satellite navigation. *John Hopkins APL Technical Digest*, 19(1). Available online at http://www.jhuapl.edu/techdi gest/td/td1901/guier.pdf.

Halpern, B.S., Walbridge, S., Selkoe, K.A., Kappel, C.V., Micheli, F., D'Agrosa, C., Bruno, J.F. et al., 2008. A global map of human impact on marine ecosystems. *Science*, 319(5865), 948–952.

Hare, R., 2015. Depth and position error budgets for mulitbeam echosounding. *The International Hydrographic Review*, 72(2).

Hayes, M.P. and Gough, P.T., 2009. Synthetic aperture sonar: A review of current status. *IEEE Journal of Oceanic Engineering*, 34(3), 207–224.

Heezen, B.C., Tharp, M. and Ewing, M., 1959. The floors of the oceans I. The North Atlantic. *Geological Society of America Special Papers*, 65, 1–126.

Hoegh-Guldberg, O. and Bruno, J.F., 2010. The impact of climate change on the world's marine ecosystems. *Science*, 328(5985), 1523–1528.

Höhler, S., 2002. Depth records and ocean volumes: Ocean profiling by sounding technology, 1850–1930. *History and Technology*, 18(2), 119–154.

INIS Hydro, 2013. INIS Hydro seabed mapping technical specification. Available online at http://www.inis-hydro.eu/INIS_HYDRO_TechSpec_v1.pdf.

International Hydrographic Organisation, 2005. *Manual on Hydrography*. 1st ed. International Hydrographic Bureau, Monaco. Available online at https://www .iho.int/iho_pubs/CB/C13_Index.htm.

International Hydrographic Organisation, 2008. IHO standards for hydrographic surveys, *Special Publication No. 44*. International Hydrographic Bureau, Monaco. Available online at https://www.iho.int/iho_pubs/standard/S-44_5E.pdf.

Joint Submission, 2006. Joint Submission to the Commission on the Limits of the Continental Shelf pursuant to Article 76, paragraph 8 of the United Nations Convention on the Law of the Sea 1982 in respect of the area of the Celtic Sea and the Bay of Biscay. Available online at http://www.un.org/depts/los/clcs_new /submissions_files/frgbires06/joint_submission_executive_summary_english.pdf.

Li, M.Z., Shaw, J., Todd, B.J., Kostylev, V.E. and Wu, Y., 2014. Sediment transport and development of banner banks and sandwaves in an extreme tidal system: Upper Bay of Fundy, Canada. *Continental Shelf Research*, 83, 86–107.

Lurton, X. and Lamarche, G., 2015. Backscatter measurements by seafloor-mapping sonars. Marine Geological and Biological Habitat Mapping. A collective report by members of the GeoHab Backscatter Working Group (May).

Lyzenga, D.R., Malinas, N.P. and Tanis, F.J., 2006. Multispectral bathymetry using a simple physically based algorithm. *IEEE Transactions on Geoscience and Remote Sensing*, 44(8), 2251.

Marine Coordination Group, 2012. *Harnessing our Ocean Wealth. An Integrated Marine Plan for Ireland*. Irish Government. Available online at http://www.ourocean wealth.ie/sites/default/files/sites/default/files/Harnessing%20Our%20 Ocean%20Wealth%20Report.pdf.

Matkin, J. and Rehbock, P.F., 1992. *At Sea with the Scientifics: The Challenger Letters of Joseph Matkin*. Honolulu, HI: University of Hawaii Press.

Monteys, X., Harris, P., Caloca, S. and Cahalane, C., 2015. Spatial prediction of coastal bathymetry based on multispectral satellite imagery and multibeam data. *Remote Sensing*, 7(10), 13782–13806.

Murray, J., 1895. *A Summary of the Scientific Results Obtained at the Sounding, Dredging and Trawling Stations of HMS Challenger*, Vol. 1. London, England: HM Stationery Office.

NPWS, 2015. Hempton's Turbot Bank SAC (Site Code: 2999) Conservation objectives supporting document – Marine habitats. Available online at http://www.npws .ie/protected-sites/sac/002999.

OSPAR Commission, 1992. *OSPAR Convention for the Protection of the Marine Environment of the North-East Atlantic*. Available online at http://www.ospar .org/site/assets/files/1290/ospar_convention_e_updated_text_in_2007_no _revs.pdf.

Pe'Eri, S., Parrish, C., Azuike, C., Alexander, L. and Armstrong, A., 2014. Satellite remote sensing as a reconnaissance tool for assessing nautical chart adequacy and completeness. *Marine Geodesy*, 37(3), 293–314.

Pinto, M. and Bellettini, A., 2007. *Shallow Water Synthetic Aperture Sonar: An Enabling Technology for NATO MCM Forces.* (Re-print series) NRC-PR-2007-10. UTD Europe, Undersea Defence Technology Europe, Naples, Italy.

Price Waterhouse Coopers, 2008. Options Appraisal Report: Final Report. Available online at http://www.infomar.ie/publications/Reports.php.

Price Waterhouse Coopers, 2013. INFOMAR External Evaluation. Available online at http://www.infomar.ie/documents/2013_PwC_Infomar_Evaluation_Final .pdf.

Reidy, M.S. and Rozwadowski, H.M., 2014. The spaces in between: Science, ocean, empire. *Isis*, 105(2), 338–351.

Sandwell, D.T., Müller, R.D., Smith, W.H., Garcia, E. and Francis, R., 2014. New global marine gravity model from CryoSat-2 and Jason-1 reveals buried tectonic structure. *Science*, 346(6205), 65–67.

Sandwell, D.T. and Smith, W.H., 1999. Bathymetric estimation. Satellite altimetry and earth sciences. In: *Satellite Altimetry and Earth Sciences: A Handbook of Techniques and Applications* (Vol. 69). Fu, L.L. and Cazenave, A. Eds. San Diego, CA: Academic Press, Chap. 12, pp. 441–459.

Smith, W.H. and Sandwell, D.T., 1997. Global sea floor topography from satellite altimetry and ship depth soundings. *Science*, 277(5334), 1956–1962.

The CSA Group, 2005. An assessment of the optimal use and application in the immediate to longterm future of the Irish National Seabed Survey deliverables to date (Phase 1) and to contribute towards the development of Phase 2. Available online at http://www.infomar.ie/documents/Assessment_of_the_INSS.pdf.

Thébaudeau, B., Monteys, X., McCarron, S., O'Toole, R. and Caloca, S., 2015. Seabed geomorphology of the Porcupine Bank, West of Ireland. *Journal of Maps*, 1–12.

Townsend, N.C. and Shenoi, R.A., 2015. Feasibility study of a new energy scavenging system for an autonomous underwater vehicle. *Autonomous Robots*, 1–13.

UN Division for Ocean Affairs and the Law of the Sea, 2015. Chronological list of ratifications of, accessions and successions to the Convention and Related Agreements. Available online at http://www.un.org/Depts/los/reference_files/chronological _lists_of_ratifications.htm [accessed 4 February 2016].

UNEP, 2006. Marine and Coastal Ecosystems and Human Wellbeing: A Synthesis Report Based on the Findings of the Millennium Ecosystem Assessment. UNEP.

UN General Assembly, Convention on the Law of the Sea, 10 December 1982. Available online at http://www.un.org/depts/los/convention_agreements /texts/unclos/unclos_e.pdf [accessed 4 February 2016].

Vilming, S., 1998. The development of the multibeam echosounder: A historical account. *J. Acoust. Soc. Am.*, 103, 2935.

3

A GIS-Based Application of Drainage Basin Analysis and Geomorphometry in the Submarine Environment: The Gollum Canyon System, Northeast Atlantic

Paul K. Murphy and Andrew J. Wheeler

CONTENTS

3.1 Introduction

The Gollum Canyon System (GCS) – more commonly referred to as the Gollum Channel System (Kenyon et al., 1978; Beyer et al., 2003, 2007; Wheeler et al., 2003; Van Rooij, 2004) – lies approximately 100 km off the southwest coast of Ireland, within the Porcupine Seabight (Figure 3.1). First identified by Berthois and Brenot (1966), the east–west orientated dendritic canyon system incises significantly into the seafloor for a distance of over 200 km. It lies in water depths of between 200 m to 3500 m, and is surrounded on four sides by continental shelf, with a restricted opening out onto the Porcupine Abyssal Plain in the southwest (Akmetzhanov et al., 2003). Downslope sedimentary processes within the canyon system were most active during glacial low-stand periods via turbidites and other mass wasting events, while interglacial periods have exhibited little or no sediment transfer activity (Van Rooij, 2004). Direct observations show contemporary sedimentary activity within the system to be minimal, with tidal currents and northward

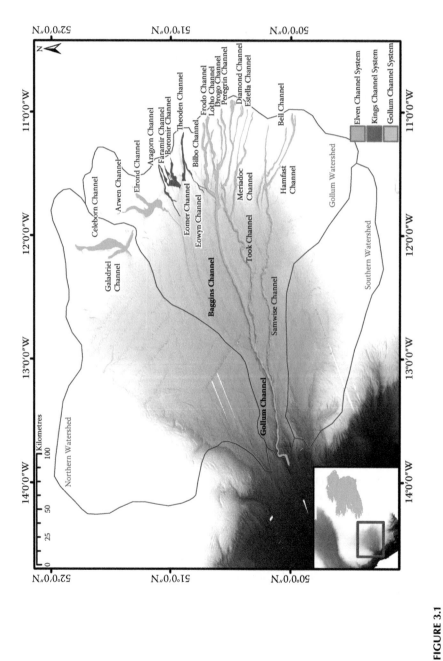

FIGURE 3.1
Channel nomenclature within the Porcupine Seabight (Murphy, 2015). (From INSS/PAD.)

flowing contourites reworking the biogenic, turbiditic and hemipelagic seabed (Tudhope and Scoffin, 1995; Wheeler et al., 2003; Van Rooij, 2004).

The GCS is composed of a series of anastomosing and individually named canyons, the dimensions of which vary significantly throughout (Figure 3.1). Previous knowledge of the system was based solely on low frequency regional side-scan sonar (see Kenyon et al., 1978), pioneering multibeam surveys (see Beyer et al., 2003) and a small-scale higher frequencey side-scan sonar survey (see Wheeler et al., 2003). These predominantly concentrated on the canyon heads of the northern-most channels (e.g. Beyer et al., 2003; Wheeler et al., 2003; Van Rooij, 2004) with fewer studies located on the deeper sections (e.g. Kenyon et al., 1978; Wheeler et al., 2003), leaving a significant proportion of the system unexplored. Between 2000 and 2002, the Irish National Seabed Survey (INSS), managed by the Geological Survey of Ireland (GSI-DCENR©), mapped 81% of Ireland's seabed territory using multibeam bathymetry (see Chapter 2 by Scott et al. of this volume), giving near-complete coverage of the GCS (Dorschel et al., 2010). The collection of these data – at a pixel resolution of approximately 25 m – provides great opportunity for scientific investigation into Ireland's offshore environment. To date, a number of projects have used the INSS data for various purposes including geomorphological (Dunlop et al., 2010, Sacchetti et al., 2011) and habitat sensitivity mapping (Guinan et al., 2009; Rengstorf et al., 2013), but its full potential has not yet been fully realized.

Previous studies into the GCS have only provided keyhole perspectives into what is potentially a vast system lying deep beneath the ocean surface. The project described here harnesses the potential of the INSS/PAD multibeam bathymetry, with the help of various toolsets within ArcGIS® (release 10.2) from Esri, to investigate the Porcupine Seabight and specifically the GCS. The concepts of geomorphometry are applied, in order to access the wealth of information contained within the data. Geomorphometry – the science of quantitative land surface analysis (Pike et al., 2009) – was originally designed for, and is most commonly applied to, terrestrial investigations, generally using digital elevation models. Many of the methods used in these investigations, however, have also proven applicable in submarine environments (Peakall et al., 2000).

Pike et al. (2009) refer to two modes of geomorphometry; namely, general and specific geomorphometry. General geomorphometry treats a continuous land surface as one entity, and in this instance it was used to define the true drainage network of the GCS. Specific geomorphometry, on the other hand, is only concerned with distinct features present on the surface and was used to define various properties of the GCS's main channels. This case study focuses on the application of GIS to marine geomorphometric studies. Analysis is undertaken on ArcMap 10.2 using the ArcGIS Spatial Analyst Tools, in addition to 3D visual analysis on ArcScene 10.2. More detailed background information, results and discussions are available in Murphy (2015) and Murphy et al. (subm.).

3.2 Data Preparation

Before analysis can be conducted, preparation and cleaning of the data is crucial to ensure the most accurate results are achieved. This investigation primarily uses the INSS dataset, although the deepest sections of the GCS are not covered. To provide complete analysis of the system, a multibeam dataset collected in 1996 by Fugro, in a project managed by the Petroleum Affairs Division (PAD), was also used. This data was collected at a much lower spatial resolution than that of the INSS, gridded at approximately 110 m. This resolution is not ideal for morphological analysis, but nevertheless provides valuable information on the deepest sections of the system which is otherwise unavailable.

Due to the size of the INSS survey area – at approximately 432,000 km^2 – the data is broken up into numerous 2° by 2° raster tiles, with 6 of these making up the GCS and its potential catchment area (Figure 3.2a). These tiles were identified and subsequently mosaicked together using the 'mosaic to new raster' tool from the ArcToolbox, in order to create a single seamless raster of the area (Figure 3.2b). The PAD dataset was received from INFOMAR's IWDDS service as one large file (Figure 3.2c). In order to minimize processing time and the amount of irrelevant data analysed, the 'Extract by Rectangle' tool was used. A rectangular vector file with vertices 51°N/15°W and 49°N/11°W was created and used as the control area for the extraction (Figure 3.2d), subsequently creating the dataset seen in Figure 3.2e which covers part of the Porcupine Seabight and Porcupine Abyssal Plain immediately to the southeast.

The 'Mosaic to New Raster' tool was then used again to combine the new INSS and PAD subsets in order to create a full dataset for the area of interest (Figure 3.2f, see also Figure 3.3 with hillshade effect). It is important to note that the 'First Mosaic Operator' was used in the workflow. This allowed higher resolution INSS pixels to retain their true values when overlapping areas were being combined in the programme, while the lower resolution PAD data filled gaps to provide full coverage of the study area.

Once combined, a visual inspection of the data identified some issues. In several locations there is no bathymetric information – indicated by white streaks – which also act as spurious drainage sinks in flow accumulation processing. In addition to this, a significant amount of noise is present over the entire study area, especially obvious when the hillshade or slope effects are used (Figure 3.4a and b, respectively). In these, the systematic and parallel tracklines of the survey vessel are quite obvious, with significant miscalculations of water depth occurring at the limits of multibeam swathe coverages.

The missing data and noise are both due to inadequate calibration of water sound velocity profiles during multibeam data acquisition leading to data rejection or retention with processing artefacts on the outer beams,

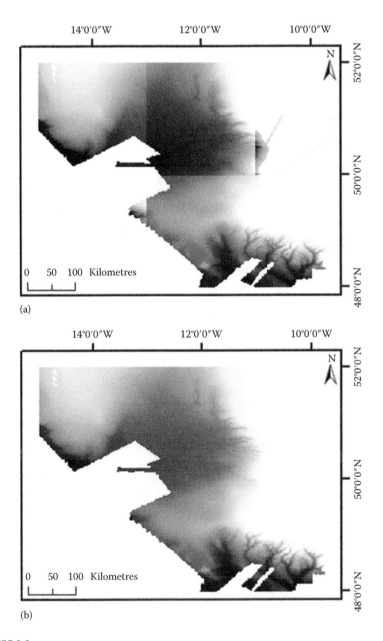

FIGURE 3.2
Steps in data preparation for drainage analysis of the Porcupine Seabight. (a) INSS data tiles at 25 m resolution. (b) Mosaicked INSS data at 25 m resolution. *(Continued)*

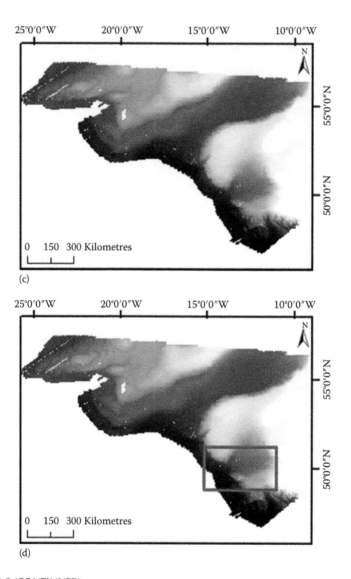

FIGURE 3.2 (CONTINUED)
Steps in data preparation for drainage analysis of the Porcupine Seabight. (c) PAD dataset at
110 m resolution. (d) PAD dataset with polygon for extraction of raster data.   (*Continued*)

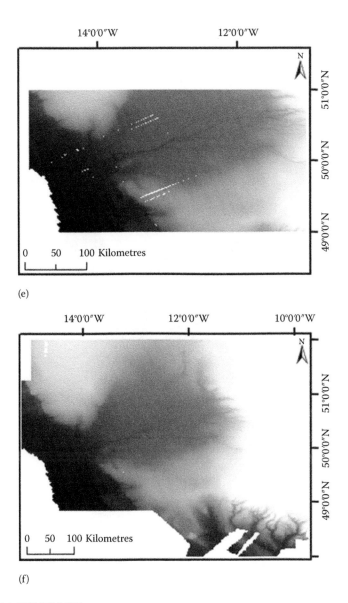

(e)

(f)

FIGURE 3.2 (CONTINUED)
Steps in data preparation for drainage analysis of the Porcupine Seabight. (e) Extracted area of PAD data, 110 m resolution. (f) Mosaicked INSS and PAD data, with the higher resolution INSS as the mosaic operator at overlap areas. (From INSS/PAD.)

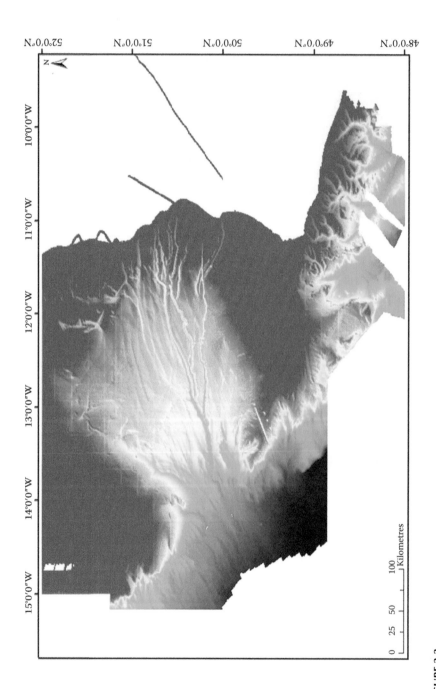

FIGURE 3.3
Mosaicked INSS and PAD dataset with hillshade effect, covering full study area. (From INSS/PAD.)

FIGURE 3.4
Hillshade (a) and slope (b) datasets of the Porcupine Seabight showing the significant noise within the data, where systematic and parallel tracklines can be seen. Scale and orientation are equal for both (a) and (b). (From INSS/PAD.) *(Continued)*

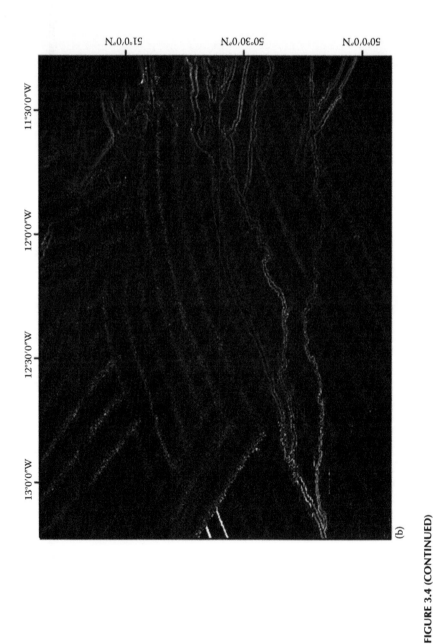

FIGURE 3.4 (CONTINUED)
Hillshade (a) and slope (b) datasets of the Porcupine Seabight showing the significant noise within the data, where systematic and parallel tracklines can be seen. Scale and orientation are equal for both (a) and (b). (From INSS/PAD.)

and should be considered within the context of the relative infancy of the data processing tools at the time of collection (F. Sacchetti, 2015, pers. comm., 9th June). The use of spatial filters reduced noise to some extent; however, not to the point that slope datasets could be used for any meaningful drainage analysis without additional processing of either the raw data or the outcome of the drainage analyses. Two methods are available to deal with this noise (B. Dorschel, 2015, pers. comm., 26th May). The first involves the complete reprocessing of the raw data files for the survey area, but was however beyond the scope of this project. The second option is manual cleaning of flow accumulation datasets, outlined in the section below.

3.2.1 General Geomorphometry

Visual inspection of the INSS and PAD datasets (Figure 3.3) provides an overview of the drainage network present, although the amount of information that can be derived is very limited. GIS packages, such as ArcGIS, provide a diverse suite of spatial analysis tools which can be harnessed to provide much more information than can be derived manually. General geomorphometry is used here for the automatic extraction of the drainage networks within the study area, similar to those done previously within terrestrial investigations (e.g. Jenson, 1991; Saraf et al., 2004; Magesh et al., 2012, 2013). The same methods as used in these land-based studies were employed for this project in order to identify any subtle drainage expressions on the seabed which may not be possible with visual analysis.

The methodology used for the extraction of channel networks is similar to that done by Tubau et al. (2013) and follows the workflow set out in Figure 3.5. The hydrology toolset in ArcMap 10.2 was applied to the multibeam data of Figure 3.3. Firstly, the 'Fill' tool was used to remove any sinks in the data and subsequently the 'Flow Direction' tool was applied, generating a new dataset which displays the direction of flow of each pixel in an eight direction pourpoint model (Magesh et al., 2012) (Figure 3.5b). The 'Flow Accumulation' tool was then used to generate a dataset exhibiting a drainage network based upon the 'Flow Direction' and the amount of pixels flowing downslope through a pixel (Figure 3.5c). Various threshold values can be set depending on the density of the drainage network believed to be present (Figure 3.5d), with higher values requiring a greater number of pixel values to be directed through a single pixel for it to register. The validity of the lower threshold values, for example 1000 (Figure 3.6a), may be questionable. The density of the network revealed shows many features not noted in previous surveys, with the danger that this low threshold may be generating drainage networks significantly influenced by data noise to the extent that they may not exist in reality. The tendency for the drainage network to show a general trend parallel to the survey tracks is especially suspicious. In contrast, the 50,000 pixel threshold value (Figure 3.6b) does not appear to extract the drainage network sufficiently enough, with many of the larger known channel heads not being identified.

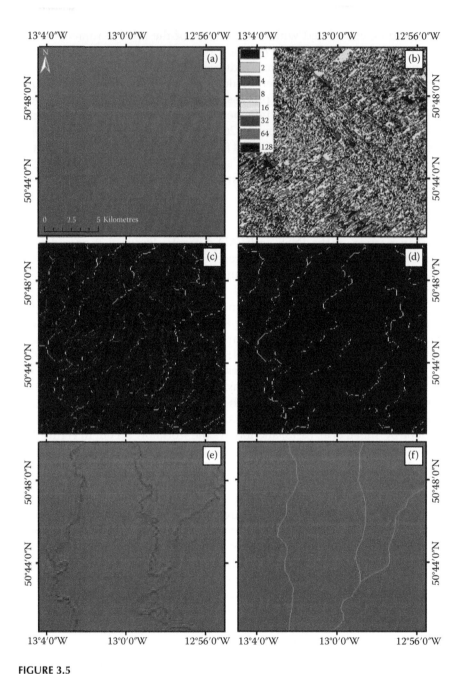

FIGURE 3.5
Automated drainage extraction workflow. (a–f) Cover the same area. (a) Fine scale area of INSS/
PAD multibeam mosaic. (b) Flow direction raster, with 8 direction pourpoint model. (c) Raw
flow accumulation raster. (d) Flow accumulation raster, selecting 10,000 pixel threshold value.
(e) Vector dataset of extracted 10,000 pixel threshold values showing raw drainage network.
(f) Manually edited drainage network to correct noise interference. (From INSS/PAD.)

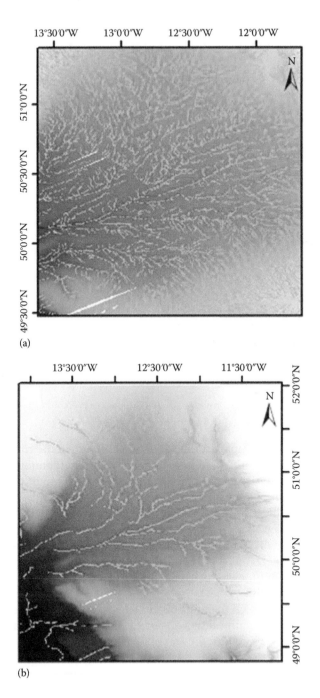

FIGURE 3.6
Flow accumulation thresholds. (a) 1000 pixels. (b) 50,000 pixels. *(Continued)*

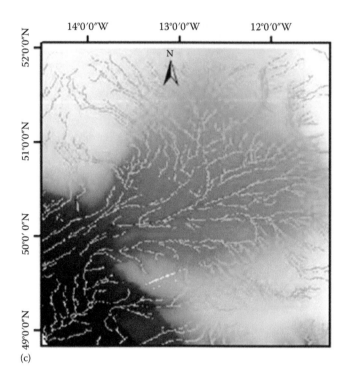

(c)

FIGURE 3.6 (CONTINUED)
Flow accumulation thresholds. (c) 10,000 pixels. (From INSS/PAD.)

The 10,000 pixel threshold value (Figure 3.6c) seems to be the most accurate when compared to the results obtained by Akhmetzhanov et al. (2003) and Van Rooij (2004). Figure 3.7 shows a larger-scale view of the mouth of the seabight at this pixel threshold, allowing a comparison between it and the channel identification done by Akhmetzhanov et al. (2003). Good correlations can be identified between both, with any differences being due to holes in the dataset on the continental rise. In addition to this, the flow accumulation dataset picks out all the channels identified by Van Rooij (2004) as well as additional channels which were identified in the preliminary manual drainage analysis of this project.

Due to inherent noise on the multibeam datasets, problems also arose with the subsequent flow accumulation raster files. It was found that at several locations, the noise created by the tracklines acted as preferential pathways for the flow network, commonly flowing either in parallel or normal to the tracklines and augmenting the flow pattern seen in the Porcupine Seabight (Figure 3.8). To remove this error in the results, manual cleaning of the data was done.

Firstly, a conditional statement was created in the 'Raster Calculator' tool and applied to the 10,000 pixel threshold values in the flow accumulation

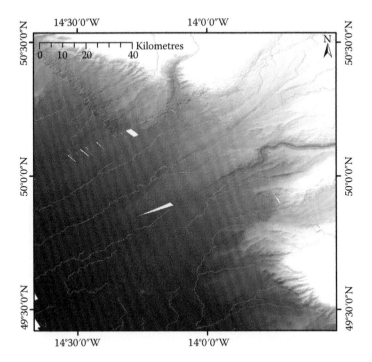

FIGURE 3.7
Verification of the 10,000 pixel threshold flow accumulation, which was compared to observations of Akhmetzhanov et al. (2003). (From INSS/PAD.)

raster. This action isolated the pixels meeting the terms of the statement, subsequently creating a new vector layer composed of a series of polylines which could then be edited as normal (Figure 3.5e,f). The editing involved the removal or altering of a channel's course, if noise in the data was determined to have affected it (Figure 3.9). The complete or partial removal of flow lines occurred where a significant proportion of the channel followed a trackline. Altering the course of a flow line was only done in locations where the source, pourpoint and evidence of channel presence could be identified; otherwise, the existence of the channel could not be verified and was therefore removed from the analysis. The result of this can be seen in Figure 3.10, representing the first ever published image of the true drainage network of the Porcupine Seabight as a whole.

Caution should be taken when analysing the results of the flow accumulation datasets presented in Figure 3.10. The use of ArcMap's hydrology tools have proven to be effective in the terrestrial environment on numerous occasions (e.g. Magesh et al., 2013) but its validity is not yet proven for submarine environments, with Tubau et al. (2013) the only previous known example issuing a similar caution. The sound velocity artefacts are significant in the data presented and therefore assessment of the network may be over- or

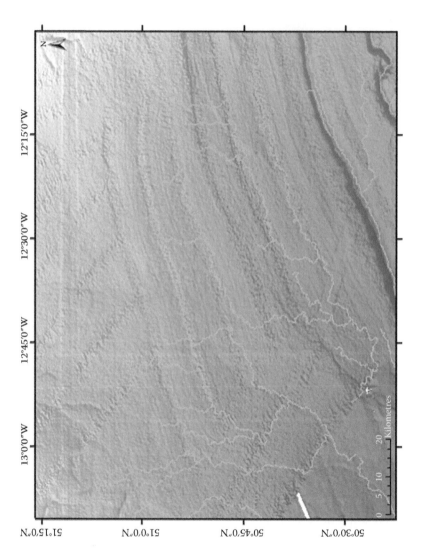

FIGURE 3.8
Interference of trackline noise with flow accumulation. Extracted channels tend to flow parallel and at right angles through the noise augmenting the true flow pattern of the seabight. (From INSS/PAD.)

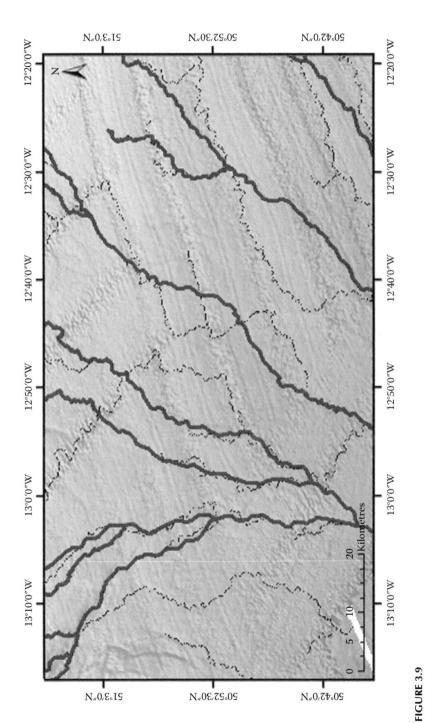

FIGURE 3.9
Editing of flow lines. Black lines represent unaltered automatic flow accumulation data, while the red represent the edited polylines. (From INSS/PAD.)

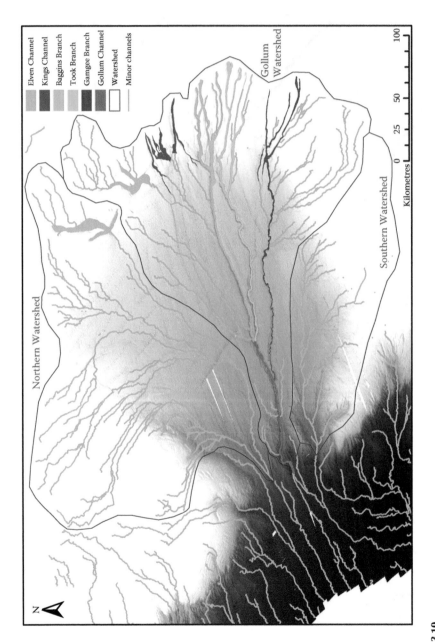

FIGURE 3.10
Completed flow accumulation for the Porcupine Seabight. (From INSS/PAD.)

underestimated. In addition, the resolution of the INSS data is 25 m and the PAD data is 110 m; therefore, features smaller than these are not detected but may potentially contribute to the drainage network.

3.2.2 Specific Geomorphometry

This section deals with the quantitative land surface analysis of the main channels within the GCS. The overall aim is to provide a mathematical description of the main channels, primarily using longitudinal and transverse profiles. This detailed analysis of the topographic features gives greater insight into the processes and controls that affect their morphology and also allows the GCS to be compared with other systems worldwide. The methodologies used are also adapted from terrestrial investigations (e.g. Strahler, 1957; Chopra et al., 2005; Pareta and Pareta, 2011) although specific geomorphometry approaches in the submarine environment are more established than the previous general investigations, with numerous examples from around the globe (e.g. Flood and Damuth, 1987; Klaucke et al., 1998; Nakajima et al., 1998; Babonneau et al., 2002; Estrada et al., 2005; Tubau et al., 2013).

A study of the literature reveals the most commonly measured variables in submarine channel analysis and those most applicable to the GCS were selected. A total of 12 variables are selected and presented in Table 3.1.

The variables in Table 3.1 were investigated using the 'Interpolate Line' tool in ArcMap's 3D Analyst extension, and applied to the mosaicked Porcupine Seabight bathymetry data (Figure 3.3). This tool produces a 2D graph, where the Y axis plots the magnitude of pixels across the transect with magnitude equated with water depth in metres. Whereas the general geomorphometic analysis is based on the WGS 1984 coordinate system, better for the representation of large areas where data covers several degrees of latitude, this is less applicable on a small scale. For small scale analysis, the Irish Transverse Mercator system was used where the x axis can be represented in metres. Data analysis of the graphs produced by ArcMap is limited to visual interpretation. For more rigorous analysis, the data were extracted from each graph and transferred to Microsoft Excel, which allows for more precise calculations to be carried out. In addition to the 3D analyst within ArcMap, ArcGlobe was also used to aid the visualization of the channel morphology and any evidence of potential processes or controls in the surrounding environment (e.g. Figure 3.11). The use of the 3D projections, as well as the 2D profiles, was extremely effective in generating a greater understanding of the study area.

Transverse profiles were taken across the main channels in the Porcupine Seabight (e.g. Figure 3.12). Each profile was taken perpendicularly to the local channel path, at the start and end of a channel as well as any location along its course where a significant change in morphology or dimension occurred. This was done in order to effectively characterize the evolution of channel morphology downstream.

TABLE 3.1

Channel Morphometric Variables

Derived Geomorphic Variable	Definition
From transverse profiles:	
Channel width	The perpendicular distance from the right to left levee, or break in slope in the cases where no levees are present
Channel depth	The average vertical distance from the base of the channel to the top of the levees or surrounding seabed
Thalweg width	The width of the thalweg defined as 'the line of greatest depth' (Whittemore Boggs, 1937) or the width of a flat aggraded floor (Babonneau et al., 2002).
Width: depth ratio	The width: depth ratio, although not accounted for in many studies, is seen to be an important factor which affects secondary circulation and overbank flow within submarine channels (Keevil et al., 2007), as well as being helpful in the differentiation of canyons from channels
From longitudinal profiles:	
Channel length	The length of the thalweg profile down the axis of the channel
Channel slope[a]	The average gradient of the channel: $\tan^{-1}\left(\dfrac{rise}{run}\right)$
Reach length	The length of the thalweg profile down the axis of the channel between significant breaks of slope
Reach slope[a]	The average gradient of the channel between significant breaks of slope: $\tan^{-1}\left(\dfrac{rise}{run}\right)$
Valley[b] length	The straight line distance between the end points of a channel or reach (Flood and Damuth, 1987)
Valley[b] slope[a]	The gradient along the straight line between the end points of a channel or reach: $\tan^{-1}\left(\dfrac{rise}{run}\right)$
Meander wavelength	The number of meanders in a reach divided by the valley length. This characteristic is important as it generally shows a significant positive correlation with channel width (Wynn et al., 2007)
Sinuosity	A measure of channel deviation from a straight line. A channel is meandering if it is greater than or equal to 1.5 (Leopold and Wolman, 1957), sinuous if it is between 1.5 and 1.15 (Keevil et al., 2007) and straight if it is less than 1.15. A perfectly straight channel has a sinuosity of 1, where the channel and valley lengths are equal (Flood and Damuth, 1987)

(Continued)

TABLE 3.1 (CONTINUED)

Channel Morphometric Variables

Derived Geomorphic Variable	Definition
Peak sinuosity	The maximum sinuosity achieved in a channel or reach (Clark et al., 1992; Peakall et al., 2011)
Thalweg and levee depth profiles	

[a] The slope of each channel, reach and valley are calculated by different methods throughout the literature. Flood and Damuth (1987) and Pirmez and Flood (1995) provide slope data in metres of rise per kilometre of run, Peakall et al. (2011) provide it as a decimal in the form rise/run, Clark et al. (1992) and Klaucke et al. (1998) provide it as a ratio in the form rise:run, while Nakajima et al. (1998), Estrada et al. (2005), Budillon et al. (2011) and Tubau et al. (2013) provide the slope in degrees. Therefore, in order to make this project compatible with previous studies, each of the formats above were calculated however degrees were chosen as the preferred unit of measurement, as it is the most common throughout the literature.

[b] The term 'valley' is also used as a morphometric parameter, despite the fact that submarine channels are not generally confined to any large physical valleys. The main purpose of this parameter is for the calculation of sinuosity, which is equal to the ratio of a reach's length to its valley length.

Longitudinal – or axial – profiles were taken along the thalweg of the main channels within the PSB (Figure 3.13). Within the investigation, the thalweg was taken as the deepest point of a V- or U-shaped channel, or the median line within channels that were determined to have a flat floor. The parameters measured from the longitudinal profiles taken include the lengths and slopes of channels and valleys, the meander wavelength, the sinuosity (Flood and Damuth, 1987) and the peak sinuosity (Clark et al., 1992) (Table 3.1). In addition to the overall measurements of the main channels, the lengths and slopes of 'channel reaches' were also calculated. Channel reaches refer to sections of a channel's course along which finer scale measurements are taken. Flood and Damuth (1987) – as well as numerous subsequent studies (e.g. Kolla et al., 2001; Wynn et al., 2007) – defined a channel reach as 'the part of the channel between two successive crossings' of the ship's nadir. This was done in order to avoid miscalculations due to distortion inherent in side-scan sonar investigations. The nadir line cannot be defined in the multibeam datasets and distortion is not generally a problem. Therefore in this study the definition of the channel reach was changed slightly, so that each reach is defined by a single slope, with a new reach beginning at any significant break in slope along a channel's course.

In addition to the distinct channel profiles above, a number of levee and thalweg profiles covering the entire length of the system were produced, in much the same way. These profiles exhibit the complete transition from the channel heads to the continental rise, effectively showing variations in slope and depth. The levee profiles were taken along the north and south levees, parallel to the axial profiles. These, in combination with the thalweg profiles,

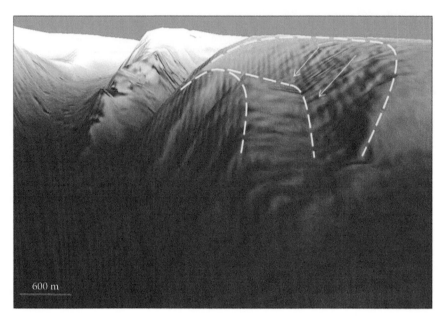

FIGURE 3.11
ArcGlobe 3D visualization showing a large slump on the southern flank of the Frodo Channel.
Image looking East, ×5 vertical exaggeration. (From INSS/PAD.)

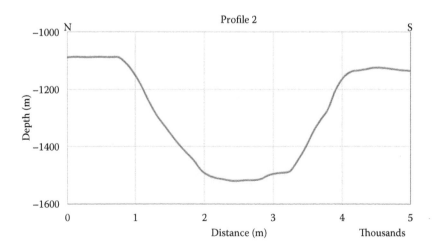

FIGURE 3.12
Example of tranverse profile, taken from Baggins Channel. (From INSS/PAD.)

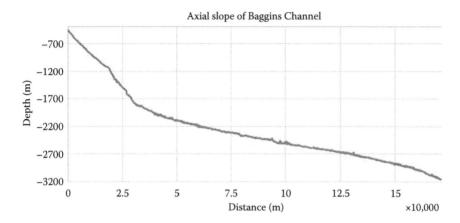

FIGURE 3.13
Example of longitudinal profile, taken from Baggins Channel. (From INSS/PAD.)

are referred to as relief profiles (Figure 3.14) and effectively show not only the changes in slope and water depth of the channel but also changes in channel depth and variations in the height between the right and left levees.

The above parameters were recorded for each of the main channels within the Gollum, Kings and Elven Channels (Van Rooij, 2004) (Figure 3.1). The parameters were collected for a total of 25 longitudinal channel

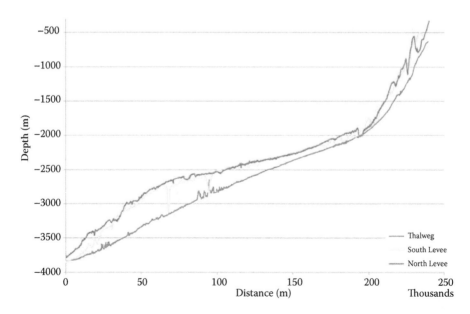

FIGURE 3.14
Example of Thalweg and Levee Profile, taken from the Baggins Branch. (From INSS/PAD.)

profiles – broken into 117 channel reaches – as well as 63 transverse profiles across the study area. This complete dataset is presented within the appendices of Murphy (2015) and Murphy et al. (subm.).

3.3 Results

The results of both the specific and general geomorphometric analysis in this chapter have added significantly to the understanding of the processes and controls on the Gollum Canyon System, as well as the extent of its drainage network. Our understanding of the extent of the drainage network has changed significantly from previous studies, due to the use of flow accumulation tools and the harnessing of the powerful GIS capabilities within ArcMap and ArcGlobe. Previously, it was thought that the GCS was the main source of downslope sediment transport within the PSB (Van Rooij, 2004) but this study has revealed that there is a far more expansive dendritic system present, which seems to drain most if not all of the area of the Seabight.

Within the study area, there are five main 'watersheds' which exit out through the mouth of the Porcupine Seabight onto the continental rise, and can be related directly to the five drainage channels identified by Akhmetzhanov et al. (2003). The three innermost channels drain the northern, Gollum and southern watersheds located within the Seabight, while the other two diffusely drain the continental margin (Figure 3.10). The drainage network analysis has revealed that previously unidentified downslope flowing portions of the Celeborn and Galadriel Channels in the Elven Channel System trace channel expressions across the Seabight. In addition, they are now revealed as hydrologically separate from one another, occurring within separate watersheds.

The Gollum watershed is the largest of the three, incorporating all of the Gollum and Kings Channel Systems as well as the Arwen and Elrond Channels of the Elven System. This is significant, in that links have been found between the previously separate channel systems. Downslope flow from the Eomer Channel joins the Kings Channels and later the Baggins Channel. In addition to this, the Eowyn Channel in the Kings Channel System flows directly into the Bilbo Channel.

Excluding the minor channels generated by the flow accumulation, the total drainage length of the Gollum watershed is approximately 1226 km, with the GCS making up 77% of this. From the channel heads to its basinward limit the GCS is on average 248 km long. The GCS is a straight to slightly sinuous dendritic system, with a maximum sinuosity of 1.45 and an average of 1.10. Slopes range from 0.2° in the mid basin to 4.5° at the channel heads. Meandering is of a low amplitude and is generally confined to the mid-slopes at about a 0.4° inclination.

From the combined efforts of the GIS-based specific and general geomorphometric analyses, several important observations and conclusions about the GCS can be made. Firstly, there is a significant geological control on channel morphology, orientation and branching patterns. This stems from observations of sudden changes in channel orientation, irregular channel floors and the restricted branching patterns of some channels. In addition, equilibrium profiles denoting the steady state conditions of a system are usually defined by a smooth concave up shape (Pirmez and Flood, 1995; Babonneau et al., 2002; Estrada et al., 2005), however, due to the geological control on the system and a consequent restriction of flow, a normal equilibrium profile cannot be applied. Instead, a two-part equilibrium profile is present (Figure 3.15). The upper section of this defines a normal canyon-channel transition, in a slightly aggraded state. The lower section begins at the start of the Gollum Channel, where the equilibrium profile diverges from the normal concave up profile. This divergence means that the lower section has been left sediment staved, and also marks the transition back into an erosional and canyonised system. This state of in-equilibrium is thought to have been significantly affected by a Late Palaeogene to Neogene tectonic inversion, where the previously rifted basin was uplifted (Van Rooij, 2004). This uplift would have affected sedimentation as well as shallowing basement geology.

The use of morphometric parameters commonly used in other submarine investigations allows the GCS to be placed within a global context. Compared to several channel systems, including the Amazon Deep-sea Fan (Flood and Damuth, 1987; Pirmez and Flood, 1995), the Bulgheria Canyon-fan (Budillon

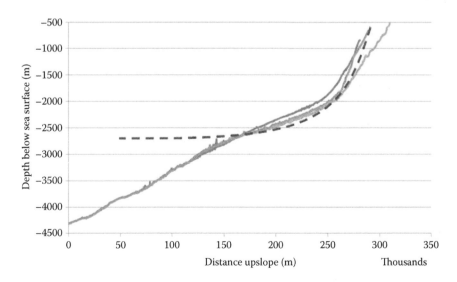

FIGURE 3.15
Individual channel profiles and approximate equilibrium profile (dashed line) for the Gollum Canyon System, due to geological influences. (From INSS/PAD.)

et al., 2011), the Magdalena Channel (Estrada et al., 2005), the Zaire Deep-sea Fan (Babonneau et al., 2002), the Trunk Channel (Klaucke et al., 1998) and the Toyama Deep Sea Channel (Nakajima et al., 1998), this analysis revealed how unique the GCS is. This is largely due to its location in the broad, geologically controlled seabight and the abundant sediment source provided by British-Irish Ice Sheet. The upper sections were correctly characterized by Van Rooij (2004) as a canyon-channel transition; however, it was found that the lower sections of the GCS do not exhibit the typical features of a channel system, as previously thought. Instead of transitioning downslope into a channel system, the geological influences on the PSB have altered the flow in the mid to lower sections, where it becomes erosive once again and has extended the canyonised system down to the mouth of the Seabight. Therefore its previous characterization as 'The Gollum Channel System' is not thought to be representative, and should instead be referred to as the Gollum Canyon System (Murphy, 2015), also making it the longest submarine canyon system in European waters.

In conclusion, this investigation has added significantly to our knowledge of the morphology and functioning of the Gollum Canyon System, via the use of both specific and general geomorphometric techniques. The hydrological toolsets within ArcMap have facilitated the derivation of the true drainage network within the Porcupine Seabight – revealing a vast dendritic system – as well as identifying hydrological connections between the previously separated Gollum, Kings and Elven Channel Systems. The 3D Analyst toolset in ArcMap, in addition to the 3D visualization of ArcGlobe, allowed for detailed analysis of the entire canyon system, for the first time. This is not only significant for the investigations into the Gollum Canyon System, but is also significant on a larger scale. There has been an apparent decrease in submarine geomorphometric investigations since the mid-2000s. This may be due to the rapid advancement of technology and switch from side-scan sonar to multibeam bathymetry as the technology of preference, leaving many of the previous techniques unsuitable. However, this project has bridged the gap between the side scan and multibeam methodologies. It has shown the power of this approach when applied to multibeam echosounder data – even when noise levels are high – potentially allowing for further comparable projects to commence in the future and therefore advancing our knowledge on the functioning of the submarine environments.

References

Akhmetzhanov, A., Kenyon, N., Ivanov, M. and Cronin, B., 2003. The continental rise west of the Porcupine Seabight, Northeast Atlantic. In: J. Meinert and P. Weaver, Eds. *European Margin Sediment Dynamics: Side-Scan Sonar and Seismic Images.* Heidelberg: Springer-Verlag, pp. 187–192.

Babonneau, N., Savoye, B., Cremer, M. and Klein, B., 2002. Morphology and architecture of the present canyon and channel system of the Zaire deep-sea fan. *Marine and Petroleum Geology*, 19, 445–467.

Berthois, L. and Brenot, R., 1966. Existence d'une flexure continentale parcourue par un réseau hydrographique, au Sud-Ouest de l'Irlande. *Comptes Rendues de l'Académie des Sciences Paris*, 263, 1297–1299.

Beyer, A., Schenke, H.W., Klenke, M. and Niederjasper, F., 2003. High resolution bathymetry of the eastern slope of the Porcupine Seabight. *Marine Geology*, 198(1–2), 27–54.

Beyer, A., Chakraborty, B. and Schenke, H.W., 2007. Seafloor classification of the mound and channel provinces of the Porcupine Seabight: An application of the multibeam angular backscatter data. *International Journal of Earth Sciences*, 96(1), 11–20.

Budillon, F., Conforti, A., Tonielli, R., De Falco, G., Di Martino, G., Innangi, S. and Marsella, E., 2011. The Bulgheria canyon-fan: A small-scale proximal system in the eastern Tyrrhenian Sea (Italy). *Marine Geophysical Research*, 32(1–2),83–97.

Chopra, R., Dhiman, R.D. and Sharma, E.K., 2005. Morphometric analysis of sub-watersheds in Gurdaspur District, Punjab using remote sensing and GIS techniques. *Journal of the Indian Society of Remote Sensing*, 2(4), 531–539.

Clark, J.D., Kenyon, N.H. and Pickering, K.T., 1992. Quantitative analysis of the geometry of submarine channels: Implications for the classification of submarine fans. *Geology*, 20(July), 633–636.

Dorschel, B., Wheeler, A.J., Monteys, X. and Verbruggen, K., 2010. *Atlas of the Deep-Water Seabed: Ireland*. New York: Springer Science & Business Media.

Dunlop, P., Shannon, R., McCabe, M., Quinn, R. and Doyle, E., 2010. Marine geophysical evidence for ice sheet extension and recession on the Malin Shelf: New evidence for the western limits of the British Irish Ice Sheet. *Marine Geology*, 276, 86–99.

Estrada, F., Ercilla, G. and Alonso, B., 2005. Quantitative study of a Magdalena submarine channel (Caribbean Sea): Implications for sedimentary dynamics. *Marine and Petroleum Geology*, 22, 623–635.

Flood, R.D. and Damuth, J.E., 1987. Quantitative characteristics of sinuous distributary channels on the Amazon Deep-Sea Fan. *Geological Society of America Bulletin*, 98(June), 728–738.

Guinan, J., Brown, C., Dolan, M.F.J. and Grehan, A.J., 2009. Ecological niche modelling of the distribution of cold-water coral habitat using underwater remote sensing data. *Ecological Informatics*, 4, 83–92.

Jenson, S., 1991. Applications of hydrologic information automatically extracted from digital elevation models. *Hydrological Processes*, 5, 31–44.

Keevil, G.M., Peakall, J. and Best, J.L., 2007. The influence of scale, slope and channel geometry on the flow dynamics of submarine channels. *Marine and Petroleum Geology*, 24(6–9), 487–503.

Kenyon, N.H., Belderson, R.H. and Stride, A.H., 1978. Channels, canyons and slump folds on the continental slope between South-West Ireland and Spain. *Oceanologica Acta*, 1, 369–380.

Klaucke, I., Hesse, R. and Ryan, W.B.F., 1998. Morphology and structure of a distal submarine trunk channel: The Northwest Atlantic Mid-Ocean Channel between lat 53°N and 44°30'N. *Bulletin of the Geological Society of America*, 110(1), 22–34.

Kolla, V., Bourges, P., Urruty, J.M. and Safa, P., 2001. Evolution of deep-water tertiary sinuous channels offshore Angola (west Africa) and implications for reservoir architecture. *American Association of Petroleum Geologists (AAPG) Bulletin*, 85(8), 1373–1405.

Leopold, L. B. and Wolman, M. G., 1957. River channel patterns; braided, meandering, and straight. *U.S. Geol. Surv. Prof. Paper* 282-B.

Magesh, N.S., Chandrasekar, N. and Kaliraj, S., 2012. A GIS based automated extraction tool for the analysis of basin morphometry. *Bonfring International Journal of Industrial Engineering and Management Science*, 2(1), 32–35.

Magesh, N.S., Jitheshlal, K.V., Chandrasekar, N. and Jini, K.V., 2013. Geographical information system-based morphometric analysis of Bharathapuzha river basin, Kerala, India. *Applied Water Science*, 3, 467–477.

Murphy, P.K., 2015. Geomorphometry of the Gollum Canyon System. Unpublished Masters Thesis, University College Cork, Ireland.

Murphy, P.K., Wheeler, A.W. and Devoy, R. J. (subm.). Geomorphometry of the Gollum Canyon System.

Nakajima, T., Satoh, M. and Okamura, Y., 1998. Channel-levee complexes, terminal deep-sea fan and sediment wave fields associated with the Toyama Deep-Sea Channel system in the Japan Sea. *Marine Geology*, 147, 25–41.

Pareta, K. and Pareta, U., 2011. Quantitative morphometric analysis of a watershed of Yamuna Basin, India using ASTER (DEM) data and GIS. *International Journal of Geomatics and Geosciences*, 2(1), 248–269.

Peakall, J., Kane, I.A., Masson, D.G., Keevil, G., McCaffrey, W. and Corney, R., 2011. Global (latitudinal) variation in submarine channel sinuosity. *Geology*, 40(1), 11–14.

Peakall, J., Mccaffrey, B. and Kneller, B., 2000. A process model for the evolution, morphology, and architecture of sinuous submarine channels. *Journal of Sedimentary Research*, 70, 434–448.

Pike, R., Evans, I. and Hengel, T., 2009. Geomorphometry: A brief guide. In: T. Hengl and H. Reuter, Eds. *Geomorphometry: Concepts, Software, Applications*. Amsterdam: Elsevier, pp. 3–30.

Pirmez, C. and Flood, R.D., 1995. Morphology and structure of Amazon Channel. *Proceedings of the Ocean Drilling Program, Initial Reports*, 155, 23–45.

Rengstorf, A., Yesson, C., Brown, C. and Grehan, A., 2013. High-resolution habitat suitability modelling can improve conservation of vulnerable marine ecosystems in the deep sea. *Journal of Biogeography*, 40(9), 1702–1714.

Sacchetti, F., Benetti, S. Georgiopoulou, A. Dunlop, P. and Quinn, R., 2011. Geomorphology of the Irish Rockall Trough, North Atlantic Ocean, mapped from multibeam bathymetric and backscatter data. *Journal of Maps*, 7(1), 60–81.

Saraf, A.K., Choudhury, P.R., Roy, B. and Sarma, B., 2004. GIS based surface hydrological modelling in identification of groundwater recharge zones. *International Journal of Remote Sensing*, 25(24), 5759–5770.

Strahler, A.N., 1957. Quantitative analysis of watershed geomorphology. *Transaction of the American Geophysical Union*, 38(6), 913–920.

Tubau, X., Lastras, G., Canals, M., Micallef, A. and Amblas, D., 2013. Significance of the fine drainage pattern for submarine canyon evolution: The Foix Canyon System, Northwestern Mediterranean Sea. *Geomorphology*, 184, 20–37.

Tudhope, A. and Scoffin, T., 1995. Processes of sedimentation in the Gollum Channel, Porcupine Seabight: Submersible observations and sediment analysis. *Transactions of the Royal Society of Edinburgh: Earth Sciences*, 86, 49–55.

Van Rooij, D., 2004. An integrated study of quaternary sedimentary processes on the eastern slope of the Porcupine Seabight, SW of Ireland, Unpublished PhD, Ghent University.

Wheeler, A. et al., 2003. Canyon heads and channel architecture of the Gollum Channel, Porcupine Seabight. In: J. Mienert and P. Weaver, Eds. *European Margin Sediment Dynamics*. Berlin: Springer, pp. 183–186.

Whittemore Boggs, S., 1937. Problems of water-boundary definition: Median lines and international boundaries through territorial waters. *Geographical Review*, 27(3), 445–456.

Wynn, R.B., Cronin, B.T. and Peakall, J., 2007. Sinuous deep-water channels: Genesis, geometry and architecture. *Marine and Petroleum Geology*, 24(6–9), 341–387.

4

Recent Developments in Remote Sensing for Coastal and Marine Applications

Melanie Lück-Vogel

CONTENTS

4.1 Introduction

The coast is the dynamic interface between land, ocean and atmosphere. Two of these environments are fluid and highly mobile, while the third, the land, is comparatively stable and enduring (Bartlett, 2000). The interactions between these involve many natural forces such as wind, river discharge, waves, salt spray and ocean currents (Bosboom and Stive, 2013). Coastal processes take place at a wide range of spatial and temporal scales (Bartlett, 1993). Algal blooms, harmful or not, can extend from several hundreds of metres up to several hundreds of kilometres across (Smayda, 1997). Weather events such as storms and flooding might affect a region, or a bay, or just the area around a particular river mouth. Cliff erosion as well as beach erosion usually occurs more locally, sometimes affecting stretches of coast only a few metres wide. Coastal vegetation succession or degradation is measurable at a sub-metre scale.

In addition to this wide range of spatial scale, coastal processes also work at a variety of temporal scales. Most dramatic are the events of erosion and land loss through single storm events, while other, slower processes such as sea level rise and related changes in nearshore sediment dynamics are progressive, and their results only become visible over decades. Thus, the only thing that is constant at the coast is that it is in a permanent state of change.

Remote sensing, whether from orbiting (space-borne) or air-borne platforms, can greatly assist in the task of monitoring coastal environments. In particular, remote sensing enables simultaneous or near-simultaneous capture of data for an extensive area of ground, which can be important for coastal management purposes given the length of many countries' coastlines (Table 4.1); while remote sensing also allows good repeat coverage and hence the acquisition of long time-series of observations, and the significance of coastal changes to be more easily evaluated.

It is striking that despite the long history of multispectral and other remote sensing techniques for land cover assessment in non-coastal, strictly terrestrial environments, a comparative scarcity of published literature suggests that very little operational remote sensing has been applied to similar requirements at the coast. Anecdotal evidence suggests that, up to and including the early years of the twenty-first century, many practitioners were sceptical of the value that remote sensing could bring to coastal management, although the author's experience suggests that this might be more due to negative experiences based on trying to apply the wrong data to the wrong purpose.

The following sections will explore the variety of established and upcoming types of remote sensing data, their spatial, temporal and spectral resolution and will give some examples of practical applications in the coastal space. The relation between remote sensing and GIS for effective marine and coastal spatial planning will also be considered, and the chapter will close with

TABLE 4.1

Examples of Coastline Length per Country

Country	Length of Coastline (km)	Country	Length of Coastline (km)
Canada	202,080	Madagascar	4828
Philippines	36,289	Thailand	3219
Japan	29,751	South Africa	2798
Australia	25,760	Germany	2389
Norway	25,148[a]	Ireland	1448
USA	19,924	Tanzania	1424
Italy	7600	Seychelles	491
Brazil	7491	The Netherlands	459[b]
India	7000	Monaco	4.1

Source: CIA (U.S. Central Intelligence Agency) 2014. *The World Factbook.* https://www.cia.gov /library/publications/the-world-factbook/ (accessed March 2016).
[a] Excludes fjords; including fjords: 100,915.
[b] Before closure of Zuiderzee, etc.: 1400.

recommendations on how to decide on which data to use for which purpose and which environmental factors are to be considered for appropriate image interpretation.

4.2 Remote Sensing

The term 'remote sensing' originated with geographers at the U.S. Office of Naval Research in the 1960s (Cracknell and Hayes, 1991), and refers to the capture of data or information about an object without being in physical contact with that object (Linz and Simonett, 1976), or 'the science and art of obtaining information about an object, area or phenomenon through the analysis of data acquired by a device that is not in contact with the object, area or phenomenon under investigation' (Lillesand et al., 2015, p. 1).

In the generic sense, the field of remote sensing covers a very wide range of methods and technologies for data capture and analysis (Lees, 2008), including digital scanning, optical (chemical) photography from satellites or aircraft, laser and radar profiling, geophysical investigations of the Earth's interior, echo-sounding of the water column and floor of the oceans and even the use of ultrasound for medical examination and X-rays for astronomical investigations or for chemical or geological spectroscopy.

The earliest form of Earth observation was through the use of aerial photography, initially from hot air balloons and later from kites, pigeons, rockets and eventually aircraft. This dates back to the middle of the nineteenth century (Anonymous, n.d.), and was based on imagery captured using light in the visible parts of the spectrum and recorded on chemically treated glass plates or film. The quality and resolution of the acquired imagery largely depended on camera and flight height, and also on the photographic process (collodion, dry plate, etc.) used. The mid-nineteenth century also saw the invention and gradual emergence of the science of photogrammetry, the use of photographs to obtain indirect measurements of building façades and other surfaces (Albertz, 2007) and, by the early 1900s, the use of photogrammetry for topographic surveying had become established (Bagley, 1917). The convergence of these two developments led to the increasing use of aerial photography for topographic mapping, reconnaissance and surveying purposes, in and subsequent to the First World War. This was followed, especially during and after World War II, by the invention and development of radar, infra-red photography and thermal imagery (Cracknell and Hayes, 1991).

From the 1960s onwards, electronic sensors started to appear, and gradually replaced photochemical imagery for collection and storage of images (Klemas, 2013). These sensors greatly expanded the scope of remote sensing to capture data at wavelengths other than those of visible light, thereby increasing by many orders of magnitude the amount, diversity and quality of information that could be derived from the data. The 1960s also saw the birth of space exploration, and this led rapidly to the use of orbiting platforms for Earth observation.

Given the wide potential range of methods, platforms and technologies involved, it is clear that many different criteria – and permutations of these – may be used to classify and examine remote sensing as a data acquisition technique. Adapted from Kramer (1994), these include:

- Measurement frequencies (the type of electromagnetic energy used to capture the data, its spectral ranges, single channel or multichannel sensor, etc.)
- Measurement platform (orbiting satellite, geostationary satellite, airborne, ship-borne, remotely piloted aircraft [drone], etc.)
- Measurement coverage (amount and frequency of repeated cover, temporal and spatial scales)
- Measurement resolution (spatial resolution, radiometric resolution, spectral resolution)
- Measurement target (Earth or ocean surface, atmospheric or water-column layer, etc.)
- Measurement pointing (nadir-looking, side-looking, forward- or backward-facing, stereoscopic viewing, etc.)
- Measurement calibration (standards adhered to, wavelengths, stability, device calibration, corrections and dynamic recalibration, data calibration, etc.)
- Measurement technique (active vs. passive remote sensing)
- Measurement data rate (temporal sampling frequency)
- Measurement target (continuously changing phenomena, once-off events, general-purpose remote sensing or specific to a particular target or set of targets)

The capture of data relating to the seabed, using multibeam sonar imagery (a form of ship-based remote sensing) has already been introduced and discussed in Chapters 2 and 3 of this volume. In this chapter, the focus of attention and discussion will be primarily on the acquisition of data from space-borne platforms.

4.3 Optical Remote Sensing

The era of space-borne remote sensing began in the late 1950s and early 1960s with the launch of the Russian Sputnik and the American Explorer 1 satellites (Melesse et al., 2007; Colwell, 1983), followed by the first meteorological satellite, TIROS-1, launched on 1 April 1960. Although TIROS-1 was only operational for 78 days, the two television cameras on board were able to capture over 20,000 images of the Earth's cloud cover (U.S. Department of Commerce, 1961) and demonstrated conclusively the potential that satellites offered for observing global weather and other phenomena from space.

Also in 1960, the U.S. military launched the first reconnaissance satellites of the Corona series which operated until 1972 (Ruffner, 1995; Lillesand et al., 2015). These satellites were able to capture grey-scale (i.e. black-and-white) imagery and, by 1963, were able to provide wide-scale, repeat coverage of the Earth, at a resolution of 10 ft (3.05 m) that was both timely and accurate (Ruffner, 1995). In 1962, the fourth version of Corona, KH-4, was developed and launched, with twin cameras that were able to provide stereoscopic photography which, by 1967, was able to provide imagery with ground resolution of 5 ft (approximately 1.5 m) (Ruffner, 1995). These early satellites did not have a digital image processing or transmission system, so capsules of film had to be returned to the Earth's surface for recovery and processing. While this imagery was classified initially, it was declassified under the terms of Executive Order 12951 of 22 February 1995, signed and ordered by then U.S. President Bill Clinton and placed in the public domain for scientific and environmental uses. Selected images can now be ordered from the U.S. Geological Survey (USGS) online. Both historic aerial imagery as well as the early reconnaissance data allow – if image quality is adequate – for the identification of small-scale land or sea features such as single trees, cars and ships, roads, buildings, as well as coastal infrastructure such as ports and moorings, and to some extent vegetation borders. They are therefore a valuable source for historical trend analysis, and can be especially useful for investigations of erosion or deposition, or changes in channel morphology. However, image acquisition intervals and image quality vary tremendously, and comparing images from one sensor with another is usually challenging.

In 1978, the US National Oceanic and Atmospheric Administration (NOAA) launched the first of a series of sensors which were dedicated particularly to ocean and weather observation on board the TIROS-N platform. Known as Advanced Very High-Resolution Radiometer (AVHRR) (Lillesand et al., 2015), these sensors offered a ground resolution of approximately 1 km, and passed over the same point on the Earth's surface at least once daily. The 'very high resolution' refers in the case of this platform to the temporal resolution, that is, how frequently images of a certain region are taken, rather than to spatial resolution. The first four AVHRR imagers had four spectral bands, and typical uses of the imagery included daytime cloud and surface mapping, snow and ice detection and sea surface temperature mapping (http://noaasis.noaa.gov/NOAASIS/ml/avhrr.html).

NOAA aims at always operating at least two identical satellites at the same time, resulting in the acquisition of at least four images per day per area. The concept of AVHRR has been very successful, and with the sensors on board the NOAA-18 and -19 and MetOp-A satellites currently operational, is still going on today. AVHRR with its spectral channels in the visible, near and thermal infrared is therefore one of the most valuable sources for historical time series of ocean and land colour, temperature and (because of its infra-red bands) vegetation processes on land and the ocean. In the ocean's context, AVHRR has been used for a diversity of applications including study of sea surface temperature applications using its thermal bands (Llewellyn-Jones et al., 1984), and for investigating sediment load (Myint and Walker, 2002) and algal blooms (using visible and near infrared bands), for example, for fishery forecasts (Barnard et al., 1997; Solanki et al., 2001; Klemas, 2012; Shen et al., 2012). A comprehensive review of ocean colour remote sensing and statistical techniques for mapping of phytoplankton blooms can be found at Blondeau-Patissier et al. (2014).

Following the AVHRR concept, in recent decades several other sensors with low spatial and high temporal resolution have been launched. Amongst those were SeaWiFS, MERIS and MODIS (Zibordi et al., 2006). Data at 250 m and 1 km resolution are available every 1 to 20 days. The European Space Agency (ESA)'s Sentinel-3A and 3B satellites, to be launched in 2016 and 2017 respectively, are designed to continue the MERIS data series with 21 (+9) spectral bands at 300 m (500 m) resolution, respectively. They also carry a radar instrument, which will be discussed separately below. All of these sensors are dedicated to monitor land cover and regional ocean colour and phytoplankton (Tassan, 1994). An example of ocean colour assessment for the Baltic Sea can be found at Kratzer et al. (2014); the usefulness of remote sensing for ocean colour monitoring, as a tool for the implementation of the European Marine Strategy Framework Directive in Portugal has been shown by Cristina et al. (2015); while Mélin and Vantrepotte (2015) used time series of SeaWIFS data to classify the regions of the coastal oceans according to their turbidity patterns. Moving closer to shore, turbidity of estuarine lagoons has also been assessed using MODIS in India (Kumar et al., 2016). Current chal-lenges and recommendations for remote sensing of coastal and inland water quality are summarized by Mouw et al. (2015), while Kachelriess et al. (2014) provide comprehensive overview of the manifold applications of optical remote sensing for management of marine protected areas. For monitoring most nearshore and terrestrial coastal processes though, the spatial resolu-tion of MODIS, MERIS, SeaWIFS and Sentinel-3 is too coarse.

In the 1970s and 1980s, another generation of 'medium resolution' Earth observation satellites emerged. The first of these was the Earth Resources Technology Satellite (ERTS-1), later renamed Landsat 1, and was launched in 1972, with the multispectral scanner (MSS) on board. This had a spatial reso-lution of about 60 m, and captured data using 4 spectral bands in the visible and infrared range (http://landsat.usgs.gov/; Lillesand et al., 2015). The suc-cessive satellites Landsat 2 and 3 carried the same sensor on board. In 1984, Landsat 4 was launched, which also had an MSS sensor for continuity with its predecessors, but which also introduced a new TM (thematic mapper) sensor that offered improved spectral and spatial properties for local, regional and global land cover and vegetation assessments. At 30 m, the spatial resolution of the TM sensor was high enough to map large coastal ecosystems such as mangroves. Therefore, mangroves are amongst the coastal ecosystems with the longest history of remote sensing mapping (e.g. Dale et al., 1996; Gao, 1998; Krause et al., 2004).

Landsat 6 failed at launch (Lillesand, 2015), but in 1999, Landsat 7 with the Enhanced Thematic Mapper plus (ETM+), including a 30 m multispectral and a new panchromatic sensor with 15 m resolution, was successfully put into orbit. This was followed, in 2013, by Landsat 8 which carries a 30 m OLI sensor (operational land imager) that captures data in the visible, near and short wave infrared wavebands, as well as a 60 m TIRS (thermal infrared sensor) (http://landsat.usgs.gov/landsat8.php). Landsat 4–8 have a repetition rate of 16 days, the older ones 18 days (van Aardt et al., 2010).

For about 40 years, Landsat has been one of the backbones of Earth observation for scientific and management applications, including marine and coastal environments. On the terrestrial side of the coast, Landsat data have only relatively limited value for mapping land cover and land use, because its medium resolution of 30 m is generally too coarse to pick up, for example, cliffline erosion, or recession of dune vegetation at a scale required by land managers. Within the nearshore zone, it has been used for a number of purposes, including mapping of tropical coral reefs, coastal water depth (bathymetry) (Johnson, 2015; Pacheco et al., 2015) and seagrass (Mumby et al., 1997; Wabnitz et al., 2008; Lyons et al., 2012) and other habitat mapping, as well as for shoreline detection (Boak and Turner, 2005; Pardo-Pascual et al., 2012), and investigation of shoreline dynamics over time (Green et al., 2000).

Moving offshore, where the spatial scale of investigation tends to be larger, numerous studies of marine processes and activities were published from the 1970s through to the early years of the present century that used Landsat data, either alone or in combination with data from other sources. Thus, as early as 1973, Maul reported on the use of ERTS imagery for mapping major ocean currents, using differences in surface chlorophyll-a concentrations (Maul, 1973), while Gagliardini et al. (1984) used Landsat MSS data along with NOAA/TIROS AVHRR and NIMBUS CZCS imagery to study how water discharging from the La Plata River in South America interacted with the receiving ocean. Elsewhere, Dwivedi and Narain (1987) investigated the potential use of Landsat TM imagery to map phytoplankton distributions in the Arabian Sea; and Forster et al. (1993) proposed a methodology for assessing sea water quality around a sewage outfall to the north of Sydney in Australia.

Despite these and other examples, the spectral and temporal resolution of Landsat data has generally been considered too low to pick up subtle spectral changes and swift temporal dynamics of currents and microorganism development, thus limiting the utility of these images for applications. The same has been true for other medium resolution sensors such as ASTER and SPOT 1-5 (van Aardt et al., 2010) (see Figure 4.1 for comparison of the level of

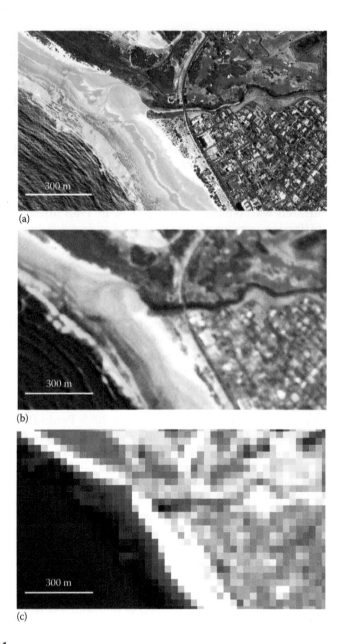

(a)

(b)

(c)

FIGURE 4.1
Effect of spatial resolution on the level of detail visible in remote sensing imagery. (a) World View-2 image from 4 April 2010 of the Lourens River mouth in False Bay, South Africa (near Cape Town). Pixel size 2 m (© DigitalGlobe). (b) RapidEye image from 2 January 2010 of the same area. Pixel size 5 m (© BlackBridge). (c) Landsat 7 image from 21 July 2002 of the same area. Pixel size 30 m (© NASA). Displayed in all three images are the near infrared, mid infrared and red spectral bands, that is, a false colour composite, in which herbaceous/grassy vegetation appears bright red and woody vegetation dark red.

spatial detail that can be derived from a 2 m resolution WorldView-2 image, a 5 m resolution RapidEye image and a 30 m resolution Landsat image). In order to provide some guidance on which spatial resolution to use for which application, Hengl (2006) suggests a simple rule of thumb: he recommends that the pixel size should optimally be half the size of the smallest object of interest. This means, if for instance the aim is to map foot paths through a dune environment which are about 1 m wide, then the best pixel size is 0.5 m. If the aim is to map coastal vegetation erosion with a footprint wider than 2 m, then the optimal resolution would be 1 m.

Because of the limitations of Landsat and other medium-resolution sensors, alternative sensors and platforms have tended to gain favour with coastal researchers in recent years, although some interest in the capabilities of Landsat data for ocean studies remain. Thus, for example, Vanhellemon and Ruddick (2015) have demonstrated the use of imagery in short-wave infrared (SWIR) wavelengths, captured by the Operational Land Imager sensor on Landsat-8 (see above), for monitoring ocean water quality through detailed examination of ocean colour data.

It should also be noted that, in 2008, the formerly significant acquisition costs for Landsat data were dropped, and since then both historic (archived) and newly acquired Landsat imagery has been made available for free download from the USGS website (http://landsat.usgs.gov/). This change in charging policies for the data has led to another burst of Landsat applications and resulting publications.

In 1996, the first of a new generation of commercially owned and operated high spatial resolution sensors and platforms, such as IKONOS, Quickbird and SPOT 4 and 5 (Fritz, 1996; van Aardt et al., 2010; Lillesand et al., 2015) appeared, which led to a major realignment and transformation of methods and practice in Earth observation (Fritz, 1996, p. 274). While the spatial resolution of these new sensors, typically between 1 m and 10–20 m, for the first time provided enough detail for mapping the coastal habitat sequence, the fact that these environments frequently occur in narrow (sometimes sub-metre) strips parallel to the coast often required a large number of separate images to be acquired and processed. This, allied to the rather high image purchase costs and data availability that was frequently on demand only, tended to prohibit application of these data in operational coastal management in many countries, particularly when extensive expanses of coastline had to be considered. The identification of the optimal remote sensing data was (and still is) frequently a trade-off between suitability and affordability (Figure 4.2).

A further bottleneck for remote sensing applications in the nearshore and dry coast environment was the relatively low repetition rate, typically around 16 days for Landsat, SPOT and ASTER (van Aardt et al., 2010). Since approximately 2007 this has been overcome with the launch of various high (spatial) resolution sensor systems and, especially, the increasing deployment of constellations of satellite, which enable images over the same area to be captured every 1 to 3 days. These modern sensors include the RapidEye

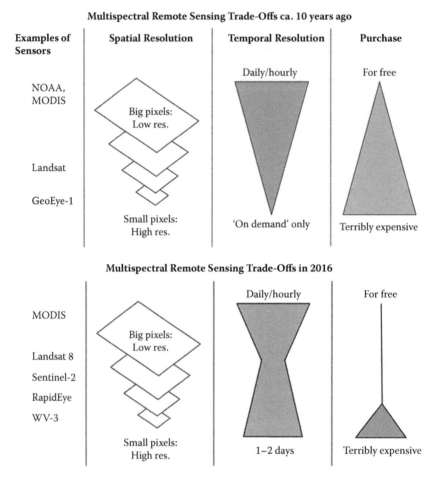

FIGURE 4.2
Change of multispectral remote sensing trade-offs over time.

constellation of five identical sensors in different orbits (http://blackbridge
.com/rapideye/), WorldView-2 and 3, SPOT-6 and 7 (http://www.geoimage
.com.au/satellite/spot-6) and the recently launched Sentinel 2A and 2B sensors (https://sentinel.esa.int/web/sentinel/missions).

RapidEye has 5 spectral bands at 5 m resolution in the visible, near and
short wave infrared parts of the spectrum (Figure 4.1b). SPOT-6 and the identical SPOT-7 have 4 spectral bands at 6 m resolution in the visible and near IR
range and a panchromatic band at 1.5 m resolution. WorldView-2 has 8 spectral bands at 2 m (Figure 4.1a) and a panchromatic band at 0.5 m resolution.
This high spatial resolution allows for the identification of individual tree
canopies and tree species, which can be useful for identification of invasive
tree and shrub species in coastal forests (Malahlela et al., 2015). WorldView-3

has 8 bands in the visible and near infrared range, 8 bands in the shortwave infrared range and 12 atmospheric bands. Its spatial resolution is between 0.31 m and 30 m, depending on the band (https://www.digitalglobe.com/). ESA's identical Sentinel-2A and 2B sensors provide 13 multispectral bands at 10–20 m resolution.

With the advent of these new sensors with better radiometric (16 bit) and spatial resolution, remote sensing of the nearshore and terrestrial coast has become much more feasible, as these sensors are now able to provide imagery detailed enough for monitoring and managing the small scale coastal environment. An increase in interest is measurable in the increase in publications on coastal remote sensing applications and an increased willingness in stakeholders to invest in this technology, after having neglected remote sensing for coastal management for so long. However, while very high resolution multispectral instruments such as that provided by WorldView-2 hold enormous potential for coastal and marine studies, Martin et al. (2016) suggest that in real scenarios, and depending on the area of coast or ocean under study, 'several portions of the WV-2 multispectral imagery acquired for coastal zones are disappointing' (Martin et al., 2016, p. 2), due to the low signal-to-noise ratio available at sensor level on shorter wavelengths. In particular, they point to the need to correct carefully for atmospheric factors, as well as the effect of sun glint (specular water reflection) on the quality of the image. Technical guidance on sun glint removal can be found at Mobley (1994), Hochberg et al. (2003) and Hedley et al. (2005).

4.3.1 Optical Remote Sensing from Unmanned Aerial Vehicles

The opening years of the twenty-first century have seen an increasing interest in the use of aerial imagery derived from high-resolution cameras mounted on unmanned aerial vehicles (UAVs) or systems (UAS). Sometimes also known as drones, these are remote controlled devices on which various sensors can be mounted. The spatial resolution of the imagery derived from these platforms can be in the centimetre range, depending on the actual camera type and flight height.

Even though UAVs have only been available to general users for little over a decade, their rapid and widespread adoption indicates that they are no longer simply the preserve of the early adopter and the enthusiast, and 'the practical application of UAVs for surveying at the coast has come of age'. (Turner et al., 2016, p. 19). In particular, with appropriate design of the flight lines, stereo-imagery can be acquired which allows the generation of high-resolution digital surface models (DSMs) through the use of digital photogrammetric software. Recent examples of this include the work of Gonçalves and Henriques (2015), who used a UAV equipped with a non-metric camera, and the Agisoft Photoscan photogrammetric software (www.agisoft.com), to monitor and analyze change in dunes and beaches on the northwest coast of Portugal, while Lim et al. (2015) used repeat UAV surveys to assess changes in coastal dune cliffs on the northeast coast of England. A comprehensive overview of UAV use for coastal surveying is provided by Turner et al. (2016).

While the spatial range of a UAV is normally limited to a few kilometres, the relative cost-effectiveness of acquisition, operation and image processing, and their ability to operate below the usual cloud level, makes this new technology increasingly popular for assessment of smaller areas of interest where high level of detail and quick updates are required. Apart from their role in supporting topographic survey, examples of the use of UAVs for coastal and marine management purposes include investigations into their potential to support search and rescue operations (Ryan and Hedrick, 2005), for marine oil spill detection (Donnay, 2009) and for identifying and mapping fish nursery areas (Ventura et al., 2016).

Besides optical sensor payloads, there has also been a rapid increase in the use of UAVs with hyperspectral sensors and with LiDAR, leading to the growing and diversifying use of these platforms for remote sensing applications (Pajares, 2015; Toth and Jóźków, 2016). Some of these sensors and their potential uses are considered in the sections below.

4.4 Hyperspectral Sensors

Hyperspectral sensors have many more and narrower spectral bands than the multispectral sensors described above. As an example, Landsat TM's spectral bands are discontinuous and between 70 and 270 nm wide, while the Hyperion satellite sensor has 196 spectral bands with a band width of 10 nm over the same spectral range which provides a continuous spectral profile (Lillesand et al., 2015); see Figure 4.3.

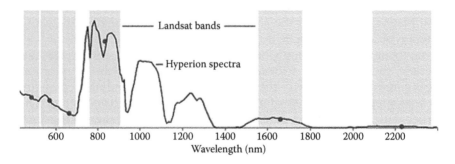

FIGURE 4.3
Comparison of continuous spectral signature from the Hyperion sensor (black line) with the Landsat TM bands (shaded blocks: respective Landsat band width; grey dots: output value per band as average over the band width). (From http://earthobservatory.nasa.gov/Features /AmazonDrought/stealing_rain2.php.)

Airborne hyperspectral sensors can have band widths of 1–2 nm. This improved sensitivity to spectral characteristics of the ground cover leads to better distinction of surface types. Hyperspectral data are very valuable for application on small areas, for plant species discrimination (e.g. detection of alien invasive species) and even detection of physiological plant stress (e.g. caused by pests or droughts). For instance, Hedley et al. (2016) used airborne hyperspectral data for the estimation of seagrass density in Florida. Le Bris et al. (2016) mapped the invasion of Pacific Oysters in European waters where they can replace the (cultivated) Portuguese Oysters. However, for operational coastal management of large areas, the costs and the analytical effort necessary for hyperspectral data analysis might be prohibitive.

4.5 LiDAR – Light Detection and Ranging

In contrast to multispectral sensors as described above, LiDAR sensors are 'active systems' and generate light pulses, whose reflection from the surface they detect themselves. The time lag between the emission of the pulse and the return can be used to derive information on the distance of the surface from the sensor and thus about surface elevation and topography (Pe'eri and Long, 2011). LiDAR, also called laser scanners, can be mounted on aircraft or vehicles, or can be ground-based (terrestrial laser scanning). There are LiDAR systems that are designed for bathymetric applications or for terrestrial (dry coast) applications only or combinations thereof (Collin et al., 2010; Pe'eri et al., 2011). LiDAR data, collected in a very accurate point cloud which can give sub-metre information on relief, are of very high value for coastal applications (Brock and Purkis, 2009). They can improve reliability of flood scenarios (Gesch, 2009) and multi-temporal datasets can give very precise information on volumetric change of sediments, for example, after storm events (Xhardé et al., 2011). Further, the combined use of multispectral data and LiDAR-derived topographic data has shown to improve accuracies of coastal vegetation maps (Elaksher, 2008; Kempeneers et al., 2009; Lück-Vogel et al., 2016). Further, more sophisticated information on ground surface properties can be derived from additional LiDAR characteristics such as the wave form (Collin et al., 2011) and signal intensities (Collin et al., 2010).

4.6 SAR – Synthetic Aperture Radar

Similar to LiDAR, radar remote sensing systems are active systems, that is, they emit the energy pulses, whose reflection from the surface they measure.

In contrast to LiDAR which uses wavelengths in the visible and near infrared range, radar uses microwaves with wavelengths in the centimetre to metre range (Mott, 2007; Lillesand et al., 2015). Frequently used space-borne SAR sensors include ESA's ERS-1 (launched 1991) and ERS-2 (launched 1995), ENVISAT (launched 2002) and the German TerraSAR-X (launched 2007). The Sentinel-3A and 3B satellites (to be launched in 2016 and 2017, respectively) will also carry radar instruments.

A typical product derived from SAR data is surface elevation (similar to LiDAR, whose main application also is topography), but usually at a coarser spatial resolution. More ocean-specific is the derivation of ocean wave spectra (Marom et al., 1990), and wind speed (Hasselmann et al., 1996) or coastal (ocean surface) altimetry (Vignudelli et al., 2011; Morrow and Le Traon, 2012).

A great advantage of SAR technology is that its wavelengths penetrate cloud cover. This is an important advantage when quick image acquisition is required, that is, in the case of a disaster like flooding, land slide or sea-based oil spills. Oil spills, either caused by ship or oil platform havoc or by illegal bilge dumping by ships at sea technically can also be detected using optical data such as Landsat (e.g. Xing et al., 2015). However, its repetition rate of 16 days (or 8 days when using several Landsat sensors in combination) frequently is not sufficient, and it seems that the culprits are frequently active when the sky is cloudy which severely impairs the usefulness of optical data. Among many examples to be found in the literature, Fan et al. (2015) successfully monitored an oil spill in China using SAR imagery; Wei et al. (2015) used SAR data and historical data to assess the economic damage after an oil spill in China. Fingas and Brown (2014) provide a comprehensive review on optical, radar and other remote sensing technologies for detection of oil spills.

A disadvantage of radar data is that it requires very specific software and specialist training. It is therefore currently much more difficult to integrate radar data into a GIS environment than it is for multispectral, hyperspectral and LiDAR data.

4.7 Remote Sensing and GIS

The sections above show that remote sensing derived data can provide information of tremendous value for various coastal and marine management aspects. However, having extracted the required information, putting it to use usually happens in a geographical information system (GIS). Here the data can be analyzed and combined with other datasets, for example, in combination with spatial planning and socioeconomic data. Krause et al.

(2004) describe the requirement of setting up a GIS for using a variety of spatial data for integrated management of mangrove systems in North Brazil. Warren et al. (2016) monitored benthic ecosystems in Qatar using remote sensing, and used GIS to help give recommendations for coastal risk assessment and management planning based on these data.

At least until the late 1990s, GIS and image processing tasks generally required different software packages and tools, although even in the 1980s some systems (mostly designed for desktop computers) were developed that were able to combine and integrate both sets of tasks within a single workflow: examples included the IDRISI system developed at Clark University (Eastman, 1990, 1995), which by the early 1990s was even offering a workbook with tutorial exercises specifically designed for training in coastal and marine applications of GIS and remote sensing (St. Martin, 1993), and the ILWIS package developed at the ITC in the Netherlands (Meijerink et al., 1988). However, most dedicated GIS software, such as Esri's ArcMap or the open source QGIS package, was primarily designed to process and analyze vector data, organized as polygons, lines and points, and integration of remote sensing data, typically raster-format images, was hardly supported. Similarly, dedicated remote sensing software (the most commonly used commercial packages internationally are ENVI, ERDAS and PCI Geomatica) had very limited vector data handling capabilities. However, in response to demands from the community, the past two decades have seen increasing interoperability and cross-over of functionality between both domains, with current versions of most mainstream GIS and image processing software offering some manipulation and joint analysis of vector and raster data.

Further, online tools such as Google Earth provide simple – but nevertheless effective – tools for a wider public to view GIS data and satellite images, without having to install specialized software packages. Google Earth also provides simple GIS functionality such as the creation of user-specific areas or points of interest, and even allows these to be linked to comments, photos or Internet links. Most remote sensing and GIS tools nowadays allow conversion of their own file formats into Google formats and vice versa. This development has greatly advanced the public's awareness and understanding of geospatial information, as do the growing number and array of web-based domain-specific geospatial information portals, many of which are specifically dedicated to the publication of data and information related to the coast. Examples of these online coastal atlases include the Coastal Viewer in South Africa (http://mapservice.environment.gov.za/Coastal Viewer/), NOAA's Digital Coast (https://coast.noaa.gov/digitalcoast/) and the Oregon Coastal Atlas (http://www.coastalatlas.net/). It also gives a very powerful opportunity for public participation in coastal and marine management (see Chapters 7 and 10 of the current volume).

4.8 Considerations for Accurate Interpretation of Coastal Remote Sensing Products

The previous sections introduced the different types of remote sensing data available for coastal management purposes, and it briefly mentioned GIS and remote sensing software packages and tools for processing, displaying and analyzing remote sensing and other geospatial information.

However, one more aspect needs some mentioning: calibration, validation and correct interpretation of remote sensing results, which create the necessity of correspondence of remote sensing derived information and field data and other ancillary data.

The dynamic nature of the coast poses a major challenge for the remote sensing practitioner, because an image taken one day can look very different, and can contain very different information, from an image taken a day later. Or a week later. Or a month earlier. The question is, do the differences matter, and are they indicative of significant actual changes at the coast? Often, answering this type of question is the specific objective of the remote sensing coastal exercise, but in other cases the timing of the data capture may lead to misinterpretations and erroneous conclusions being drawn. Lück-Vogel et al. (2011), for example, used Landsat imagery to map variations on the soft sandy coast off Maputo (Mozambique) in terms of width of beaches and visibility of partly submerged sand spits and the position of the surf zone. Through analysis of tidal data at the respective image acquisition dates, they were able to show that most of the observed variations were simply related to tidal differences. Without relating those land cover products to tidal data, the observed changes could have been misinterpreted as coastal erosion or accretion processes – with potentially disastrous consequences in an operational coastal management context.

In a separate study, Lück-Vogel et al. (2016) mapped estuarine vegetation in St. Lucia, South Africa, using high resolution multispectral imagery from 2010, 2011, 2012 and 2014, based on a field survey-based GIS map of vegetation from 2008. Analysis of the classification accuracies showed that misclassifications could almost entirely be related to the highly dynamic environmental conditions such as water level variations in the estuary, eradication of vegetation patches through single flood events and even extreme wind conditions at the image acquisition time, leading to turbid water and wind-wave disturbed surfaces of the water bodies. These and similar experiences lead to the conclusion that field data should match the date, or at least season and tidal condition of the remote sensing acquisition time as closely as possible; and that the interpretation of remote sensing products considers the respective relevant environmental conditions at the time of data acquisition. It is particularly essential that these precautions be taken when remote sensing is being used for operational coastal management purposes, on which people's lives might depend.

4.9 Conclusions and Recommendations

This chapter gave a brief overview of different types of remote sensing imagery and their applications. Multispectral and hyperspectral sensors and data characteristics were described, as well as active LiDAR and SAR sensor systems. It was explained that the choice of the appropriate satellite data must consider technical details such as spatial and temporal resolution, number of spectral bands and acquisition costs, and that frequently trade-offs between these factors need to be made. Further, it was recommended that additional environmental data, such as water levels, tidal data and weather data are being used for accurate interpretation of remote sensing products.

At the root of decision making for the acquisition of remote sensing data must stand the following questions:

- Which land cover feature or which coastal/marine process do I want to assess?

- What is the spatial extent of that feature, phenomenon? For example, size of vegetation patches, width of coastal vegetation bands, ocean wave length. This will determine the optimal pixel size (spatial resolution).

- What is the temporal dimension of the target? For example, long-term change, seasonal patterns, once-off storm event. This will determine the optimal acquisition date or acquisition time window. If long-term monitoring is envisaged, this will also determine the optimal repetition rate, for example, seasonal or annual. This will also inform if historical availability of the remote sensing data is required. For the assessment of seasonal dynamics a 3–5 year observation period might be sufficient, while for assessment of long-term trends, for example, climate change effects, a 10–30 year observation period might be the bare minimum requirement.

- Do I use a combination of remote sensing and field-based or other ancillary data? If yes:

  - To what extent is my target feature influenced by differences in the acquisition dates of the field data and the satellite data? The more dynamic the target, the closer all datasets should match temporally.

  - Do I have ancillary data, for example, tide calendars, weather data to interpret conditions at the respective acquisition dates appropriately? Without such additional information, remote sensing observations might be misinterpreted and lead to unsubstantiated coastal management decisions.

As long as attention is paid to these questions, remote sensing products can add great value to many environmental or other coastal and marine management exercises.

Acknowledgement

The author thanks Dr. Fiona Cawkwell, University College Cork (Ireland) for helpful comments on an earlier draft of this chapter.

References

Albertz, J. 2007. 140 years of "photogrammetry" some remarks on the history of photogrammetry. *Photogrammetric Engineering and Remote Sensing*, May 2007.

Anonymous. n.d. History of Aerial Photography. http://professionalaerialphotographers.com/content.aspx?page_id=22&club_id=808138&module_id=158950 (accessed 3 May 2016).

Bagley, J.W. 1917. The use of the panoramic camera in topographic surveying. *U.S. Geological Survey Bulletin 657*. Washington, DC: Government Printing Office, p. 88.

Barnard, A.H., Stegmann, P.M., and Yoder, J.A. 1997. Seasonal surface ocean variability in the South Atlantic Bight derived from CZCS and AVHRR imagery. *Continental Shelf Research* 17(10), 1181–1206.

Bartlett, D.J. 1993. Space, time, chaos and coastal GIS. In *Proceedings, Sixteenth International Cartographic Congress*, Cologne, Germany, 3–9 May 1993.

Bartlett, D.J. 2000. Working on the frontiers of science: Applying GIS to the coastal zone. In *Marine and Coastal Geographical Information Systems*, D. Wright and D.J. Bartlett (Eds.), Chapter 2. London, England: Taylor & Francis.

Blondeau-Pattissier, D., Gower, J.F.R., Dekker, A.G., Phinn, S.R. and Brando, V.E. 2014. A review of ocean color remote sensing methods and statistical techniques for the detection, mapping and analysis of phytoplankton blooms in coastal and open oceans. *Progress in Oceanography 123*, 123–144. http://dx.doi.org/10.1016/j.pocean.2013.12.008.

Boak, E.H., and Turner, I.L. 2005. Shoreline definition and detection: A review. *Journal of Coastal Research* 21, 688–703.

Bosboom, J., and Stive, M.J.F. 2013. Coastal dynamics I – Lecture notes CIE4305. Version 0.4 (January 2013), Delft University, the Netherlands.

Brock, J.C., and Purkis, S.J. 2009. The emerging role of LIDAR remote sensing in coastal research and resource management. *Journal of Coastal Research* 53, 1–5.

CIA ([U.S.] Central Intelligence Agency) 2014. *The World Factbook*. Available online at https://www.cia.gov/library/publications/the-world-factbook/ (accessed March 2016).

Collin, A., Long, B., and Archambault, P. 2010. Salt-marsh characterization, zonation assessment and mapping through a dual-wavelength LiDAR. *Remote Sensing of Environment* 114, 520–530.

Collin, A., Long, B., and Archambault, P. 2011. Benthic classifications using bathymetric LIDAR waveforms and integration of local spatial statistics and textural features. *Journal of Coastal Research* 62, 86–98.

Colwell, R.N. (Ed.). 1983. *Manual of Remote Sensing*. Falls Church, VA: American Society of Photogrammetry.

Cracknell, A.P. and Hayes, L.W.B. 1991. *Introduction to Remote Sensing*. London, New York, Philadelphia: Taylor & Francis, p. 293.

Cristina, S, Icely, J., Costa Goela, P., Angel DelValls, T., and Newton, A. 2015. Using remote sensing as a support to the implementation of the European marine strategy framework directive in SW Portugal. *Continental Shelf Research* 108, 169–177. doi:10.1016/j.csr.2015.03.011.

Dale, P.E.R., Chandica, A.L., and Evans, M. 1996. Using image subtraction and classification to evaluate change in subtropical intertidal wetlands. *International Journal of Remote Sensing* 17, 703–719.

Donnay, E. 2009. Use of unmanned aerial vehicle (UAV) for the detection and surveillance of marine oil spills in the Belgian part of the North Sea. *Proceedings of the 32 AMOP Technical Seminar on Environmental Contamination and Response*, Volume 2. Canada: Environment Canada, p. 940.

Dwivedi, R.M. and Narain, A. 1987. Remote sensing of phytoplankton. An attempt from the Landsat thematic mapper. *International Journal of Remote Sensing* 8, 1563–1569.

Eastman, J.R. 1990. *IDRISI: A Grid-Based Geographic Analysis System*. Worcester, MA: Clark University, Graduate School of Geography.

Eastman, R. 1995. *IDRISI for Windows: User's Guide-Version 1.0*. Worcester, MA: Clark University, Graduate School of Geography.

Elaksher, A.F. 2008. Fusion of hyperspectral images and LIDAR-based DEMs for coastal mapping – Review. *Optics and Lasers in Engineering* 46, 493–498.

ESA (European Space Agency). n.d. Observing the Earth. http://www.esa.int/Our_Activities/Observing_the_Earth/How_does_Earth_Observation_work (accessed 30 April 2016).

Fan, J., Zhang, F., Zhao, D., and Wang, J. 2015. Oil spill monitoring based on SAR remote sensing imagery. *Aquatic Procedia* 3, 112–118.

Fingas, M., and Brown, C. 2014. Review of oil spill remote sensing. *Marine Pollution Bulletin* 83, 9–23.

Forster, B.C., Xingwei, S., and Baide, X. 1993. Remote sensing of sea water parameters using Landsat-TM. *International Journal of Remote Sensing* 14, 2759–2771.

Fritz, L.W., 1996. The era of commercial earth observation satellites. *Photogrammetric Engineering and Remote Sensing* 62(1), 39–45.

Gagliardini, D.A., Karszenbaum, H., Legeckis, R., and Klemas, V. 1984. Application of Landsat MSS, NOAA/TIROS AVHRR, and Nimbus CZCS to study the La Plata River and its interaction with the ocean. *Remote Sensing of Environment* 15, 21–36.

Gao, J. 1998. A hybrid method toward accurate mapping of mangroves in a marginal habitat from SPOT multispectral data. *International Journal of Remote Sensing* 19, 1887–1899.

Gesch, D.B. 2009. Analysis of Lidar elevation data for improved identification and delineation of lands vulnerable to sea-level rise. *Journal of Coastal Research* 53, 49–58.

Gonçalves, J.A., and Henriques, R. 2015. UAV photogrammetry for topographic monitoring of coastal areas. *ISPRS Journal of Photogrammetry and Remote Sensing* 104, 101–111.

Green, E.P., Mumby, P.J., Edwards, A.J., and Clark, C.D. 2000. *Remote Sensing Handbook for Tropical Coastal Management*. Paris, France: UNESCO Publishing.

Hasselmann, S., Brüning, C., Hasselmann, K., and Heimbach, P. 1996. An improved algorithm for the retrieval of ocean wave spectra from synthetic aperture radar image spectra. *Journal of Geophysical Research: Oceans* 101(C7), 16615–16629.

Hedley, J.D., Harborne, A.R., and Mumby, P.J. (2005). Simple and robust removal of sun glint for mapping shallow water benthos. *International Journal of Remote Sensing* 26, 2107–2112.

Hedley, J., Russell, B., Randolph, K., and Dierssen, H. 2016. A physics-based method for the remote sensing of seagrasses. *Remote Sensing of Environment* 174, 134–147.

Hengl, T. 2006. Finding the right pixel size. *Computers & Geosciences* 32, 1283–1298.

Hochberg, E.J., Andréfouët, S., and Tyler, M.R. 2003. Sea surface correction of high spatial resolution INKONOS images to improve bottom mapping in nearshore environments. *IEEE Transactions on Geoscience and Remote Sensing* 41, 1724–1729.

Johnson, M. 2015. Deriving bathymetry from multispectral Landsat 8 satellite imagery in South Africa. Unpublished BSc Hons thesis, Department of Geography and Environmental Studies, Stellenbosch University, South Africa.

Kachelriess, D., Wegmann, M., Gollock, M., and Pettorelli, N. 2014. The application of remote sensing for marine protected area management. *Ecological Indicators* 36, 169–177. doi:10.1016/j.ecolind.2013.07.003.

Kempeneers, P., Deronde, B., Provoost, S., and Houthuys, R. 2009. Synergy of airborne digital camera and Lidar data to map coastal dune vegetation. *Journal of Coastal Research* 53, 73–82.

Klemas, V. 2012. Remote sensing of algal blooms: An overview with case studies. *Journal of Coastal Research* 28, 34–43.

Klemas, V., 2013. Airborne remote sensing of coastal features and processes: An overview. *Journal of Coastal Research* 29(2), 239–255.

Kramer, H. 1994. *Earth Observation Remote Sensing. Survey of Missions and Sensors,* 2nd ed. Berlin: Springer Science and Business Media.

Kratzer, S.E., Harvey, T., and Philipson, P. 2014. The use of ocean color remote sensing in integrated coastal zone management – A case study from Himmerfjärden, Sweden. *Marine Policy* 43 (January 2014), 29–39. doi:10.1016/j.marpol.2013.03.023.

Krause, G., Bock, M., Weiers, S., and Braun, G. 2004. Mapping land-cover and mangrove structures with remote sensing techniques – A contribution to a synoptic GIS in support of coastal management in North Brazil. *Environmental Management* 34, 429–440.

Kumar, A., Equeenuddin, S.M., Mishra, D.R., and Acharya, B.C. 2016. Remote monitoring of sediment dynamics in a coastal lagoon: Long-term spatio-temporal variability of suspended sediment in Chilika. *Estuarine, Coastal and Shelf Science* 170 (5 March 2016), 155–172. doi:10.1016/j.ecss.2016.01.018.

Le Bris, A., Rosa, P., Lerouxel, A., Cognie, B., Gernez, P., Launeau, P., Robin, M., and Barillé, L. 2016. Hyperspectral remote sensing of wild oyster reefs. *Estuarine, Coastal and Shelf Science.* doi:10.1016/j.ecss.2016.01.039.

Lees, B.G. 2008. Remote sensing. In *The Handbook of Geographic Information Science.* Wilson, J.P. and Fotheringham, A.S. (Eds.). Malden, MA: Blackwell Publishing Ltd., pp. 49–60.

Lillesand, T.M., Kiefer, R.W., and Chipman, J.W. 2015. *Remote Sensing and Image Interpretation*, 7th ed. New York: Wiley.

Lim, M., Dunning, S.A., Burke, M., King, H., and King, N. 2015. Quantification and implications of change in organic carbon bearing coastal dune cliffs: A multiscale analysis from the Northumberland coast, UK. *Remote Sensing of Environment* 163, 1–12.

Linz, J., and Simonett, D.S. 1976. *Remote Sensing of Environment*. Reading, MA: Addison-Wesley.

Llewellyn-Jones, D.T., Minnett, P.J., Saunders, R.W., Zavody, A.M. 1984. Satellite multichannel infrared measurements of sea surface temperature of the N.E. Atlantic Ocean using AVRHH/2. *Quarterly Journal of the Royal Meteorological Society* 110, 613–631.

Lück-Vogel, M., Barwell, L., and Theron, A. 2011. Transferability of a remote sensing approach for coastal land cover classification. *Proceedings of the CoastGIS 2011 Symposium*, Oostende, Belgium, 6–8 September 2011, 3, 71–78.

Lück-Vogel, M., Mbolambi, C., Rautenbach, K., Adams, J., and van Niekerk, L. 2016. Vegetation mapping in the St. Lucia estuary using very high resolution multispectral imagery and LiDAR. *South African Journal of Botany*. Available online at doi:10.1016/j.sajb.2016.04.010.

Lyons, M.B., Phinn, S.R., and Roelfsema, C.M. 2012. Long term land cover and seagrass mapping using Landsat and object-based image analysis from 1972 to 2010 in the coastal environment of South East Queensland, Australia. *ISPRS Journal of Photogrammetry and Remote Sensing* 71, 34–46.

Malahlela, O.E., Cho, M.A., and Mutanga, O. 2015. Mapping the occurrence of Chromolaena odorata (L.) in subtropical forest gaps using environmental and remote sensing data. *Biol. Invasions* 17, 2027–2042.

Marom, M., Goldstein, R.M., Thornton, E.B., and Shemer, L. 1990. Remote sensing of ocean wave spectra by interferometric synthetic aperture radar. *Nature* 345, 793–795 (28 June 1990). doi:10.1038/345793a0.

Martin, J., Eugenio, F., Marcello, J., and Medina, A. 2016. Automatic sun glint removal of multispectral high-resolution worldview-2 imagery for retrieving coastal shallow water parameters. *Remote Sensing* 8, 37.

Maul, G.A. 1973. Mapping ocean currents using ERTS imagery. In NASA. *Goddard Space Flight Center Symposium on Significant Results Obtained from the ERTS-1*, Vol. 1, Sect. A and B, 1365–1375.

Meijerink, A.M.J., Valenzuela, C.R., and Stewart, A. (Eds.). 1988. *ILWIS: The Integrated Land and Watershed Management Information System*. Enschede, the Netherlands: International Inst. for Aerospace Survey and Earth Sciences.

Melesse A.M., Weng, Q., Thenkabail, P.S., and Senay, G.B. 2007. Remote sensing sensors and applications in environmental resources mapping and modelling. *Sensors* 7, 3209–3241.

Mélin, F., and V. Vantrepotte 2015. How optically diverse is the coastal ocean? *Remote Sensing of Environment* 160, 235–251.

Mobley, C.D. 1994. *Light and Water: Radiative Transfer in Natural Waters*. San Diego: Academic Press.

Morrow, R., and Le Traon, P.-Y. 2012. Recent advances in observing mesoscale ocean dynamics with satellite altimetry. *Advances in Space Research* 50(8), 1062–1076.

Mott, H. 2007. *Remote Sensing with Polarimetric Radar*. New York: Wiley-IEEE Press.

Mouw, C.B., Greb, S., Aurin, D., DiGiacomo, P.M., Lee, Z., Twardowski, M., Binding, C. et al. 2015. Aquatic color radiometry remote sensing of coastal and inland waters: Challenges and recommendations for future satellite missions. *Remote Sensing of Environment* 160 (April 2015), 15–30. doi:10.1016/j.rse.2015.02.001.

Mumby, P.J., Green, E.P., Edwards, A.J., and Clark, C.D. 1997. Measurement of seagrass standing crop using satellite and digital airborne remote sensing. *Marine Ecology Progress Series* 159, 51–60.

Myint, S.W., and Walker, N.D. 2002. Quantification of surface suspended sediments along a river dominated coast with NOAA AVHRR and SeaWiFS measurements: Louisiana, USA. *International Journal of Remote Sensing* 23(16), 3229–3249.

Pacheco, A., Horta, J., Loureiro, C., and Ferreira, O. 2015. Retrieval of nearshore bathymetry from Landsat 8 images: A tool for coastal monitoring in shallow waters. *Remote Sensing of Environment* 159, 102–116.

Pajares, G. 2015. Overview and current status of remote sensing applications based on unmanned aerial vehicles (UAVs). *Photogrammetric Engineering & Remote Sensing* 81(4), 281–329.

Pardo-Pascual, J.E., Almonacid-Caballer, J., Ruiz, L.A., and Palomar-Vázquez, J. 2012. Automatic extraction of shorelines from Landsat TM and ETM+ multi-temporal images with subpixel precision. *Remote Sensing of Environment* 123, 1–11.

Pe'eri, S., and Long, B. 2011. LIDAR technology applied in coastal studies and management. *Journal of Coastal Research* 62, 1–5.

Pe'eri, S., Morgan, L.V., Philpot, W.D., and Armstrong, A.A. 2011. Land-water interface resolved from airborne LIDAR bathymetry (ALB) waveforms. *Journal of Coastal Research* 62, 75–85.

Ruffner, K.C. 1995. Preface. In *CORONA: America's First Satellite Program*, K.C. Ruffner (Ed.). Washington, DC: CIA Cold War Records Series, Center for the Study of Intelligence, Central Intelligence Agency.

Ryan, A., and Hedrick, J.K. 2005. A mode-switching path planner for UAV-assisted search and rescue. *Proceedings of the 44th IEEE Conference on Decision and Control.* 1471–1476. doi: 10.1109/CDC.2005.1582366.

Shen, L., Xu, H., and Gui, X. 2012. Satellite remote sensing of harmful algal blooms (HABs) and a potential synthesized Framework. *Sensors* 12(6), 7778–7803.

Smayda, T.J. 1997. What is a bloom? – A commentary. *Limnol. Oceanogr* 42(5, part 2), 1132–1136.

Solanki, H.U., Dwivedi, R.M., Nayak, S.R., Jadeja, J.D., Thakar, D.B., Dave, H.B., and Patel, M.I. 2001. Application of ocean colour monitor chlorophyll and AVHRR SST for fishery forecast: Preliminary validation results off Gujarat coast, northwest coast of India. *Indian Journal of Marine Sciences* 30, 132–138.

St Martin, K. (Ed.). 1993. *Explorations in Geographic Information Technology Volume 3: Applications in Coastal Zone Research and Management.* Worcester, MA: Clark University, Graduate School of Geography and Geneva, Switzerland: United Nations Institute for Training and Research (UNITAR).

Tassan, S. 1994. Local algorithms using SeaWiFS data for the retrieval of phytoplankton, pigments, suspended sediment, and yellow substance in coastal waters. *Applied Optics* 33(12), 2369–2378.

Toth, C., and Jóźków, G. 2016. Remote sensing platforms and sensors: A survey, *ISPRS Journal of Photogrammetry and Remote Sensing*, 115, 22–36. http://0-dx.doi.org.library.ucc.ie/10.1016/j.isprsjprs.2015.10.004.

Turner, I.L., Harley, M.D. and Drummond, C.D. 2016. UAVs for coastal surveying. *Coastal Engineering* 114, 19–24. http://dx.doi.org/10.1016/j.coastaleng.2016.03.011.

U.S. Department of Commerce. 1961. *Catalogue of Meteorological Satellite Data – TIROS I Television Cloud Photography.* Washington, DC: U.S. Dept. of Commerce, Weather Bureau. Available online at http://docs.lib.noaa.gov/rescue/TIROS /QC8795C381961TIROS1.pdf (accessed 7 April 2016).

van Aardt, J.A.N., Vogel, M., Lück, W., and Althausen, J.D. 2010. Remote sensing systems for operational and research use. In *Manual of Geospatial Science and Technology,* 2nd ed. London, England: Taylor & Francis, pp. 319–361.

Vanhellemont, Q., and Ruddick, K. 2015. Advantages of high quality SWIR bands for ocean colour processing: Examples from Landsat-8. *Remote Sensing of Environment* 161, 89–106.

Ventura, D., Bruno, M., Lasinio, G.J., Belluscio, A., and Ardizzone, G. 2016. A low-cost drone based application for identifying and mapping of coastal fish nursery grounds. *Estuarine, Coastal and Shelf Science* 171, 85–98.

Vignudelli, S., Kostianoy, A.G., Cipollini, P., and Benveiste, J. (Eds.). 2011. *Coastal Altimetry.* Berlin: Springer. doi: 10.1007/978-3-642-12796-0.

Wabnitz, C.C., Andréfouët, S., Torres-Pulliza, D., Müller-Karger, F.E., and Kramer, P.A. 2008. Regional-scale seagrass habitat mapping in the Wider Caribbean region using Landsat sensors: Applications to conservation and ecology. *Remote Sensing of Environment* 112, 3455–3467.

Warren, C., Dupont, J., Abdel-Moati, M., Hobeichi, S., Palandro, D., and Purkis, S. 2016. Remote sensing of Qatar nearshore habitats with perspectives for coastal management. *Marine Pollution Bulletin.* doi:10.1016/j.marpolbul.2015.11.036.

Wei, L., Hu, Z., Dong, L., and Zhao, W. 2015. A damage assessment model of oil spill accident combining historical data and satellite remote sensing information: A case study in Penglai 19-3 oil spill accident of China. *Marine Pollution Bulletin* 91, 258–271.

Xhardé, R., Long, B.F., and Forbes, D.L. 2011. Short-term beach and shoreface evolution on a cuspate foreland observed with airborne topographic and bathymetric LIDAR. *Journal of Coastal Research* 62, 50–61.

Xing, Q., Ruolin, M., Mingjing, L., Lei, B., and Xin, L. 2015. Remote sensing of ships and offshore oil platforms and mapping the marine oil spill risk source in the Bohai Sea. *Aquatic Procedia.* Maritime Oil Spill Response, 3 (March 2015), 127–32. doi:10.1016/j.aqpro.2015.02.236.

Zibordi, G.M., Melin, F., and Berthon, J.-F. 2006. Comparison of SeaWifs, MODIS and MERIS radiometric products at a coastal site. *Geophysical Research Letters* 33(6), L0661.

5

Current and Future Information and Communication Technology (ICT) Trends in Coastal and Marine Management

Dian Zhang, Noel E. O'Connor and Fiona Regan

CONTENTS

5.1 Introduction

Monitoring of water, globally and within Europe, had been increased in the past decade and will continue to grow in the coming years to comply with legislative requirements such as the EU Water Framework Directive (EC, 2000), the Bathing Water Directive (EC, 2006), the Water Floods Directive (EC, 2007), the Marine Strategy Framework Directive (EC, 2008), and in response to the pressures of climate change, which will lead to resource scarcity, ocean acidification and water quality changes. Traditionally, monitoring of water relies on field studies using conventional manual sampling and subsequent laboratory analysis. These traditional methods are unlikely to provide reasonable estimates of the true maximum or mean concentrations for the physico-chemical variables in a water body to help scientists to better understand the natural environment. The use of a network of sensors (Sensor Web) in conjunction with advanced geographic information technologies offers the potential to reduce costs considerably, providing more useful, continuous monitoring capabilities to give an accurate idea of changing environment and water quality. In addition, Sensor Web can provide real-time information and contribute to a greater representation of long-term trends in aquatic environments, which can help operators better understand and subsequently better manage these resources (Greenwood et al., 2007).

The Sensor Web, a term first used by Kevin Delin in the National Aeronautics and Space Administration (NASA) New Technology Report on Sensor Webs, was originally defined as a web of wireless, intra-communicating, spatially distributed sensor devices that are deployed to monitor and explore environments (Delin et al., 1999). Each sensor node in a Sensor Web comprises a transducer that physically interacts with the environment and converts environmental parameters into electrical signals, along with telecommunication capabilities, power sources, energy harvesting devices and computation devices (Delin and Jackson, 2001). A Sensor Web with wireless communication capability is sometimes referred to as a Wireless Sensor Network (WSN). Modern aquatic environmental monitoring systems often require sensing at high spatial and temporal scales. A Sensor Web (or WSN) can provide the capability to meet these challenges. A wireless aquatic environmental monitoring sensor network typically consists of a number (a few to thousands) of *in situ* sensors, or a combination of multimodal sensors, working together to obtain data about the environment (Yick et al., 2008). In the context of environmental sensing, *in situ* means sensors in direct contact with the medium of interest, as opposed to methods such as remote sensing, where no contact is made between the sensor and the analyte. Sensor Web can provide the fundamental components of monitoring programmes that involve observation, understanding and controlling of the physical world. The advancements in Sensor Web have resulted in the development of affordable, low-power, environmentally friendly, diminutive and multifunctional devices that consist of sensing, data pre-processing and transmitting components (Yick et al., 2008). These low-cost units have been successfully employed in many exciting application domains, such as surveillance, sports, health, automobile industry, etc. and are the key components in the next generation of coastal and marine monitoring and management systems.

In recent years, the concept of Sensor Web has been extended to a whole new scale, the Internet of Things (IoT), primarily due to the exponentially increasing number of connected devices deployed (Weber and Weber, 2010; Kopetz, 2011). Generally speaking, IoT refers to the infrastructure of ubiquitous wireless sensing and identification system interconnection of anything from anywhere at any time (Xia et al., 2012). The world-leading networking technology provider Cisco predicts that 50 billion 'things' will be connected to the Internet by the year 2020, each one sensing, controlling or providing information about the physical world (Figure 5.1). IoT provides the fundamental infrastructure of future global-scale environmental monitoring systems, which fuse information captured from a vast number of multimodal smart devices all over the world.

The rapid growth of connected devices brings numerous challenges, such as those posed by Big Data and the associated analytics, in future environmental monitoring programmes. The term 'Big Data' refers to the collected datasets that are too large for traditional applications to store or process. Thus, intelligent methods must be created to solve these challenges. A possible solution would be distributing the data processing, by converting traditional dumb units into smart devices that perform data analytics on chip. Geoinformatics, which already played an important role in current large-scale coastal and marine monitoring systems, is likely to remain the key component in the future, especially when modelling natural phenomena.

Early stage works in both academia and industry have already shown the potential of such smart systems in terms of better environmental understanding, cost reduction, natural resource saving and improved environmental management. However, ongoing research is still needed in many directions to further advance current findings.

5.2 Aquatic Environmental Management Systems – The Components

Large-scale aquatic environmental monitoring systems are complicated and often require multidisciplinary knowledge. Typically, a monitoring system consists of sensing, data transmission and data mining. In the following section, each of these components is discussed in detail, which provides an overall picture of the requirements to develop such a system.

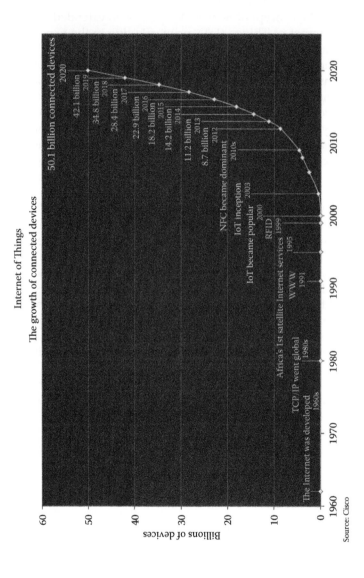

FIGURE 5.1
The exponential growth of connected devices brings new chances and challenges that connect people to the environment. (From Cisco.)

5.2.1 Sensing

Sensing is used to detect and respond to some kind of input from the physical environment using a sensor device. In the past decade, sensing technology has been significantly enhanced in the environmental monitoring domain. New portable and *in situ* sensing devices have been introduced. Remote sensing, satellite or camera-based, has been used to perform standalone large-scale monitoring, or to complement the *in situ* sensing. Depending on the sensing methods used, sensors can also be categorized as reagent-based chemical analysers or electrical-optical sensors. A reagent-based chemical analyzer is based on discrete injection of the reagent and aqueous sample by metering pumps into a microfluidic chip. The reacted sample is presented to the analyzer's spectrophotometer, which converts the reflected light into digital signals. An electrical-optical sensor determines the characteristics of a sample by measuring the current flow between positive and negative electrodes or the reflected light from the analyte.

Portable sensing is defined as acquiring information from the physical environment using a portable device. Portable sensing is a semi-automatic process, which generally requires some kind of manual operations, for example, collecting and preparing water samples, mixing with reagent, reading and logging outputs. Portable sensing is also commonly used to calibrate *in situ* sensors on site. Depending on the parameter it measures, a portable device may give an instant output or a result after a short period. Figure 5.2a shows a YSI Pro DSS multi-parameter portable sensor device that can provide instant measurements of many water quality parameters such as dissolved oxygen, conductivity and total dissolved solids and so on. A ColiSense portable sensor, which is a rapid on-site unit for the detection of the faecal pollution indicator *Escherichia coli* (*E. coli*) in environmental waters, is also shown in Figure 5.2b. It may require up to 75 minutes to deliver the results of analysis. However, it has significantly reduced the complexity and time required to gather data compared to traditional laboratory-based methods (Heery et al., 2016). Although portable sensing is convenient, it still requires trained personnel to operate at the observation site.

In situ sensing is defined as a technology used to acquire information about an object or system when the distance between the object and the sensor is comparable to or smaller than any linear dimension of the sensor (Teillet et al., 2002). The key differences between *in situ* sensors and portable sensors are that *in situ* sensors can be deployed over a long period of time and can perform autonomous measurements. Moreover, *in situ* sensor units are often equipped with, or can be connected to, an external data transmission unit that can send real-time data to a base station. *In situ* sensors can be powered by mains electricity, battery or renewable energy such as solar or wind. Figure 5.3 shows two examples of *in situ* sensors that can monitor multiple water quality parameters over long periods. A list of some currently available commercial multi-parameter aquatic environmental monitoring sondes

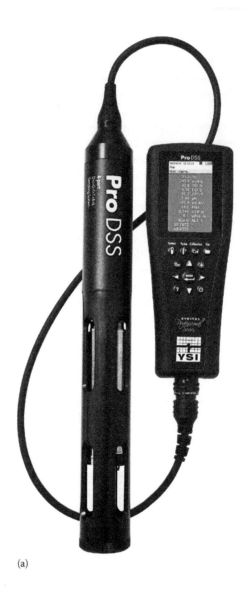

(a)

FIGURE 5.2

(a) YSI Pro DSS, a multi-parameter portable sensor that provides instant measurements of a variety of combinations for dissolved oxygen, conductivity, total dissolved solids, and so on. (From www.ysi.com.) *(Continued)*

with their manufacturers and the parameters that they are reported to measure is provided in Table 5.1.

Remote sensing, in contrast to *in situ* sensing, represents the case when the distance between the object and the sensor is much greater than any linear dimension of the sensor (Teillet et al., 2002). Recently, visual sensing using

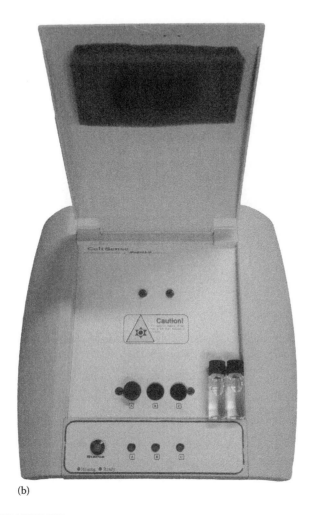

(b)

FIGURE 5.2 (CONTINUED)
(b) ColiSense is a field portable fluorimeter, designed for on-site enzyme fluorescence assay for *E. coli*. (From Heery, B., Briciu-Burghina, C., Zhang, D., Duffy, G., Brabazon, D., O'Connor, N. and Regan, F., 2016. *Talanta*, 148, 75–83.)

cameras has been used to complement *in situ* sensing. Zhang et al. (2014), for example, describe how a camera was used to monitor surface activities at an estuary, which provided additional context information of the surroundings in conjunction with a multi-parameter *in situ* sensor. The information that the camera provided was used to contextualize the abnormal events detected by the *in situ* sensor. Elsewhere, satellite remote sensing has been used to monitor inland and marine environments at regional and global scales. Sentinel-3A, the first newly launched satellite in the three-satellite constellation (including

FIGURE 5.3
YSI 6600 V2 and Hydrolab DS5X multi-parameter water quality sondes that are designed for long-term monitoring and profiling real-time multiple water quality parameters, such as turbidity, dissolved oxygen, algae, conductivity and so on *in situ*. (From www.mestech.ie.)

TABLE 5.1

A List of Some Commercially Available Multi-Parameter Water Quality Monitoring Sondes – The Manufacturer and the Available Sensor Probes for Multiple Water Parameters Are Also Listed

Sensor	Parameters	Manufacturer
TROLL® 9500	Ammonia, barometric pressure, chloride, conductivity, dissolved oxygen, level, nitrate, pH, turbidity	In-Situ Inc. https://in-situ.com
YSI 6600 V2	Ammonia, biochemical oxygen demand, blue-green algae, chloride, chlorophyll, conductivity, depth, dissolved oxygen, hydrocarbon, level, nitrate, pH, photosynthetic active radiation, rhodamine, temperature, turbidity	YSI https://www.ysi.com
Hydrolab DS5X	Ambient light, ammonium, blue-green algae, chlorophyll, chloride, conductivity, depth, dissolved oxygen, nitrate, oxygen reduction potential, pH, rhodamine, temperature, total dissolved gas, turbidity	Hydrolab http://hydrolab.com
Sea-Bird Coastal HydroCAT	Conductivity, dissolved oxygen pressure	Sea-Bird Coastal http://www.ott.com
Eureka 2 multiprobe	Ammonium, blue-green algae, chloride, chlorophyll, conductivity, crude oil, depth, dissolved oxygen, nitrate, pH, rhodamine, salinity, sodium, temperature, turbidity	Eureka Environmental http://www.waterprobes.com
AP-7000	Conductivity, depth, dissolved oxygen, oxygen reduction potential, resistivity, salinity, SSG, temperature, total dissolved solids	Aquaread http://www.aquaread.com

Sentinel-3B and Sentinel-3C) developed by the European Space Agency (ESA), aims at achieving an autonomous, multi-level operational Earth observation capacity. The main objectives of the Sentinel-3 satellite constellation include measuring sea-surface topography, sea-surface temperature, ocean and land-surface colour, sea-water quality and pollution, inland water monitoring and the Earth's thermal radiation (Donlon et al., 2012). The role of satellite remote sensing in coastal and marine management is discussed in greater detail by Lück-Vogel in Chapter 4 of the present volume.

5.2.2 Data Communication

In order to communicate to the base station, sensor nodes in a Sensor Web incorporate a radio module. There are various data transmission technologies applied in coast and marine environmental monitoring systems. The optimal data transmission method for a Sensor Web is task-dependent, but is normally based on three main factors: bandwidth, distance and power consumption. Some commonly used wireless data transfer technologies are listed below and discussed.

Satellite Communication: A communication satellite functions as a relay station, to provide a communication link between a source and a receiver. Since the communication satellite operates at the geostationary orbit, it can cover a wide area over the surface of the Earth. Satellite communication has a number of drawbacks, such as high cost, long delay, special hardware requirement and so on.

WiMax: Worldwide Interoperability for Microwave Access (WiMax) is a wide-range wireless communications standard designed to provide high-speed connection over a long distance (Eklund et al., 2002). It can reach up to 10 Mbps at 10 km with line-of-site (optical visibility). However, it is a power-hungry technology and requires significant electrical support. In addition, the installation and operational costs are very high. Weather conditions, such as rain, could also affect the signal. WiMax has been used for data communication between near-coast buoys and base stations.

Wi-Fi: It is defined as any wireless local area network (WLAN) that is based on IEEE 802.11 standards (IEEE Computer Society LAN MAN Standards Committee, 1997). Wi-Fi is an ad-hoc network, where devices equipped with a wireless network interface controller connect to a hotspot and the data is transferred to the Internet or other devices on the local network. Deploying and maintaining a Wi-Fi network involves relatively low cost and effort. Depending on the standards that the devices support, the data communication speed can reach 300 Mpbs over a 100 m range. The disadvantage of a Wi-Fi enabled sensor is that the power consumption is relatively high compared to other technologies, such as Bluetooth and ZigBee.

Mobile Network: In recent years, mobile network operators have started providing fast mobile broadband connection services, over 3G and 4G technologies, along with existing voice and text services based on GSM (Global System for Mobile Communications) and GPRS (General Packet Radio Service) technologies. Speed, coverage and cost are the main advantages of mobile networks. The throughput of mobile broadband connection is between a few hundred kilobits and tens of megabits per second. According to the Ericsson Mobility Report 2014 (Ericsson, 2014), more than 85% of the world's population is expected to have 3G coverage by the year 2017. Mobile network technology is also commonly used in Sensor Web for data transmission. Many sensors have a GPRS module and can send data to another GSM enabled device or an online data portal for further processing. GPRS has great coverage throughout the world and the cost of purchase, deployment, maintenance and operation is low. 3G or 4G mobile networks are also used for transmitting image data from camera-based visual sensors. One example of the use of mobile network-enabled sensors comes from Del Villar-Guerra et al. (2012), who used telemetry-enabled tags to monitor the movement and behaviour of seals in Irish coastal waters. Data from these tags were transmitted via mobile networks to databases onshore from which they could be retrieved, processed and analyzed.

Bluetooth: It is designed for data communication over short distances for fixed and mobile devices. The new Bluetooth 4.0 includes Classic Bluetooth, Bluetooth high speed and Bluetooth Low Energy (BLE, marketed as Bluetooth Smart). Bluetooth Low Energy (Decuir, 2010), the key new feature in Bluetooth 4.0, is aimed at very low power applications running off small cell batteries, while still providing long range and high-speed data exchanging. Depending on use cases, the power consumption of BLE varies between 0.01 and 0.5 Watt. BLE can cover up to 100 m and the application throughput is 0.27 Mbps. Bluetooth is often used for inner Sensor Web data communication.

ZigBee: Similar to Bluetooth, ZigBee is designed for low power and short range data communication. It is typically used in low data rate applications that require long battery life. The data rates of ZigBee standard vary from 20 Kbit/s to 250 Kbit/s and the range is between 10 and 100 meters. The power consumption of ZigBee standard is 0.1 mW. The ZigBee high power can reach up to 1500 m; however, the power consumption is much higher (60 times more than the standard version). The cost of deploying and maintaining a ZigBee network is relatively low. Similar to Bluetooth Low Energy, Zigbee is commonly used for inner Sensor Web data transmission among sensor nodes or sensor nodes to a centre node. However, data communication between centre node and base stations may still require other types of technologies.

Low Power Wide Area Networks: The emergence of IoT has put focused attention on *low power wide area networks* (LPWAN), a specialized long range, low cost, low bit rate and low power wireless data communication technology. Albeit relatively new and not yet formalized, LPWAN is ideally suited for low-intensity information exchange among a large number of connected devices. Different companies or organizations have their own implementations and protocols, such as SIGFOX, LoRa, NWave, OnRamp and so on. Depending on the implementation, the end-devices can be located up to tens of kilometres from a base station. In addition, due to better power efficiency, the battery on a sensor device may last for many years. SIGFOX, a French company founded in 2005 (www.sigfox.com), is one of the most widely known LPWAN service providers. By early 2016, SIGFOX's service covered 16 countries across Europe and North America. Compared to other relatively high cost technologies, such as mobile network or satellite communication, SIGFOX charges can be as low as $1 per device per year. However, it should be noted that LPWAN is not suitable for large quantities of data transmission such as visual data across a network.

5.2.3 Data Mining

The huge volumes of data (Big Data), collected from monitoring systems, can only be utilized after being converted to meaningful information. The potential benefit of machine learning has already been demonstrated in the literature in adjacent research domains, such as smart cities air pollution forecasting models (www.siemens.com) where a number of artificial intelligence (AI) methods, such as artificial neural networks (Aleksander and Morton, 1990), deep learning (Ngiam et al., 2011) and support vector machines (Cortez and Vapnik, 1995), have already been employed for event detection (Sadlier and O'Connor 2005; Sakaki et al., 2010). In aquatic management systems, similar applications map to use cases around real- or near real-time alerting systems. Outputs from the machine learning algorithms can provide information to notify relevant personnel (aquaculture farmers or local authorities, etc.) in relation to water quality in the case of toxic algal blooms, or water level in the case of flooding, for example, thus appropriate action can be taken to avoid or limit negative impacts. In addition, with the support of data mining techniques, multiple complementary data sources can be organized into a more understandable structure and combined to provide high-level information to support coastal zone management. Used in conjunction with geographic information systems (GIS) technologies, this high-level information can be mapped back to the environment for communication to decision makers, as well as for further analysis and modelling.

Data mining in the context of environmental monitoring can be decomposed into data storage, data processing and information management.

5.2.3.1 Data Storage

Traditionally, data are often stored on a standard desktop PC. However, in recent years, it has been recognized that there are significant advantages to store data on cloud-based storage provided by hosting companies, such as Dropbox, Google Drive, Amazon S3 (Amazon Simple Storage Service) and Microsoft Azure Storage. Collecting a set of environmental data often needs a lot of effort, and is time consuming, while some natural phenomena being monitored may only occur over time periods measured in years or longer. In these and other situations, cloud storage is often considered more reliable compared to traditional local storage. Service providers normally have data protection schemes, in which the data are stored on multiple drives at multiple locations to provide redundancy and guard against hardware failure or natural disaster. Cloud storage is also more flexible. The size of the data may increase over time, while on the cloud storage the storage size can be expanded or shrink dynamically by just changing a few settings.

5.2.3.2 Data Processing

Data processing is the procedure of producing useful information from a dataset. An example of such a data processing flow would be data cleaning, which validates inputs and removes invalid sensor measurements from the dataset; abnormal detection, which isolates unusual readings from the data stream; event creation that groups abnormal measurements into events and clustering/classification, which converts data into meaningful information. Various statistical and machine learning techniques can be applied at each step of the process.

Depending on the scale of the dataset generated from a monitoring system, data processing can be performed on a local machine, on super computers or on a computing cloud. A local machine may be suitable for proof-of-concept, in which new algorithms can be developed, evaluated and tested. Computationally expensive tasks such as modelling or simulation, particularly when geotechnology is applied, may need to run on supercomputers, which normally have several hundred thousand CPU cores and multi-terabytes of random access memory (RAM). These supercomputers are generally run by nationally supported high performance computing platforms, to which researchers can apply for computing resources and the time frames to run processes. Some supercomputers in Europe include ICHEC (Ireland), JUGENE (Germany), Tera 100 (France) and so on. Moreover, some research centres such as NASA may have their own computer clusters (NASA's Pleiades is one of the world's most powerful supercomputers). Similarly, computationally expensive tasks can be performed on computing cloud platforms as well, such as Amazon Web Services (AWS), Microsoft Azure, Google Cloud Dataproc and so on.

5.2.3.3 Information Management

To make large-scale monitoring systems more efficient and effective, the analyzed data need to be presented in a convenient manner that can be easily accessed and interpolated by operators. Interactive data visualization (Murray, 2013) such as Ireland's Digital Ocean (www.digitalocean.ie) provides the infrastructure to access and manage this information more efficiently (Ward, 2010). Moving beyond the traditional static graphics and spreadsheets, which allow the user to browse the final output, an interactive graphical user interface enables users to gain more details and interactively changes the information to be shown and how it is processed.

Leveraging the ubiquity of smart devices, the visualization tools can be web-based or mobile app-based, which allow end users to gain access to information on their mobile devices at any time and any location. Due to the obvious spatial and topographic attributes in location-enabled aquatic monitoring systems (all the data are geotagged), GIS are often used to reflect directly the spatial distribution of environmental elements. GIS can also be of use in water quality prediction, water environmental capacity assessment, pollution prediction and understanding the distribution of the amount of pollutant reduction. It can provide multi-faceted and multi-form support with table and graphics for the water environmental management decisions (Yang et al., 2011).

5.3 Current Coastal and Marine Water Quality Monitoring Systems

A number of environmental monitoring systems have been developed globally and within Europe. In this section, some of the state-of-the-art coastal and marine monitoring systems are explored, and their details are summarized in Table 5.2.

5.3.1 Argo

The Argo project (a global array of temperature/salinity profiling floats) provides that the physical state of the global upper ocean is being systematically measured and the data assimilated in near real-time into computer models

TABLE 5.2

Current State-of-the-Art Coast and Marine Environmental Monitoring Programmes

System Name	Sensor and Platform	Scale	Operator	Data Transmission	Parameters	Open Data
Argo	Argo float	Global	CLS	Satellite	Water (temperature/salinity)	No
NDBC	Moored buoys, ground stations	Global	NOAA	Satellite	Water, atmosphere	Yes
GOS	Ground stations, radiosondes (air-based), ships, drifting buoys, aircraft, satellite	Global	World Meteorological Organization	Satellite	Water, atmosphere	Unknown
IMOS	Vessel, satellite, buoys, ground stations, AUV	Global	University of Tasmania, Australian Marine and Climate Science Community	Satellite	Water, atmosphere, Earth surface	Yes
GOOS	Volunteer ship and yachts, drifting buoys, tide gauges	Global	Intergovernmental Oceanographic Commission	Satellite	Water, atmosphere	
Copernicus	Satellites, ground stations, airborne and sea-borne sensors	Global	European Commission	Satellite	Land, water, atmosphere	Yes
SmartBay	Buoys and cabled observatory	National	SmartBay Ltd.	Cable, WiMax	Water, atmosphere	Yes

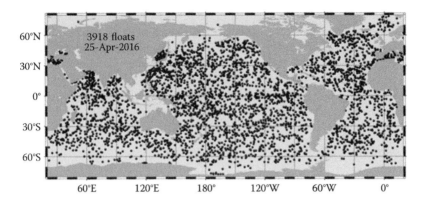

FIGURE 5.4
Argo free-drifting profiling floats distribution map shows the location of all the Argo floats worldwide. (From www.argo.ucsd.edu.)

(www.argo.ucsd.edu). Argo has already grown to be a major component of the ocean observing system, and is a standard to which other developing ocean observing systems can look to. Deployments began in 2000 and continue today at the rate of about 800 per year.

To date, almost 4000 Argo units, as shown in Figure 5.4, are floating on and below the ocean surface worldwide. Each unit operates autonomously, and contains sensors that collect data on a number of marine environmental parameters including sea water temperature, pressure, salinity, electrical conductivity, dissolved oxygen and chlorophyll. The data are collected and transmitted in real time via satellite to regional base stations. At the two major Argo data processing facilities, in France and the United States, the data quality is first verified and then measurements, along with the time-stamp, geolocation, device ID number and so on, are extracted and made available to users. In conjunction with geoinformatics technologies, the data collected from the Argo system have already been used for applications such as ocean surface current mapping, marine biological production (Bushinsky and Emerson, 2015), analyzing world ocean barotropic circulation (Colin de Verdière and Ollitrault, 2016) and seasonal variability of nutrient concentrations (Pasqueron de Fommervault et al., 2015).

5.3.2 National Data Buoy Center

The National Data Buoy Center (NDBC), a subdivision of the U.S. National Oceanic and Atmospheric Administration (NOAA), aims to design, develop, operate and maintain a network of coastal stations and data collecting buoys, which are used for weather forecasting.

Currently, more than 100 moored buoys, 55 Tropical Atmosphere Ocean (TAO) (TAO) stations, 50 Coastal-Marine Automated Network (C-MAN) stations and 39 Deep-ocean Reporting and Assessment of Tsunamis (DART®) tsunameter stations are operating, as shown in Figure 5.5 (Evans et al., 2003). All stations measure wind speed, direction and gusting, atmospheric pressure and air temperature. Some C-MAN stations also measure sea surface temperature, wave height and period. Conductivity and water current are measured at selected stations. Data are primarily transmitted from the platform via a Geostationary Operational Environmental Satellite (GOES) to the NESDIS Data Acquisition Processing System (DAPS) at Wallops Island on the coast of Virginia.

5.3.3 Global Ocean Observing System (GOOS)

GOOS is a permanent global marine and ocean variables observing, modelling and analyzing system that supports operational ocean services worldwide. GOOS provides accurate information of the present state of the oceans, continuous forecasts of the future conditions of the sea and the basis for the analysis of climate change. The primary targets of GOOS include monitoring, understanding and predicting global weather and climate; improving management of marine and coastal ecosystems and assets; minimizing the damage from nature disasters and protecting life and property on coasts and at sea.

As shown in Figure 5.6, GOOS contains 1424 drifting buoys, which record the sea surface current and temperature and the atmospheric pressure. Over 300 embarked systems are on commercial or cruising yachts, which measure temperature, salinity, oxygen and the carbon dioxide in the ocean and the atmosphere and the atmospheric pressure are also included. GOOS also consists of 200 marigraphs and holographs, which transmit information in near real time, providing the possibility of detecting tsunamis. Long-term observations of weather, chemical and biological parameters between sea surface and bottom are also recorded through 482 moorings in open sea.

5.3.4 Global Observing System (GOS)

The purpose of GOS is to monitor the state of the atmosphere and ocean surface from the Earth and space for the preparation of weather analyses, forecasts and warning for climate change. In GOS, there are six types of observation methods: surface observation (11,000 land stations); upper-air observation (1300 radiosondes attached to free-rising balloons from just above ground to heights of up to 30 km); marine observation (4000 ships, 1200 drifting buoys); aircraft-based observation (3000 aircrafts); satellite observation (3 operational near polar-orbiting satellites, 6 operational geostationary environmental observation satellites and several R&D satellites) and weather radar observation (in 76 countries). Figure 5.7 illustrates the architecture of the GOS monitoring system.

5.3.5 Integrated Marine Observing System (IMOS)

IMOS has been regularly operating a wide range of observing systems throughout Australia's coast and oceans since 2006. IMOS is designed to be

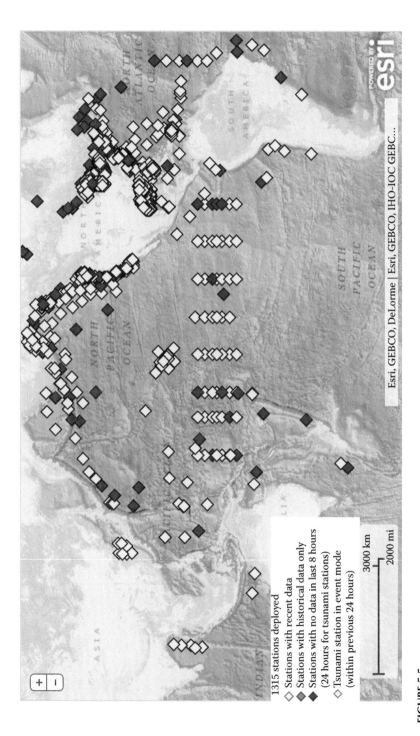

FIGURE 5.5
The distribution of NDBC's global marine and coastal monitoring stations. All the data are transmitted to the NESDIS Data Acquisition Processing System via GOES satellite. (From www.ndbc.noaa.gov.)

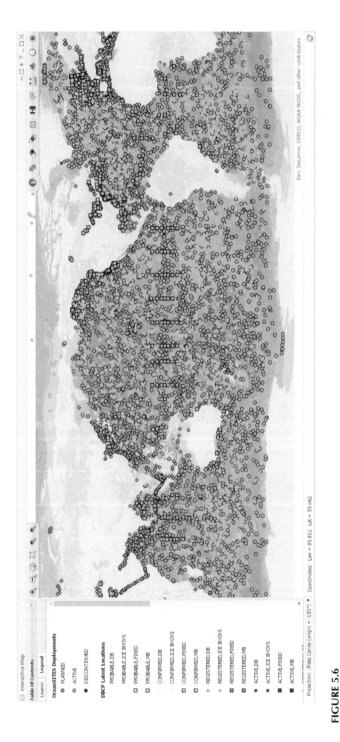

FIGURE 5.6
GOOS monitoring network includes 350 surface measurements from volunteer ships, 1424 global drifting surface buoys, 300 tide gauges, and so on. (From www.jcommops.org.)

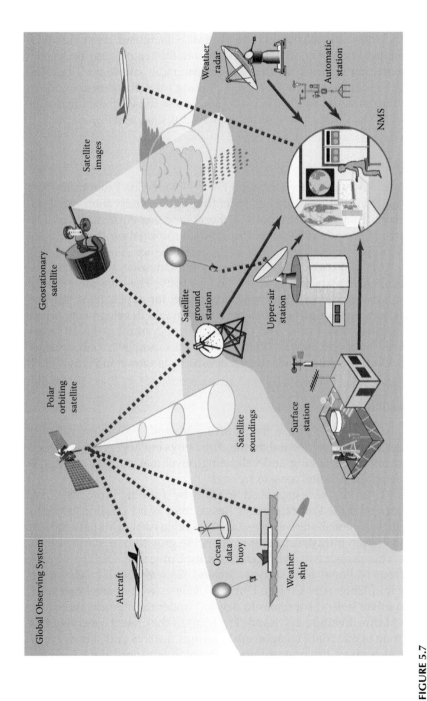

FIGURE 5.7

The Global Observing System includes surface observation, upper-air observation, marine observation, aircraft-based observation, satellite observation and weather radar observation. (From www.wmo.int.)

a fully integrated, national system, observing at ocean-basin and regional scales, and measuring biological, chemical and physical variables (Hill et al., 2009). The aim of IMOS is to observe the oceans around Australia to meet the national and international research needs. IMOS facilitates many forms of observation including deep water moorings, satellite remote sensing, autonomous underwater vehicles, ocean gliders, ocean radar and so on. The value of the IMOS system is to create a robust working infrastructure for end-to-end data management, search, discovery and access to IMOS data that support research. The IMOS open data access portal is shown in Figure 5.8.

5.3.6 Copernicus

Copernicus is a revolutionary European Earth Observation Programme that collects data from multiple sources including Earth observation satellites, *in situ* sensors from ground stations, airborne and sea-borne sensors. The system aims to provide reliable and up-to-date information through a set of services related to environmental and climate issues. The services that Copernicus provides cover six thematic areas: land, marine, atmosphere, climate change, emergency management and security. The data are publicly available at www.copernicus.eu/project-database, for applications such as civil protection, environment protection, sustainable development, urban area management and so on. The Copernicus programme is coordinated and managed by the European Commission. An example visualization of sea-surface elevation data provided by Copernicus is shown in Figure 5.9.

5.3.7 SmartBay

In contrast to the systems above, the SmartBay Ireland project does not only operate a series of environmental monitoring buoys but also provides a subsea infrastructure for testing and validating novel marine technologies in Galway Bay on the Atlantic coast of Ireland.

SmartBay Ireland was established in 2010 by funding from the Irish Government under the Programme for Research in Third Level Institutions (PRTLI V). It was a collaborative project that aimed to establish a national test and demonstration infrastructure to support the growth of a marine sector in Ireland. This project was led by Dublin City University (DCU) and with partners from National University of Ireland Galway (NUIG), Maynooth University and University College Dublin (UCD) as well as Intel and IBM. In 2016 SmartBay Ireland continues to develop under the support and direction of the Marine Institute in Ireland. The focus of the project covers the test and validation of materials, components, coatings, sensors and ocean energy devices. It also offers the opportunity to develop advanced data analytics using the multiple data streams available. Collaborating with the world's largest information and communications technology (ICT) companies gives

FIGURE 5.8
The open access data portal for IMOS data. (From imos.aodn.org.au/imos123/home.)

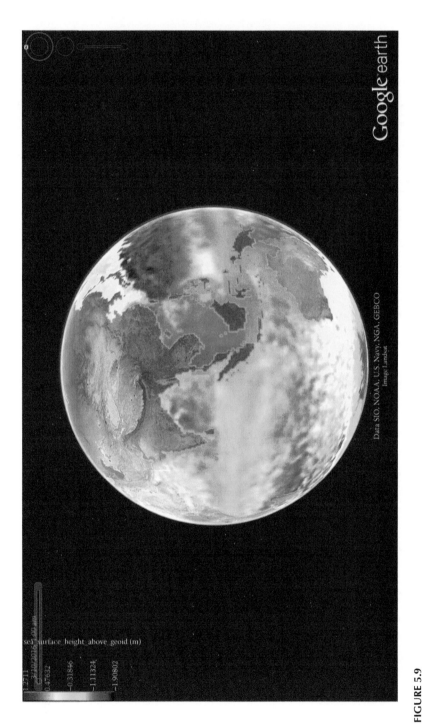

FIGURE 5.9
The visualization of sea surface height (2016-03-10) using open data provided by Copernicus and Google Earth software.

SmartBay the potential to generate projects for the facility, aimed at developing new ICT software products and services for various marine sector applications. Figure 5.10 illustrates the current and projected SmartBay infrastructure, including a fleet of Mobilis DB5800 buoys, sensors, high-speed optical fibre communication systems, power source and a range of transport vessels.

Numerous research and development projects have already been supported in SmartBay, where technologies have been developed, tested and validated (Hayes et al., 2009; Vega et al., 2013; Barattini et al., 2015) in Galway Bay. The future of SmartBay is to continue to grow and support ocean energy developers in a 1/4 scale site in the Atlantic. This phase is a preparation for scaling up of ocean energy devices as well as projects involving marine sensing and communication technologies or translation of technologies to the marine environment.

5.4 The Road Ahead: The Future Ideal Monitoring System

The ideal coastal or marine monitoring system of the future might consist of a network of multimodal interconnected sensors, deployed globally, capable of autonomous operation in the field for long periods (annual to decade time scales). Such a system would enable the possibility of collecting rich and varied information, with high spatial and temporal scale datasets that allow users to build, evaluate, test and interpolate environmental models, thus providing greater confidence in the information provided to the user.

Figure 5.11 shows a possible architecture of a future large-scale multimodal smart environmental monitoring system. The IoT-scale multimodal smart Sensor Web performs autonomous sensing, data collection and data pre-processing, before uploading measurements to a centralized cloud-based data repository. The smart system then loads and processes the data from the data repository, perhaps along with other information gathered from external data sources such as social media, World Wide Web, government agencies and so on to provide high level information to the operator and decision maker. With the support from GIS technology, natural phenomena can be simulated and represented to the end user, based on these sensed data, which provide a richer content-based knowledge to help operators better understand and manage the environment.

Although it is evident that some elements of this ideal monitoring system are already in place, ongoing research and development is required in several areas relating to sensor development, data processing and information access/representing and management. In addition, ensuring the efficient interoperability of these different components presents several challenges, many of which are also the subject of current research (see Chapter 6 of the

FIGURE 5.10
The SmartBay infrastructure includes a fleet of Mobilis DB5800 buoys, sensors, high-speed optical fibre communication systems, power source and a range of transport vessels. (Courtesy of SmartBay Ireland, www.smartbay.ie.)

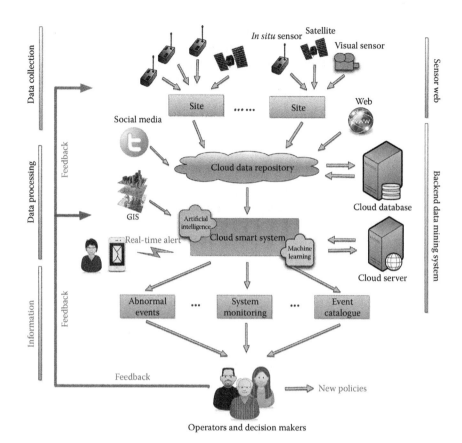

FIGURE 5.11
A system architecture of future global scale, multimodal smart environment monitoring system.

present volume, by Lassoued and Leadbetter, which discusses the specific issue of ontologies and their role in facilitating semantic interoperability between technologies and between the datasets obtained from them). Data communication is generally being developed on its own thread by some professional associations, such as the Institute of Electrical and Electronics Engineers (IEEE), through standardization processes.

5.4.1 Sensor Development

Current *in situ* sensors in aquatic environmental monitoring applications are passive systems (all the settings are pre-configured and do not vary during the period of deployment), which are relatively easy to develop, configure and deploy. However, pre-configured sensors do not adapt based on the occurrences at the scene. In contrast, *smart sensors* can adapt, according to the dynamics on site, which can provide richer details of environmental phenomena, for example by increasing sampling rate when a particular event of interest is occurring, or by reducing sampling rate when no events arise so as to expand the sensor's operating time, especially in the case of reagent-based analyzers that contain a limited amount of reagent, or battery powered systems that need to conserve power.

To date, approximately 20 water quality parameters can be monitored using *in situ* and portable sensors. However, there are more than 100 principal parameters in water quality standards (Flanagan, 1986). Sensors for accurate measurement of these parameters are simply not yet available. There are still challenges to develop, evaluate and field test new reliable, affordable, environment-friendly and long life sensors for these remaining water quality parameters. Early research may focus on development of portable devices and later convert to *in situ* sensors. This development often requires a team of multidisciplinary experts including chemists, microelectronic engineers and software developers.

In addition, future sensors are not limited to physical devices. The vast increasing of online resources, such as social media, can act as 'virtual sensors', which can be used to provide useful information of nature events. An example of this could be the estimation of an area affected by a natural disaster. These virtual sensors can also be used to estimate the public engagement. For example, Figure 5.12 shows the number of Tweets that contain the word 'climate change' during COP21 Paris, France, Sustainable Innovation Forum 2015. It can be clearly seen that the public is more aware of the environment during COP21 (see Chapter 7 by Goldberg, D'Iorio and McClintock, this volume, for further discussion of volunteered geographic information [VGI], crowdsourcing and the use of social media as virtual sensors of coastal and marine information).

5.4.2 Data Processing

Data processing is the core element of future global scale monitoring systems. Many machine learning and data mining techniques have been developed in the computer science research domain. However, they are not tailored for analyzing environmental data. In addition, there is a gap between environmental scientists and computer science experts. Computer science researchers are generally not familiar with all the real problems and subtle requirements of aquatic monitoring systems. On the other hand, environmental scientists are usually not, or not completely, aware of the potential that computer science technologies can offer. Future research may focus on bringing these two fields together, specifically to adopt and adapt the state-of-the-art machine learning and artificial intelligence techniques from computer science, to automate the processing of raw sensor data from multiple sensing modalities and to create a rich content-based information repository that is more suitable for management. Also, lightweight data processing

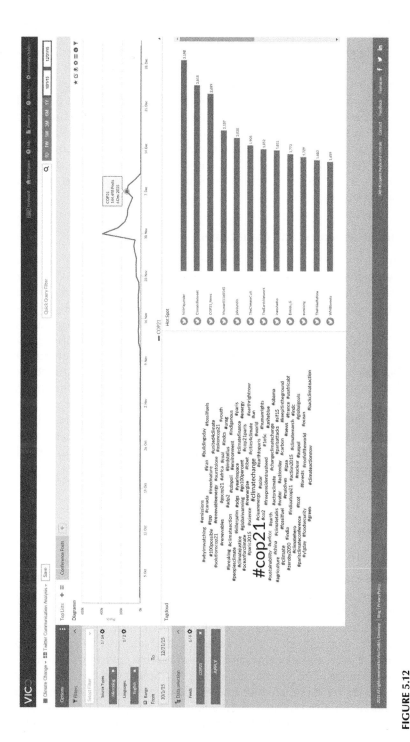

FIGURE 5.12

The number of Tweets (act as 'virtual sensors') that mentioned the words 'climate change' during the 2015 United Nations Climate Change Conference (COP21) increased significantly during the conference, which shows the public awareness of environmental health.

methods need to be developed to perform data pre-processing on-chip, to convert a sensor to a smart sensor.

Smart sensing enables more accurate and automated collection of environmental data by including additional functionalities into the sensor such as self-calibration, self-diagnosis, self-validation and multisensing capabilities. A smart sensor can also adapt itself according to the variation of the physical and chemical parameter that it is measuring. Smart sensing can distribute the workload on the server side, reduce the amount of data needed to be transmitted and make the Sensor Web more efficient. Information produced on one smart device can be used to validate the information generated by nearby sensors or used to wake up nearby sensors to gather a richer dataset of a phenomenon.

5.4.3 Information Access and Management

Information can only be accessed, managed and interpolated by system operators efficiently through the medium of an interactive graphical user interface (GUI). Users can drag, click or move the mouse cursor over to the components in the GUI to access additional information or control the outputs of the interface. Generally, environmental monitoring system operators or environmental scientists may not have the IT skills needed to efficiently manage large scale digital content. Creation of interactive content management tools, which focus on this requirement, is another research direction. The research in this domain may require the development of user-friendly, large scale, non-expert operative and content-rich front end systems, perhaps based on alternatives to the classic GUI such as tactile or gesture-base manipulation within some form of virtual reality environment. Such a system may also be able to control and send feedback to the physical/virtual sensor networks themselves.

5.5 Conclusion

The challenges of coastal and marine monitoring are many. Spatial and temporal variations mean that individual sensors or samples are not providing representative information for management of the aquatic environment. The technology to integrate information sources (*in situ* sensing, remote sensing, online resources, meteorological, etc.) is available and addresses a critical step in advancing national and global scale coast and marine environmental management programmes. The shortfall in terms of complexity is the availability of specific sensors for parameters of interest. The development of sensors to the specifications required for marine monitoring is costly and time consuming. While there are initiatives like the X-prize (Wendy Schmidt Ocean Health,

2015) and the nutrient challenge (Nutrient Sensor Challenge, 2015) to help drive these sensor developments, there is still a lag in terms of availability of specific sensors. Meanwhile, there are also opportunities to use existing technologies (sensors, datasets, network technology) to develop valuable management tools or decision support tools for marine and coastal monitoring. In this chapter, the components required to create a water monitoring system have been discussed, and examples of some current state-of-the-art coastal and marine water quality monitoring systems have been given. Based on our knowledge in this area, we also illustrate our vision of the future ideal aquatic monitoring systems, and the future research directions to achieve this ideal scenario are outlined.

References

Aleksander, I., and Morton, H., 1990. *An Introduction to Neural Computing* (Vol. 240). London: Chapman and Hall.

Barattini, P., Garcés, E., Bonasso, M., Mier, S., Philimis, P., Giusti, A., Thomas, K. et al., 2015. MARIABOX an autonomous monitoring device for marine pollution: From the laboratory to a product: Design challenges and real world trade-off. *Instrumentation Viewpoint* (18), 56–56.

Bushinsky, S.M., and Emerson, S., 2015. Marine biological production from in situ oxygen measurements on a profiling float in the subarctic Pacific Ocean. *Global Biogeochemical Cycles*, 29(12), 2050–2060.

Colin de Verdière, A., and Ollitrault, M., 2016. A direct determination of the world ocean barotropic circulation. *Journal of Physical Oceanography*, 46(1), 255–273.

Cortez, C., and Vapnik, V., 1995, Support vector networks. *Machine Learning*, 20, 273–279.

Decuir, J., 2010. *Bluetooth 4.0: Low Energy*. Cambridge, UK: Cambridge Silicon Radio SR plc, p. 16.

Delin, K.A., Jackson, S.P., and Some, R.R., 1999. Sensor webs. *NASA Tech Briefs*, 23(80), 27.

Delin, K.A., and Jackson, S.P., 2001, May. Sensor web: A new instrument concept. In *Symposium on Integrated Optics*, International Society for Optics and Photonics, pp. 1–9.

Del Villar-Guerra, D., Cronin, M., Dabrowski, T., and Bartlett, D., 2012. Seals as collectors of oceanographic data in the coastal zone. *Estuarine, Coastal, and Shelf Science*, 115, 272–281.

Donlon, C., Berruti, B., Buongiorno, A., Ferreira, M.H., Féménias, P., Frerick, J., Goryl, P. et al., 2012. The global monitoring for environment and security (GMES) sentinel-3 mission. *Remote Sensing of Environment*, 120, 37–57.

EC, Directive 2000/60/EC of the European Parliament and of the Council of 23 October 2000 establishing a framework for community action in the field of water policy. OJ L 327, 22.12.2000, 1–73.

EC, Directive 2006/7/EC of the European Parliament and of the Council of 15 February 2006 concerning the management of bathing water quality and repealing Directive 76/160/EEC. OJ L 64, 4.3.2006, 37–51.

EC, Directive 2007/60/EC of the European Parliament and of the Council of 23 October 2007 on the assessment and management of flood risks. OJ L 288, 6.11.2007, 27–34.

EC, Directive 2008/56/EC of the European Parliament and of the Council of 17 June 2008 establishing a framework for community action in the field of marine environmental policy. OJ L 164, 25.6.2008, 19–40.

Eklund, C., Marks, R.B., Stanwood, K.L., and Wang, S., 2002. IEEE standard 802.16: A technical overview of the WirelessMAN™ air interface for broadband wireless access. *IEEE Communications Magazine*, 40(6), 98–107.

Ericsson, A., 2014. Ericsson mobility report, on the pulse of the networked society. *Ericsson, Sweden, Tech. Rep. EAB-14*, 61078.

Evans, D., Conrad, C.L., and Paul, F.M., 2003. *Handbook of Automated Data Quality Control Checks and Procedures of the National Data Buoy Center*. NOAA National Data Buoy Center Tech. Doc, 3(2).

Flanagan, P.J., 1986. Parameters of water quality: Interpretation and standards (Vol. 6). An Foras Forbartha. Information and Training Centre.

Greenwood, R., Webster, J., and Regan, F., 2007. Royal Society of Chemistry Report.

Hayes, J., O'Hare, G.M., Kolar, H., and Diamond, D., 2009. Building an adaptive environmental monitoring system using Sensor Web. *ERCIM NEWS* (76), 38–39.

Heery, B., Briciu-Burghina, C., Zhang, D., Duffy, G., Brabazon, D., O'Connor, N., and Regan, F., 2016. ColiSense, today's sample today: A rapid on-site detection of β-d-Glucuronidase activity in surface water as a surrogate for E. coli. *Talanta*, 148, 75–83.

Hill, K., Moltmann, T., Meyers, G., and Proctor, R., 2009. The Australian Integrated Marine Observing System (IMOS). *Proceedings of OceanObs'09: Sustained Ocean Observations and Information for Society* (Vol. 1), Venice, Italy, 21–25 September 2009, Hall, J., Harrison, D.E. and Stammer, D., Eds., ESA Publication WPP-306. Available online at http://www.oceanobs09.net/proceedings/ac/FCXNL-09A02-1656575-1-IMOS_Oceanobs_final.pdf.

IEEE Computer Society LAN MAN Standards Committee, 1997. Wireless LAN medium access control (MAC) and physical layer (PHY) specifications.

Kopetz, H., 2011. *Real-Time Systems: Design Principles for Distributed Embedded Applications*. Springer Science & Business Media.

Murray, S., 2013. *Interactive Data Visualization for the Web*. O'Reilly Media, Inc.

Ngiam, J., Khosla, A., Kim, M., Nam, J., Lee, H., and Ng, A.Y., 2011. Multimodal deep learning. In *Proceedings of the 28th International Conference on Machine Learning* (ICML-11) pp. 689–696.

Nutrient Sensor Challenge, 2015. Alliance for Coastal Technologies, accessed 21 March 2016, http://www.act-us.info/nutrients-challenge/.

Pasqueron de Fommervault, O., D'Ortenzio, F., Mangin, A., Serra, R., Migon, C., Claustre, H., Lavigne, H. et al., 2015. Seasonal variability of nutrient concentrations in the Mediterranean Sea: Contribution of Bio-Argo floats. *Journal of Geophysical Research: Oceans*, 120, 8528–8550, doi:10.1002/2015JC011103.

Sadlier, D.A., and O'Connor, N.E., 2005. Event detection based on generic characteristics of field-sports. In: *ICME 2005 – Proceedings of the IEEE International Conference on Multimedia and Expo*, 6–8 July 2005. Amsterdam, the Netherlands.

Sakaki, T., Okazaki, M., and Matsuo, Y., 2010, April. Earthquake shakes Twitter users: Real-time event detection by social sensors. In *Proceedings of the 19th International Conference on World Wide Web*, ACM, pp. 851–860.

Teillet, P.M., Gauthier, R.P., Chichagov, A., and Fedosejevs, G., 2002. Towards integrated earth sensing: The role of in situ sensing. *International Archives of Photogrammetry Remote Sensing and Spatial Information Sciences*, 34(1), 249–254.

Vega, A., Corless, R., and Hynes, S., 2013. *Ireland's Ocean Economy, Reference Year: 2010.* SEMRU, NUI Galway.

Ward, M.O., Grinstein, G., and Keim, D., 2010. *Interactive Data Visualization: Foundations, Techniques and Applications.* Boca Raton, FL: CRC Press.

Weber, R.H., and Weber, R., 2010. *Internet of Things: Legal Perspectives* (Vol. 49). Springer Science & Business Media.

Wendy Schmidt Ocean Health 2015, accessed 21 March 2016, http://oceanhealth .xprize.org/.

Xia, F., Yang, L.T., Wang, L., and Vinel, A., 2012. Internet of Things. *International Journal of Communication Systems*, 25(9), 1101.

Yang, F., Liang, F., and Shi, J., 2011. Applications of GIS in environment monitoring. Available online at http://www.seiofbluemountain.com/upload/product /201105/2011fzjz31.pdf.

Yick, J., Mukherjee, B., and Ghosal, D., 2008. Wireless sensor network survey. *Computer Networks*, 52(12), 2292–2330.

Zhang, D., Sullivan, T., Briciu Burghina, C.C., Murphy, K., McGuinness, K., O'Connor, N.E., Smeaton, A.F., and Regan, F., 2014. Detection and classification of anomalous events in water quality datasets within a smart city–smart bay project. *International Journal on Advances in Intelligent Systems*, 7(1&2), 167–178.

6

Ontologies and Their Contribution to Marine and Coastal Geoinformatics Interoperability

Yassine Lassoued and Adam Leadbetter

CONTENTS

6.1 Introduction

Over recent years, the idea of publishing structured data on the World Wide Web has gained much traction, through high profile applications such as Google's Rich Snippets, which add encyclopaedic information, location details and product data to search results. Facebook's Graph schema is another such high profile application allowing media and previews of websites to enter the social networking platform, and providing recommendations of pages and sites to users. Structured data is also at the heart of the World Wide Web Consortium's (W3C) best practices for publishing data on the Web. When structured data are made available online using the W3C's Resource Description Framework standard, and web addresses are used to identify both the datasets being published and the definitions of their contents and related datasets, it is Linked Data as defined by Tim Berners-Lee and conforms to his vision of Web 3.0 being a web of interconnected data queryable as a single, huge database. Ontologies are the most formal way of structuring data, information and knowledge in a Web-based context, and this chapter explores their application to the geoinformatics communities within the marine and coastal domains.

6.2 Ontologies

Ontologies are part of a wider family of systems called knowledge organization systems (KOS).

Hodge, in *Systems of Knowledge Organization for Digital Libraries: Beyond Traditional Authority Files* (Hodge, 2000), defines KOS as 'all types of schemes for organizing information and promoting knowledge management'. The author provides a taxonomy of KOS, with the following three top-level groups, listed in order of complexity and expressiveness:

1. *Term lists*, which include controlled vocabularies, glossaries, dictionaries and gazetteers, are lists of terms, often with definitions.

2. *Classifications and categories*, such as subject headings, classification schemes and taxonomies, go a step further than term lists by providing a (typically shallow) hierarchical categorization of terms.

3. *Relationship lists*, which include thesauri, semantic networks and ontologies, go another step further by providing semantic relationships between terms (concepts).

In this context, a thesaurus (pl. 'thesauri') defines concepts and the semantic relationships between them. Relationships commonly expressed in a thesaurus are: hierarchy (narrower/broader), synonymy and relatedness. Semantic networks organize concepts as a network rather than a hierarchy. Nodes of the network are concepts, and edges are semantic relationships between concepts. Relationships in a semantic network go beyond the common hierarchy, synonymy and relatedness relationships of thesauri. For instance, you may express that an **instrument** *measures* a **parameter**, in which case '**instrument**' and '**parameter**' are nodes and '*measures*' is a relationship (edge).

Ontologies can be thought of as a generalization of thesauri and semantic networks. As described in the next section, ontologies give the ability to standardize the syntax and semantics of data descriptions; allow browsing through different aspects of a collection of data; aid the discovery of data and allow data to be combined more easily through greater interoperability. In addition to semantic relationships, ontologies introduce logical rules and axioms for reasoning with concepts and relationships. In general, as defined by Marine Metadata Interoperability (MMI), an ontology consists of one or many (but not necessarily all) of the following:

- Classes: Types of 'things', which may be organized hierarchically,
- Individuals: Instances of classes, that is, 'things',
- Relationships between individuals,

- Properties: Data associated with individuals,
- Constraints, and logical rules and axioms over the above.

Figure 6.1 shows the Stimulus-Sensor-Observation ontology (Janowicz, 2010), which defines a model for sensor observations. In this figure, classes are represented with boxes, for example, Sensor, Stimulus, ObservedProperty and so on. Relationships are links (oriented arrows) between classes. For example, a Sensor *detects* a Stimulus, which *is a proxy for* an Observed Property. A Sensor instance may be a thermistor temperature sensor, which detects changes in electrical resistance (Stimulus instance). A change in electrical resistance is a proxy for temperature (instance of ObservedProperty).

Ontologies are mostly useful and powerful when used with a (semantic) reasoner, which can infer new knowledge from asserted facts. For instance, using the above ontology, if you assert that 'change in electrical resistance is a proxy for temperature', a reasoner will infer that change in electrical resistance is a Stimulus, and that temperature is an Observed Property. Reasoners can further deal with the properties of relationships (inverse relationship, transitive relationship, etc.), logical rules and constraints over instance types (classes).

One of the main benefits of ontologies is knowledge reuse, as typically an ontology may refer to resources (classes, relationships or instances) in another ontology.

Ontologies are commonly expressed in ontology languages, such as RDF, the Resource Description Framework (Cyganiak, 2014), or OWL, the Web Ontology Language (W3C OWL, 2012). These provide standard formalisms for expressing and exchanging ontologies.

In this chapter, conforming to the MMI definition, we consider all forms of KOS as ontologies provided that they are expressed in an ontology language.

6.3 Ontology Use Cases

Ontologies are rather general-purpose tools, which may be used for various objectives and in different more or less sophisticated ways. In this section, we present the most common use cases of ontologies in coastal and marine applications.

Use Case 1: Standard Vocabularies

The most basic use case of ontologies is to provide reusable standard domain vocabularies, similar to a data dictionary for a field but with the addition of defined relationships between the terms.

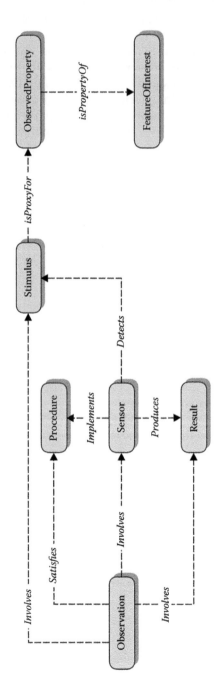

FIGURE 6.1
Stimulus-sensor-observation ontology.

The terms of such vocabularies may be used as standard field values in databases, or metadata (catalogues). Such vocabularies may be interrelated to provide semantic relationships between their terms. And thus can help ensure the interoperability of datasets by providing common terminology, and providing links between data values (cf. Use Case 4). Examples of such vocabularies are the SeaDataNet vocabularies, which provide linked terms with definitions, pertaining to several aspects of marine data, such as parameters, instruments, platforms, themes, places and so on. These constitute a valuable asset for standardizing marine data and metadata field names and values. The relationships between terms are the basis for reasoning with these vocabularies.

Use Case 2: Browsing

Ontology browsing is the ability to graphically navigate an ontology or a thesaurus in order to understand the meaning of the concepts (ideas represented by terms) defined therein, and to find out how these relate to each other semantically. Ontology browsing is useful as a way to provide educational information about a given domain (domain knowledge). It is commonly used in discovery interfaces (cf. Use Case 3) as a way to browse products/resources by topic (e.g. multifaceted product browsing). This is typically of interest to coastal and marine web atlases, where topic (educational) information is crucial for the general users to allow them to understand the provided data.

Use Case 3: Discovery

Ontologies may be used by data discovery services (e.g. search engines and catalogue services) as a means to improve the pertinence of their search results, by exploiting the semantic relationships between terms (narrower, related, same as, etc.), and/or interpreting the meaning of a user free-text keyword according to a given thesaurus. For instance, if you search for datasets matching the term 'seabed', you would be able to get those tagged with the keyword 'seafloor' (synonym), or if you search for 'CTD' (i.e. Conductivity, Temperature, Depth), you would be able to get 'Sea Surface Salinity' datasets, as salinity is related to conductivity.

Use Case 4: Interoperability

Ontologies benefit interoperability in two ways:

1. Conforming to Use Case 1, ontologies may be used to define standard vocabularies for use across different systems. This, in itself, facilitates interoperability.

2. Ontologies can be used as a mechanism for linking terms across different vocabularies, thus facilitating translation from one model (or vocabulary) into another.

In the second case, ontologies may be used as a mapping mechanism between

- Two data structures/schemas (structural interoperability): This is useful when two information systems, with structurally heterogeneous backends, need to interoperate with each other or with a third-party system, for example, mediator, broker, extract transform and load (ETL) tool and so on.
- The values of similar properties (attributes) in different databases, using different representations (semantic interoperability): In this case, we assume that structural interoperability has already been achieved and that actual data values need to be mapped or translated from one model, classification scheme or terminology, to another. This is the typical case of distributed catalogue services using different vocabularies, possibly from different domains or in different natural languages, for metadata values, for example, descriptive keywords, units of measure, parameter names, organization names, and so on.

6.4 State of Ontologies in Marine and Coastal Applications

In the 1980s, UNESCO's International Oceanographic Data and Information Exchange (IODE) Group of Experts on the Technical Aspects of Data Exchange (GETADE) group undertook some extremely high quality content governance on controlled vocabularies in order to underpin the General Formatting (GF3) data exchange format. The result of their work, known as the GF3 code tables, was published as a printed document in 1987 (IOC, 1987). Of particular note from this work was the extension of the code table entry attribute set to include a detailed description of what the code was to be used for as well as a text-based label, which went a long way to improving the understanding of code semantics. Extensive, well thought out knowledge was used to populate the descriptions; the code list entries covering ocean waves remain as exemplars to this day.

Around the turn of the last century, the European Commission SEASEARCH project set out to produce a federated metadata catalogue from across the European oceanographic data centres to allow the integration of marine data created across the continent without any replication of data and information. Within this project, it was quickly realized that controlled vocabularies were essential to such an enterprise and so SEASEARCH worked to set up some 20 vocabularies under the label of common data libraries. However, SEASEARCH failed to fully address the issues associated with vocabulary management. Content governance for the common data libraries was the responsibility of individuals leading to decisions based upon restricted knowledge and opinion. However, technical governance presented an even more serious issue as no authoritative master copies were maintained and so multiple local copies sprang up and started to evolve in different directions.

SEASEARCH demonstrated the need for quality vocabulary management in European oceanographic data management. Vocabulary Management for the Oceanographic Domain was delivered by three projects between 2001 and 2004: NERC DataGrid in the United Kingdom, Enabling Parameter Discovery and the Marine XML Study Group. NERC DataGrid delivered version 0 of the NERC Vocabulary Server that was able to deliver authoritative master copies of controlled vocabularies through universally accessible Web Services. Enabling Parameter Discovery delivered the British Oceanographic Data Centre Parameter Usage Vocabulary, which contains detailed concepts for annotating scientific measurements made in the marine environment to identify what was measured, with the text labels for concepts within the vocabulary systematically built from an underpinning semantic model. Marine XML provided the foundation for vocabulary content governance by committee such as the International Council for the Exploration of the Seas Platforms list and governance body that are still active today.

Parallel developments, particularly spurred by the Marine XML work, led to the creation of the Marine Metadata Interoperability (MMI) project, with the aim of easing the discovery, access and use of marine science datasets. The goal of MMI is to promote collaborative research in the marine science domain, by simplifying the incredibly complex world of metadata into specific, straightforward guidance. MMI hopes to encourage scientists and data managers at all levels to apply good metadata practices from the start of a project, by providing the best advice and resources for data management. A key outcome to date of the MMI project is its semantic framework, which is a set of guidance documents, created in collaboration with members of the marine science community to establish a set of best practices, and a suite of tools to allow users to work with semantic technologies without becoming bogged down in the intricacies of the semantic web.

At a European Commission level, the INSPIRE Directive obliges national authorities of the EU member states to contribute their spatial data according to over 30 harmonized themes (e.g. Hydrography, Protected Sites or Elevation), to make them accessible and to have them described via standardized Geospatial Web Services. These datasets are considered to be up-to-date, quite reliable, EU-wide and mostly freely available, forming a very impressive data source for multi-thematic information retrieval. Tschirner, Scherp and Staab (2011) have used the basic conversion rules to take the original INSPIRE data models (targeted at the Geography Markup Language, and written using XML Schema Definitions) to create focussed INSPIRE ontologies in the Web Ontology Language (OWL). This activity began an ongoing effort to create semantic web representations of both the underlying Open Geospatial Consortium standards used in INSPIRE (e.g. in the Observations and Measurements ontology of Cox, 2013) and the INSPIRE data models (e.g. as in Leadbetter and Vodden, 2015).

Finally, in the computer science sphere, there has been an increasing demand for patterns around ontology development to allow data publishers to better supply compatible resources. The EarthCube cyberinfrastructure programme in the United States has funded the GeoLink project (http://www.geolink.org/), which brings together experts from geosciences, computer science and library science in an effort to develop Semantic Web components that support discovery and reuse of data and knowledge. GeoLink's participating repositories include content from field expeditions, laboratory analyses, journal publications, conference presentations, theses/reports and funding awards that span scientific studies from marine geology to marine ecosystems and biogeochemistry to paleoclimatology. GeoLink is building a set of reusable ontology design patterns (ODPs) that describe core geoscience concepts, a network of Linked Data published by participating repositories using those ODPs and tools to facilitate discovery of related content in multiple repositories.

6.5 Ontology Resources

Taking this review of ontology development in the marine and coastal domain as the state-of-the-art, web-based resources which fall into each of the categories described above will be discussed in detail below.

The most populous resource of relevance to the marine and coastal domain is the NERC Vocabulary Server (http://vocab.nerc.ac.uk; Leadbetter, Lowry and Clements, 2014). This resource provides access to over 98,000 terms registered in approximately 200 vocabularies. Due to it being maintained by, and hosted at, the British Oceanographic Data Centre, the content of the NERC Vocabulary Server is mainly focussed on the marine domain, but it has also been used by the International Coastal Atlas Network (http://www.iode.org/ican) to host some of its ontology resources (cf. Section 6.6.2). The content of the NERC Vocabulary Server is on the less formal end of the ontology spectrum, as it has no formal declarations of terms belonging to a specific ontology class; but it does use narrower-broader relationships to informally declare concepts to be classes or instances of those classes. Ongoing work with the National Science Foundation funded GeoLink project may in due course formalize these class-instance definitions in some way. The vocabulary terms of the NERC Vocabulary Server have been used to provide interoperability in the European Commission SeaDataNet (http://www.seadatanet.org/) and NETMAR (https://netmar.nersc.no/) projects.

A related resource to the NERC Vocabulary Server is the Marine Metadata Interoperability Ontology Registry and Repository (MMI-ORR, http://

mmisw.org/; Graybeal, Isenor and Rueda, 2011). The MMI-ORR expands the NERC Vocabulary Server model in that it can be used to provide access to, and maintain, vocabulary content and formal ontology resources. In contrast to the NERC Vocabulary Server, the MMI-ORR also provides maintenance interfaces for users to be able to register new and update existing content on the server.

Similarly, the Spatial Information Services Stack Vocabulary Service developed by the Commonwealth Scientific and Industrial Research Organisation in Australia, and the UK Government Linked Data Register are both services used by government agencies within those countries to publish and maintain controlled vocabularies relating to the environment, including many coastal and marine resources. In a parallel to the IODE activity described earlier, the Open Source nature of the UK Government Register has seen the register software redeployed by the World Meteorological Organizations as they seek to move their code lists away from paper and PDF to the semantic web (see http://codes.wmo.int/).

Thus far, the reader may feel that no formal ontologies exist for use in the coastal or marine domain as only informal ontologies have been described. However, there are several resources which are formal ontologies, but which also may make use of the resources described above. One example is the Observations & Measurements ontology (Cox, 2013), which allows the description using a formal data structure both of values obtained from the environment and of estimates created by models, which is both an Open Geospatial Consortium and ISO standard. Within O&M, the key definition is that an **observation** is an act that results in the estimation of the **value** of a **feature property**, and involves application of a specified **procedure**, such as a sensor, instrument, algorithm or process chain. The procedure may be applied in situ, remotely or ex-situ with respect to the sampling location. The feature property has been specifically modelled within the INSPIRE Spatial Data Infrastructure, and the Complex Property Model ontology builds upon that work (Leadbetter and Vodden, 2015) to allow a rich semantic description of concepts from controlled vocabularies in the NERC Vocabulary Server and MMI-ORR, and to use those concepts as O&M feature properties. A further ontology resource of note is the Semantic Web for Earth and Environmental Terminology (SWEET) published by NASA, which contains over 6000 concepts in 200 small ontologies, at a level that allows users to plug their domain specific ontologies into the mid-level SWEET ontology. This will be revisited in one of the case studies below.

Perhaps one of the greatest strengths of the Semantic Web and Linked Data approaches (Berners-Lee, Bizer and Heath, 2009) is the fact that they demand, particularly in the case of Linked Data, linkages and connections between resources to be found and exploited. The interconnections between the resources described above have been developed over a number of years, and their state in October 2015 is visualized in Figure 6.2. These are shown as

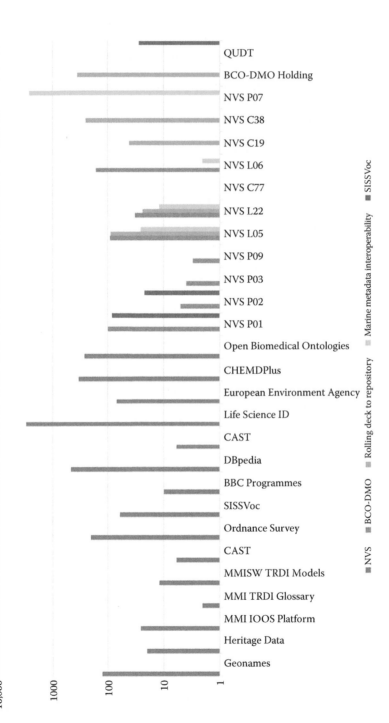

FIGURE 6.2
A frequency graph showing the number of incoming connections to various ontology services used in the marine and coastal domains from other relevant ontology services.

a frequency graph, following the findings of Leadbetter et al. (2016) in which traditional network diagrams are shown to be of limited use for visualizing these data.

6.6 Case Studies and Projects

6.6.1 The Climate and Forecast Conventions

One important use of ontologies in cross-disciplinary research, such as marine and coastal studies, is the integration of data, which have been created by disciplines that may not traditionally have a 'common' language. One particular vocabulary resource, which has been developed in order to broach this particular problem, is the Climate and Forecast Standard Names, served from the NERC Vocabulary Server at http://vocab.nerc.ac.uk /standard_name. While primarily designed for data exchange within the confines of the NetCDF format, Climate and Forecast conventions are increasingly gaining acceptance and have been adopted by a number of projects and groups as a primary standard. The conventions define metadata that provide a definitive description of what the data value in each variable represents, and the spatial and temporal properties of the data. This enables users of data from different sources to decide which quantities are comparable, and facilitates building applications with powerful extraction, regridding and display capabilities. One weakness of the standard names (e.g. 'air_temperature') is that they are used to create scalar coordinate variables (e.g. 'air_temperature_at_2m'). There is, however, no standard 'grammar' within the Climate and Forecast conventions. So this example can proliferate as '2m_air_temperature', 'air_temperature_at_2m' or 'airTemp2m' (this last using the official Climate and Forecast standard name abbreviation). Since 2008, this issue has been raised within the Climate and Forecast community, but has never been resolved.

Following the development of the Complex Properties Ontology, which is specifically designed to address the definition of exactly what has been measured in the environment, it is possible to model the grammar of the Climate and Forecast conventions from the perspective of the Observations and Measurements (O&M) data model. The ObservedProperty of an O&M Observation is the property of the FeatureOfInterest that is observed during the act of observation. For example, if the temperature of the FeatureOfInterest is measured, then the ObservedProperty is 'temperature'. The observedProperty in O&M must be a reference to some definition of that property. So typically this would be to an item in a published vocabulary. However, it is quite common in practice that definitions of observed properties in published vocabularies are not specific enough to allow end users to interpret exactly which property was observed. For example, the ObservedProperty may be 'radiance', and this may be defined in a vocabulary. However, the actual property observed is radiance at a particular wavelength or wavelength range, for example, between 300 and 400 nm. Another example is temperature. It may not be sufficient to simply state that an Observation has the ObservedProperty 'temperature'. It may be important to know that it is 'Air Temperature at 2 metres above the ground'. The Complex Properties Model ontology extends the INSPIRE Observable Property model to provide a framework for extending a predefined term in a vocabulary with additional information, such as constraints (e.g. the earlier wavelength example) or statistical measures (e.g. the earlier temperature example).

It has therefore been proposed that the Complex Properties Model may provide a solution to the Common Concepts problem – and also bridge the Climate and Forecast Standard names from a simple vocabulary to being instances of an ontology with mappings into the Observations and Measurements domain. Considering the '2m_air_temperature' example, the core property is that of 'temperature', which (as seen in Figure 6.3) is well defined in the SWEET ontology as a property. The measurement matrix is 'atmosphere', which is modelled within the NERC Vocabulary Server, as is the canonical unit for the standard name of Kelvin, which is also well defined in the Quantities, Units, Dimensions and Data Types (QUDT) ontologies (also published by NASA). The Complex Properties Model also defines restrictions which can be placed on an observable property. In this case, the restriction is that the measurement is at a height (defined by the Climate and Forecast standard name 'altitude') of 2 m, with the units of measure again defined in the NERC Vocabulary Server and the QUDT ontologies. In this way it should be seen that the labels of '2m_air_temperature', 'air_temperature_at_2m' and 'airTemp2m' are now redundant, and the data model gives the missing grammar for Common Concepts. Finally, both the instance of '2m air temperature' and the constraint of 'Height: 2m' are linked back to the standard name 'air_temperature'.

6.6.2 International Coastal Atlas Network (ICAN)

ICAN (http://www.iode.org/ican) is a community of more than 50 organizations led by the International Oceanographic Data Exchange (IODE) programme of the Intergovernmental Oceanographic Commission of UNESCO. The strategic aim of ICAN is to share experiences and common solutions to Coastal Web Atlas (CWA) development, and to promote the interoperability of CWAs. The ICAN community has developed an ontology-based mediation prototype for providing transparent access to information from different CWAs (Lassoued et al., 2008). The architecture of the ICAN mediator is shown in Figure 6.4.

Atlases in ICAN share metadata through standard catalogue services (OGC CSW) and metadata encodings (ISO-19139). However, they use

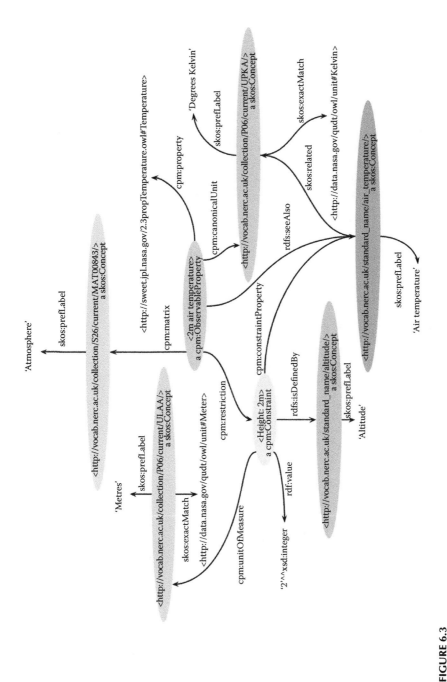

FIGURE 6.3
The 'grammar' of the Climate and Forecast conventions cast in the formal data model of the Complex Properties Ontology.

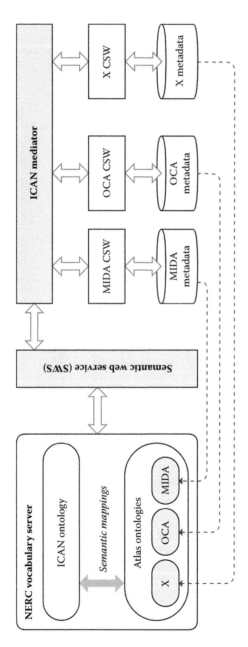

FIGURE 6.4
ICAN semantic mediation architecture.

different vocabularies and languages in metadata field values. For instance, while the Irish Marine Digital Atlas (MIDA) uses the keyword 'coastline', or 'seabed' to describe a given dataset, the Oregon Coastal Atlas (OCA) uses 'shoreline' or 'seafloor' to describe the same type of data. Atlas vocabularies in ICAN are modelled as multilingual ontologies. A common (also multi-lingual) ontology, called the ICAN ontology, is provided to the ICAN users and hosted on the NERC Vocabulary Server. Semantic mappings (relation-ships) between the atlas ontologies and the ICAN ontology are provided as a way to facilitate translation from one vocabulary into another. The ICAN ontologies are accessible through a semantic web service (SWS), which pro-vides a high-level application programming interface (API) for querying the ontology concepts and their relationships. The ICAN mediator provides a common point of access to the atlas catalogues. Provided with a user query formulated using the ICAN ontology terms, the mediator translates the terms within the query into terms supported by the atlases using the SWS. It, then, integrates responses from the atlas catalogues and returns them to the user.

The ICAN semantic mediation approach helps address the atlas interoper-ability issue (cf. Use Case 4, Section 6.3) by allowing metadata distributed across semantically heterogeneous atlases to be queried seamlessly regard-less of the vocabularies or languages they use. It further improves search results as it exploits the semantic relationships between multilingual terms both within and across ontologies (cf. Use Case 3, Section 6.3).

6.6.3 Marine Debris Data Portal (MDDP)

The Marine Debris Data Portal project (Lassoued et al., 2015) developed an ontology-based mediation approach for the semantic integration of the West Coast Marine Debris Database (MDDB, http://debris.westcoastoceans.org). The goal is to help marine debris experts visualize spatial patterns, and inform marine debris policies and prevention on the U.S. West Coast using the West Coast Ocean Data Portal (WCODP, http://portal.westcoastoceans .org).

The MDDB stores data from beach clean-up and monitoring and der-elict gear removal events from 12 organizations. During these clean-up events, debris or derelict gears are counted or weighed by type. Data from a clean-up event are collected in a data sheet, which records event informa-tion (date, location, organization, etc.) and the details of the debris or der-elict gear quantities (weight or count) removed or observed by type (e.g. beverage cans count, glass bottles count, trash weight, etc.). Typically, due to historic circumstances, data sheets from different organizations use dif-ferent classifications of debris and derelict gears, according to whatever is relevant to the different clean-up communities. As a first attempt to inte-grate data properly, debris and derelict gear types were standardized in the MDDB using a common code list known as the master list. This is a flat list of debris and gear types that directly map to those of the data cards.

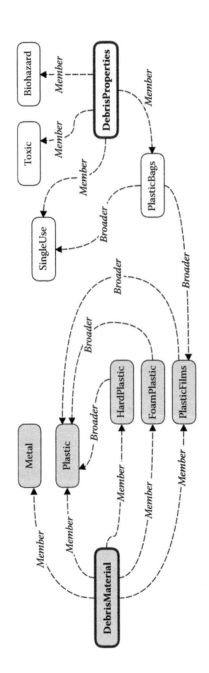

FIGURE 6.5
Extract of the WCODP Debris Categories Ontology: Grey boxes pertain to debris materials, while white ones pertain to debris properties. Labelled arcs represent semantic relationships.

While this unifies the debris types vocabulary, it still doesn't enable data aggregation and analysis across clean-up events, as no semantic relationships between the debris types are provided. For example, 'toxic', 'car batteries' and 'spray paint cans' all appear in the master list as separate debris types, while in fact the last two are clearly subtypes of the first. Therefore, in order to get an estimate of toxic debris along a given coastline or in a given area, you need to manually figure out what debris types, other than 'toxic', should be taken into account. Moreover, the master debris types, which simply reflect the debris classifications of the clean-up organizations, are not necessarily relevant to the WCODP end users (marine debris experts, decision makers, etc.).

The MDDP project addressed the difficulties outlined above through the development of an ontology-based approach. The approach relies on the provision of two ontologies: (1) the MDDB debris types ontology, which captures the debris types represented in the MDDB, and (2) the WCODP debris categories ontology, which defines new high-level and general-purpose debris types, called categories, requested by the WCODP user community. Semantic mappings between the newly developed WCODP debris categories and the MDDB debris types are provided as part of these ontologies. The WCODP ontology classifies debris from different perspectives: by material (e.g. plastic, metal, wood, etc.), by property (e.g. toxic, biodegradable, bio-hazard, etc.) and by usage or industry (e.g. home, automotive, etc.). These may be used as different debris filtering facets. Semantic relationships amongst the WCODP categories are provided, which facilitates data aggregation. For instance, if you request plastic debris, all subtypes of plastic (hard plastic, foam, plastic films, etc.) will be automatically included. An extract of the WCODP ontology is illustrated and explained in Figure 6.5.

The MDDP architecture, which is illustrated in Figure 6.6, is similar in principle to the ICAN semantic mediation architecture. The main difference is that it is applied to data stored in a single database rather than metadata delivered through distributed catalogues.

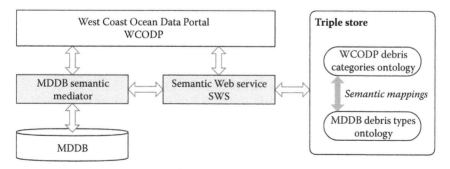

FIGURE 6.6
MDDP semantic mediation architecture.

Using such ontology-based architecture, the MDDP project succeeded in facilitating debris data integration and aggregation, but also enabled delivering data to end users in a representation (debris classification) that is more suitable to their needs than that of the original data.

6.7 Conclusion

In conclusion, we have seen that the emergence of structured data publication on the World Wide Web has also given geoinformatics practitioners new methods of structuring their understanding of the marine and coastal domains, and advanced techniques to integrate data from various sources. While much of the groundwork has been done for integrating metadata and information using these techniques, there still remains work to be done to see a wide-scale adoption of them and the challenge remains to apply these ideas to the data themselves. The 'Born Connected' concept advanced by Fredericks (2015) and Leadbetter et al. (2016) has applied the ideas of ontologies and structured data to the streams of data produced by Internet of Things devices meaning that data from oceanographic sensors arrive from the marine environment fully structured and annotated. It is hoped that in the future this research and development activity will become operational and supported by instrument manufacturers such that it proliferates.

References

Berners-Lee, T., Bizer, C., and Heath, T. (2009). Linked data – The story so far. *International Journal on Semantic Web and Information Systems*, 5(3), 1–22.

Cox, S. (2013, October). An explicit OWL representation of ISO/OGC Observations and Measurements. In *SSN@ ISWC*, pp. 1–18.

Cyganiak, R., Wood, D., and Lanthaler, M. (2014). RDF 1.1 Concepts and Abstract Syntax. W3C Recommendation, 25 February 2014.

Fredericks, J. (2015). Persistence of knowledge across layered architectures. In P. Diviacco, P. Fox, C. Pshenichny, and A. Leadbetter (Eds.), *Collaborative Knowledge in Scientific Research Networks*. Hershey, PA: IGI Global.

Graybeal, J., Isenor, A. W., and Rueda, C. (2012). Semantic mediation of vocabularies for ocean observing systems. *Computers & Geosciences*, 40, 120–131.

Hodge, G. (2000, April). *Systems of Knowledge Organization for Digital Libraries: Beyond Traditional Authority Files*. Washington, DC: Council on Library and Information Resources, Digital Library Federation.

IOC (1987). A General Formatting System for Geo-Referenced Data, Volume 2: Technical description of the GF3 format and code tables. Intergovernmental Oceanographic Commission, UNESCO, Paris.

Janowicz, K., and Compton C. (2010). The stimulus-sensor-observation ontology design pattern and its integration into the semantic sensor network ontology. *3rd International Workshop on Semantic Sensor Networks*, vol. 668, CEUR-WS, 2010.

Lassoued, Y., Haddad, T., and Hallenbeck, T. (2015). Semantic integration of marine debris data from multiple cleanup organizations. In *12th International Symposium for GIS and Computer Cartography for Coastal Zone Management CoastGIS'15*, Cape Town, South Africa, 2015.

Lassoued, Y., Wright, D., Bermudez, L., and Boucelma, O. (2008). Ontology-based mediation of OGC catalogue service for the web: A virtual solution for integrating coastal web atlases. In *3rd International Conference on Software and Data Technologies ICSOFT'08*, Porto, Portugal, 2008.

Leadbetter, A. M., Lowry, R. K., and Clements, D. O. (2014). Putting meaning into NETMAR – The open service network for marine environmental data. *International Journal of Digital Earth*, 7(10), 811–828.

Leadbetter, A. M., and Vodden, P. N. (2015). Semantic linking of complex properties, monitoring processes and facilities in web-based representations of the environment. *International Journal of Digital Earth* (ahead-of-print), 1–25.

Leadbetter, A., Cheatham, M., Shepherd, A., and Thomas, R. (2016) Linked Ocean Data 2.0, in Diviacco, P., Leadbetter, A., and Glaves, H. (Eds.), *Oceanographic and Marine Cross-Domain Data Management for Sustainable Development*, Hershey, PA: IGI Global.

Tschirner, S., Scherp, A., and Staab, S. (2011, October). Semantic access to INSPIRE. In *Terra Cognita 2011 Workshop Foundations, Technologies and Application*.

W3C OWL Working Group (2012). *OWL 2 Web Ontology Language, Document Overview*, 2nd ed. W3C Recommendation, 11 December 2012.

7

Applied Marine Management with Volunteered Geographic Information

Grace Goldberg, Mimi D'Iorio and Will McClintock

CONTENTS

7.1 Introduction

The term 'volunteered geographic information' (VGI), first suggested by Goodchild (2007), refers to the gathering and dissemination of geographic information provided voluntarily by the public, using emerging Web 2.0 technologies along with cyberinfrastructure and crowdsourcing methods (Elwood, 2008; Elwood et al., 2012; Sui et al., 2013). Increasingly, the general public contributes this georeferenced information using their web-enabled devices. This includes contributions through their participation in social media or citizen science initiatives (Pimm et al., 2015; Wood et al., 2015). Equally important is information collected from non-expert audiences that have specific place-based knowledge, using digital mapping tools and targeted methods that may be administered in-person or remotely (Corbett and Rambaldi, 2009; Taylor and Lauriault, 2014).

These data play an increasingly important role in informing coastal and marine resource management. VGI methods may be used to produce a range of spatial datasets (Table 7.1) that allow resource managers to map biodiversity and human activities. These maps help answer key questions relevant to designing effective spatial management plans, such as: 'What areas should be highest priority for protection?'; 'What areas are best suited for new development?' and 'What areas contain multiple conflicting uses and may require better management?' Collecting these data with VGI methods may be cheaper and quicker to implement than more traditional empirical survey methods performed by expert researchers. In many cases, VGI data are of equal or higher quality than data collected by other means, although this is not necessarily the case and data quality inevitably remains a key consideration (Goodchild and Li, 2012; Hyder et al., 2015; Tulloch et al., 2013). In the case of mapping human dimensions, non-traditional data contributors are valuable sources for providing reliable and appropriate information

TABLE 7.1

Key Types of Volunteered Geographic Information (VGI)

Types of VGI Information	Knowledge Contributions
Distribution of human activities and uses	These data can represent locations, intensity of use and extents of fishing grounds, recreational dive spots, traditional use areas, etc. Attributes associated with the spatial information may include the importance or value of these areas to the user, temporal variation in use of that area or more detail about the specific uses.
Distribution of natural resources	These data are used to identify and demarcate user and stakeholder knowledge of habitats, species occurrence and movements, key spawning aggregation locations for fished species or the extent of specific habitat types.
Management scenarios	In addition to the above VGI data that may inform planning and decision making, stakeholders or the general public may be asked to contribute specific ideas and opinions on the details of area-based management, resource use planning, policy, etc.
Perceptions of environment condition	Non-expert data contributors may be asked to report on the environmental health of an area. These may be perceptions of change over time, the presence of marine debris, etc. These data may be combined with information about the character and frequency of public use of that area.
Citizen science programmes	Observational data (photographs, species occurrences and lists) collected by the general public through citizen science programmes can be used to inform real-time decision making, contribute to baseline understanding of species diversity and help map temporal population shifts. This is particularly useful for collecting biological and oceanographic data. VGI may be integrated with expert data to increase the extent of scientific knowledge without the cost of expanded expert data collection.

(Brown et al., 2015; Hyder et al., 2015). In many cases, they are also the most qualified to contribute key data about community perceptions and values of coastal and marine resources.

This chapter outlines the actual and potential role for crowdsourced information and VGI (Goodchild, 2007) to contribute to the coastal and marine management. This chapter considers the range of data types involved (Table 7.1), collection methods and ultimate uses of VGI. A broad definition of VGI applies, in which crowdsourced data and completely open web platforms are examined alongside more targeted methods of capturing data derived from local experience and expertise. We also share insights on how processes and procedures of data collection may have consequences for the usefulness of VGI.

7.2 Informing Marine Resource Management

Marine spatial planning (MSP) is widely recognized as an effective method for implementing ecosystem-based management (Katsanevakis et al., 2011; Douvere, 2008). This requires the integration of best available science and data into decision making through the development of measurable indicators and planning guidelines (Ehler and Douvere, 2009). By definition, MSP results in geospatially explicit plans, and as such requires spatial data and information as key inputs. A variety of resource management and policy initiatives, including MSP, are increasingly making use of information provided by citizen science endeavours, and can make important contributions to environmental monitoring and mapping biodiversity (Hyder et al., 2015; Wood et al., 2015).

In the case of the United States, Executive Order 13547, signed by President Barack Obama on July 19, 2010, Coastal and Marine Spatial Planning (CMSP, equivalent to MSP elsewhere) refers to a process that helps reduce conflicts among uses, reduces environmental impacts, facilitates compatible uses and preserves critical ecosystem services, while also meeting economic, environmental, security and social objectives. Integrating human-dimension information into CMSP is essential to delivering effective, long-term solutions that benefit both marine systems and the coastal communities (Ehler and Douvere, 2009; Kittenger, 2014; Sullivan et al., 2015). VGI offers a means to capture spatial intelligence and harness the power of collective human knowledge to help unravel the complexity of the human dimensions of marine planning strategies ('human dimensions' refers here to the many ways in which people and societies interact with, affect and are affected by natural ecosystems and environmental change through time). Human dimension data and applied social research are increasingly recognized as essential to management and policy around the globe (see also Chapter 10 by Aswani of the present volume). However, effective means to incorporate human dimensions data are limited, due, in part, to the complexity of human relationships with the ocean and coasts, and a dearth of tools to effectively characterize these linkages (Ehler and Douvere, 2009; Hall-Arber et al., 2009; Kittenger, 2014; Koehn et al., 2013; Sullivan et al., 2015). The process of engaging stakeholders and the general public in collecting and revising these data is more than just a one-way information flow: it also promotes inclusive policy and decision making, builds trust and increases buy-in for effective implementation of policies (Black et al., 2015; Jarvis et al., 2015; Pomeroy and Douvere, 2008).

7.3 Data Collection Methods and Case Studies

Historically, VGI was often collected by encouraging stakeholders to participate in a mapping process by, for example, using a stick and a beach-sand canvas. Contemporary practitioners have access to a wide range of digital tools for the collection of VGI. The process is supported by sophisticated geographic information systems (GIS) platforms that allow users to contribute more detailed spatial information and associated attributes (Carver et al., 2001; Corbett and Rambaldi, 2009; Dunn, 2007; Geertman and Stillwell, 2009). These tools add greater precision to the volunteered information when compared to data collection methods with *post hoc* transcription steps (Mehdipoor et al., 2015). In addition to what we will refer to as participatory methods, relevant data are being contributed through a suite of web-based and mobile applications (more commonly referred to as 'apps') without the contributor being embedded in a participatory mapping process. This allows for the creation and collection of new types of datasets not previously available to or used in resource management.

Notably, non-experts can contribute to crowdsourced datasets with a range of tools such as simple mobile apps that allow users to record species sightings in real-time (Pimm et al., 2015; Wood et al., 2015). This is also increasingly used to map and understand human dimensions. Other datasets may be opportunistically generated, using geotagged web content posted to open social media platforms. In such cases, users may not necessarily intend or be aware that their contribution or content is being used to inform and aid research, decision making and policy development, and this may raise ethical or other issues that need to be resolved. At present few general guidelines exist to cover such eventualities. Nonetheless, appropriate use of digital tools adds efficiency to data gathering efforts, and can allow for data collection by a greater number, wider geographical spread and broader diversity of contributors. Participants can often see and follow the information they have contributed upon submission. Some tools even allow them to edit spatial data and associated attributes through a verification process.

This chapter describes many methods and programmes as forms of VGI, categorized into three groups: crowdsourcing, geotagged web content and participatory mapping. Where Heipke (2010) uses the terms VGI and 'crowdsourcing' interchangeably, we will use the term 'crowdsourcing' to define citizen science and web-based survey methods with a geographic component. The typology of VGI suggested by Lauriault and Mooney (2014) takes a different approach, treating VGI, citizen science and participatory mapping as three distinct methods of crowdsourcing. This typology did not suit the wide range of programmes we examine that fulfil the definition of VGI put forth by Goodchild and Li (2012), as any contributed information containing a geographic location and one or more attributes. The nature of emerging technologies is to blur these classifications (Lauriault and Mooney, 2014).

7.3.1 Crowdsourcing Information

Crowdsourcing utilizes knowledge from the public to inform a particular topic, often involving an open call for voluntary contributions from a large group of unknown individuals ('the crowd'; Heipke, 2010). A good example is the development of map-based mobile apps for navigation, which has shifted the user's role from that of a passive receiver of information to being an active contributor and editor of map features (Buganza et al., 2015). For example, within the recreational boating community, mobile apps provide navigational charts and also offer integration with chart plotters, though this also raises issues of safety and liability if these crowdsourced data are used for navigation. These apps also provide the means for users and subscribers to contribute points of interest, geotagged photos and place-based commentary (e.g. 'Community Edits' by Navionics & ActiveCaptain integrated with Garmin BlueChart). These contributed data are then accessible to all users via chart plotter interfaces and mobile apps, as well as to online web-viewers. Recent and likely future trends in electronic navigation charting (ENC) include the role of volunteered and 'unofficial' information as inputs, though charts based on such information will lack the legal authority of those created and disseminated officially: this is further discussed by Weintrit in Chapter 14 of the present volume.

Monitoring programmes that assess the state of the marine environment may use 'humans as sensors' (Goodchild, 2007, 2010) to inform science and policy decisions. Many of these programmes refer to the use of 'citizen science' and involve the use of volunteers to collect or process scientific information (Haklay, 2013; Silvertown, 2009). Thiel et al. (2014) provide a comprehensive review of marine-based research projects that leverages volunteer participation. Volunteered information may include georeferenced species observations, recorded by volunteers with a hand-held GPS unit, or photographs that can be used for the verification of remotely sensed habitat maps (Kordi et al., 2016). Smartphone devices, integrated GPS and mobile apps allow non-traditional data contributors to accurately collect point data of relevance. In addition, some mainstream GIS software vendors now also offer apps and other tools that may be downloaded and installed on mobile devices to facilitate field data collection and upload to geodatabases: One example of such tools is the ArcCollector app provided by Esri (http://doc .arcgis.com/en/collector/).

In some cases volunteers may also enter attributes associated with data points. These attributes may include time and date of the observation or sample, or more detailed information by way of data input forms, or with photo and video upload tools. Through the review of photos and videos, both experts and non-experts (i.e. 'the crowd') may verify visual information collected in the field. Some examples of species identification by way of mobile apps include iNaturalist (California Academy of Sciences, 2014), iSeaHorse (Jeffrey, 2015) and Eye on the Reef (GBRMPA, 2014). The iNaturalist platform encourages the 'community of users' to agree, comment or propose names for recorded species observation. Elsewhere, an edited volume by Taylor and Lauriault (2014, 2nd ed.) documents a number of projects and initiatives that link crowd sourcing and the use of VGI to community participation and the collection of traditional ecological knowledge among indigenous peoples of the Canadian Arctic.

Other web-based applications for data entry are usually employed after returning home, capturing location with either a text entry of latitude and longitude, dropping a pin on a map or sketching an area (Town et al., 2013; Marshall et al., 2012; McClintock, 2013). MantaMatcher (www.mantamatcher .org) – a visual database of manta rays (*Manta birostris* and *M. alfredi*) – allows for the upload of underwater photography of manta ray sightings. Each photograph uploaded is associated with trip metadata and the contributor. Photographs often capture markings that are unique to individual rays, allowing scientists to better study animal movement and population dynamics (Town et al., 2013).

In order to improve data quality, scientists and software developers have identified some best practices for the development of these applications. The use of drop-down menus and compulsory input fields promote accurate data and complete and consistent contributions. A common limitation of submitted volunteer sightings is the lack of or limited effort data. This information, typically recorded by scientists, is used to make interpretations of absence and presence of species, that is, spatial data and time spent in the field with no animals observed are equally valuable (Wood et al., 2015).

Web-based surveys are frequently used for data collection to answer specific questions in a discrete timeframe. SeaSketch (www.seasketch.org), a marine planning decision-support tool, includes a survey builder for the development of web-based surveys for gathering spatial and non-spatial information, usually by marine resource managers, to inform a specific policy process (see Box 7.1). Contributors normally receive a survey URL link

BOX 7.1   SEASKETCH

SeaSketch (www.seasketch.org) is a web-based mapping platform for collaborative, science-based marine spatial planning (MSP). It supports an iterative geodesign workflow (Goodchild, 2010), in which users can sketch out their ideas on the map and get immediate analytical feedback that informs iterative design. They can then share those sketched and edited plans with other users, editing and iterating on each other's plans. This process of collaboratively expressing ideas on the web-mapping interface allows users to better understand the marine planning process they are engaged in, and supports joint problem solving (Cravens, 2014).

The SeaSketch user interface allows for the exploration of spatial data and information in the 'Data Layers' tab. The 'My Plans' tab is a private sandbox for users to sketch out and analyze their ideas, which they can then share in the 'Participate' tab through map-based discussion forums and surveys (Figure 7.1). Out of the box features include a map viewer ready to display web-hosted map services and their associated attributes, legends and metadata, standard map tools such as location search and measurement tools, sketching tools for points, lines and polygons and the ability to create forums and design surveys with specific privacy permissions, among others. In addition, you may link your project to geoprocessing services, cloud-hosted python scripts that allow users to get immediate analytical reports about their sketched plans.

Each of these features is configured to meet the specific process needs of any given project in the Administrative Dashboard, where project leads use form builders and upload tools to create surveys, designate sketch types with set attributes and link to data and analytical services. This is possible because SeaSketch is a software service (SaaS), a single cloud-based application that can support an unlimited number of uniquely configured projects. The SaaS architecture allows for rapid deployment at a reduced total cost of ownership to any given project, as a new fully featured site can be created with the click of a button.

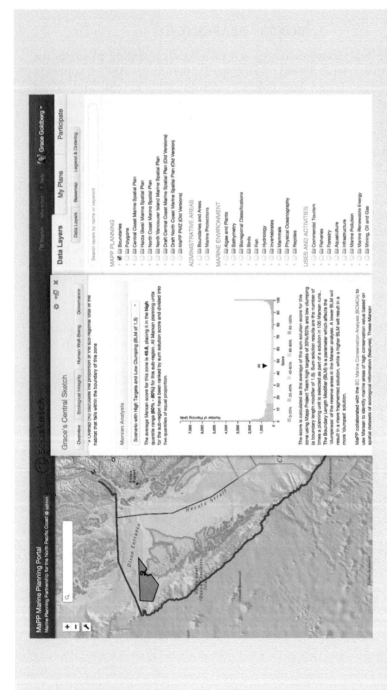

FIGURE 7.1

The SeaSketch user interface, displaying an analytical report with feedback about Marxan analyses related to the example sketch outlined in blue on the map viewer, with available data in the Data Layers tab at right. This example sketch was created and analyzed in the mapp.seasketch .org project. Not all features shown are publicly accessible.

via email invitation or, if crowdsourced, the link may be shared over social media and by other outreach channels. Survey respondents answer questions and contribute spatial responses using a computer mouse or track pad. These spatial responses may include graphic objects (points, lines or polygons), and may be associated with a range of attribute types. Although these surveys may be used in citizen science programmes, flexible survey builder tools allow the development of survey instruments for a wide variety of contexts (Goldberg and McClintock, 2015).

Mobile and web technology plays to the strength of this type of data collection. Crowdsourced observations can be hosted on the Internet, and can be subject to expert or crowdsourced verification. Verification may be associated with the geographic location of the sighting, as in the case of eBird (www.ebird.org), where observations made outside of the normal spatial range of the observed species are automatically flagged for additional review (Sullivan et al., 2009). However, the focus is on verifying the species, rather than emphasizing verification of the geographic location itself.

7.3.1.1 Case Studies: Using Web-Based Surveys at Different Scales

Barbuda is a small island in the Eastern Caribbean, part of the state of Antigua and Barbuda, with fewer than 1900 residents. The Blue Halo Initiative is a collaboration between the Waitt Institute (San Diego, California – see http://http://waittinstitute.org/) and partner local governments at multiple sites across the Caribbean to envision, create and implement comprehensive sustainable ocean policies. In Barbuda, they focused the project on planning sanctuaries and multiple-use zones in the coastal waters out to a distance of three nautical miles from the coastline. In order to negotiate ocean zones by way of community participation (Waitt, 2013), stakeholder groups and a proportion of Barbuda's population were requested to complete a survey (DeGraff and Ramlal, 2015; Johnson and McClintock, 2013; Pomeroy et al., 2014). At the start of the planning process, there was little spatially explicit information about the distribution of human activities in the ocean. Interviewers used an Internet-connected laptop, and SeaSketch (Box 7.1; Figure 7.2) to capture stakeholder information about ocean use within the state jurisdiction, defined as the area between the shoreline and three nautical miles to sea. Paper maps were also available to provide context and orient stakeholders.

Fishermen were first asked to map or outline their fishing grounds using the SeaSketch tool (Figure 7.3). Once these were mapped, they were asked to identify the most valuable fishing grounds by distributing 100 points to their most productive fishing grounds, in an adaptation of a social science research exercise known as the '100 pennies exercise' (Scholz et al., 2006). In this instance, the operation benefitted from the use of a web-based platform, adding flexibility for scientists and data contributors. The resulting polygons were readily available to be summarized into key data products such

FIGURE 7.2
Waitt Institute and associated contractors interview a group of fishermen, interrupting their domino game at the docks to collect information about ocean use and perceived value. (From McClintock, W.J., *Coast Guard Journal of Safety & Security of Sea, Proceedings of the Marine Safety & Security Council*, 70(3), 63–67, 2013.)

as fishing intensity or value heat map (Figure 7.4) that were then available for use to inform planning and trade-off analyses in the process of designing the ocean zoning scheme. There are also other areas of the Caribbean where the Blue Halo Initiative approach is used. In some cases, local experts can and want to contribute additional data and information and can do so on their own time. In those cases, these local experts may return to SeaSketch to adjust or add to their responses after the initial interview. One-on-one consultation is time consuming and thus expensive. It is usually unfeasible to collect very large samples using this approach and the use of a combination of individual interviews, focus group meetings and online data submission should be considered as part of the information collection strategy.

The New Zealand Department of Conservation (DOC) has required a similar widely accessible web-based survey to reach the general public in and around the Hauraki Gulf and Port of Auckland, an area with a population of over 1 million people. The 2013 initiative entitled 'Sea Change Tai Timu Tai Pari' (www.seachange.org.nz) is an ongoing large-scale multisector marine spatial planning scheme that includes the use of crowdsourced information about public use and perceptions of ocean space as inputs for consideration by a diverse stakeholder committee. Data were collected over a two-month period during 2014 (Jarvis et al., 2015, 2016). The public was requested to share views (or information) regarding points of importance

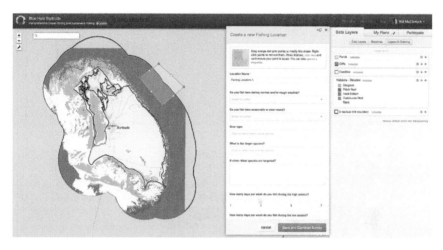

FIGURE 7.3
The SeaSketch survey tool interface, as used in the Barbuda Blue Halo Initiative. In the screen-shot above, users are asked to indicate where and how they fish, and to place a value on each fishing area identified.

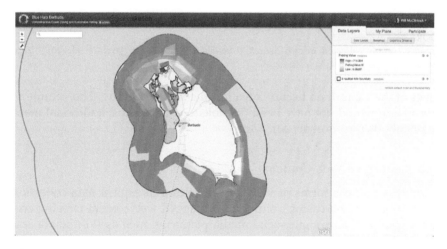

FIGURE 7.4
A heat map showing the distribution of valued fishing grounds as indicated by fishermen using SeaSketch (see Figure 7.3).

within the study area, using a map interface provided by SeaSketch, and by adding attribute information to those points. These attributes ranged from the nature and frequency of their visits to a point of interest, to their perceptions of environmental health and degradation. Using this approach, it was possible to successfully collect 4495 points of interest (Figure 7.5). These points were used by scientists and managers to characterize public use and

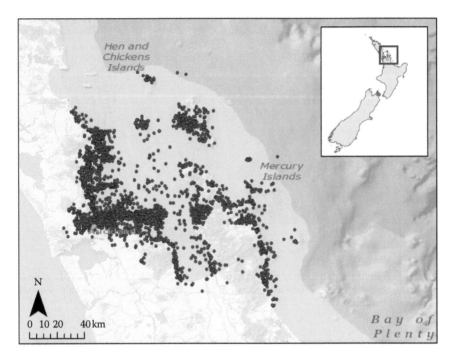

FIGURE 7.5
A map of points collected in crowdsourced SeaSketch surveys by Jarvis et al. (2015, 2016) around the Hauraki Gulf, New Zealand.

values of the 1.2 million hectare area for fine-scale planning. The volume of information would not have been possible to collect using traditional mapping methods (Jarvis et al., 2015, 2016).

7.3.2 Geotagged Web Content

In addition to these forms of VGI acquired through explicit data collection programmes, an increasing volume of geotagged web content is being created by the public through social media platforms such as Wikimapia and Flickr (Goodchild, 2007). This content may also provide valuable data for use in marine and coastal policy formulation, although in most cases it is unlikely that the creators of this information are aware of how their contributions are being used or even that they are being used at all.

Data contribution can occur with mobile devices *in situ*, or uploaded and tagged later via a web browser. Part of the strength of this type of dataset is that the sheer volume of contributors increases the reliability of the data. This may be caused simply by, for example, the millions of travel photos in which outliers are drowned out. Alternatively, a community coalescing around and contributing to initiatives such as Wikipedia actively edits contributions to the point where content converges on accuracy (Elwood et al.,

2012). For example, Flickr (www.flickr.com) is a photo sharing website identified by Elwood et al. (2012) as a rich source of VGI. Many digital cameras, including those available on current generations of smartphones, have the ability to automatically geotag photographs with coordinates acquired via in-built GPS chips. This locational information is typically stored as part of the metadata associated with the JPEG file format used as a common standard for digital photographs (CITT, 1992). Flickr encourages tagging of geographic location when users contribute media to the public database, and these media may subsequently be used by scientists to map ecosystem services and inform policy decision making (Wood et al., 2013). In addition to using the geotagged photos to map the value of nature-based tourism and other cultural ecosystem services, photographs sourced from Flickr have contributed to projects that map habitat extent, specific wildlife populations and for planning physical offshore infrastructure (Davies et al., 2012; Griffin et al., 2015; Richards and Friess, 2015).

When examining human dimensions and natural resource, not all social media platforms are created equal. Content privacy and the nature of the content contributed to these online communities determine how they may potentially contribute valuable VGI. For example, although researchers in Singapore identified Instagram (www.instagram.com) as the more widely used platform to post geotagged photographs in their study area, providing a larger dataset than Flickr, Instagram's strict privacy policy makes this type of research infeasible. Flickr, however, is much more accessible for the harvesting of content and the application programming interface (API) facilitates analysis of uploaded content (Richards and Friess, 2015).

Posts to the Twitter platform (www.twitter.com) are a very different form of geotagged web-content. In this case, the postings are limited to 140 character text posts, although they may also include photos or other media. These brief and high frequency 'Tweets' have been identified as valuable sources of information useful for disaster response and management (e.g. Hurricane Sandy in the U.S. Northeast; Goodchild, 2007; Guan and Chen, 2014). To date, there has been relatively limited use of this type of geotagged content to inform our understanding of offshore areas, or to determine real-time decision making by ocean users and coastal emergency response, and it is an area where future research would be both timely and appropriate.

7.3.3 Participatory Mapping with Web-Based Tools

Participatory mapping has evolved greatly in recent decades both as a concept and as a tool to create more socially inclusive GIS (Dunn, 2007). To support this evolution, many participatory mapping strategies have been designed and employed to collect VGI. They tend to differ in scope by focusing on smaller target audiences and on specific management or planning objectives. However, they share a similar purpose in transforming community knowledge into useable spatial data (Corbett and Rambaldi, 2009; Taylor and Lauriault, 2014). These participatory processes may employ a range of formal and informal engagement strategies (NOAA, 2014; Pomeroy and Douvere, 2008). Web-based mapping applications support this collection of stakeholder-contributed information and spatial planning ideas through the involvement of wider audiences in iterative science-based decision making (Carver et al., 2001; Geertman and Stillwell, 2009; Pettit and Nelson, 2004).

Personal or one-on-one interviews focus on topical data collection through structured conversation between an investigator and an informant. Interview methods can vary between open questions, where the respondents contribute an objective answer, and close-ended questions where they select from a list of predetermined responses (Berg and Lune, 2012). The interview method is flexible and can be adapted to the respondent's needs. The personal face-to-face nature of an interview may add accountability that increases the reliability of information gathered. Answers, however, may not be completely objective, conducting interviews is time consuming, and responses received are often anecdotal so that the accuracy of data cannot be assumed. Alternatively, focus groups may be used to target groups of stakeholders or other experts, who then collaborate and provide a collective response or submission on a particular topic (Kaplowitz and Hoehn, 2001).

Web-GIS decision-support tools can support and facilitate interviews, and also allow for meaningful remote participation and contribution (Garau, 2012; Pettit and Nelson, 2004). The SeaSketch software service, like its predecessor MarineMap (Merrifield et al., 2013), enables the contribution of spatial information, plan ideas and other information for marine planning processes (Goldberg and McClintock, 2015). Participants in a one-on-one interview or a focus group can use web-applications such as SeaSketch to:

1. View available spatial data.
2. Discuss additional information needs in map-based forums.
3. Sketch out spatial plan ideas.
4. Evaluate and edit those sketches with sophisticated GIS tools.
5. Contribute and debate alternatives in online forums.

For capturing VGI in a group consultation setting, a web-based tool may be projected for easy viewing. Spatial data and information can be entered in real time by a technician, guided by the participants. Participants may also have the platform open on their own laptop or tablet device, and add spatial information interactively.

The verifiability of data collected by crowdsourcing and participatory methods may differ. Typically, data collected by participatory GIS methods require face-to-face meetings where individuals may scrutinize each other's input which may, in turn, lead to some degree of verification by peers. Crowdsourced information, while often not subject to the same level of verification, might allow researchers to collect larger datasets, which ultimately represent a wider range of opinion and/or knowledge, with a higher confidence in the mean.

7.3.3.1 Case Study: Developing the Pacific Regional Ocean Uses Atlas

To incorporate human dimensions information into marine planning, NOAA's Marine Protected Areas Center designed a participatory mapping process that combines traditional stakeholder engagement methods with simple GIS techniques to gather spatial knowledge from community experts (NOAA, 2014). From MPA management planning in the U.S. Virgin Islands to hazard response planning in New England, to coral reef management in Hawaii, the NOAA participatory process has been used at various scales and with many different target communities to integrate the spatial human dimension into planning strategies.

With funding from the U.S. Bureau of Ocean Energy Management (BOEM), the process was also applied as a fundamental component of the Pacific Regional Ocean Uses Atlas (PROUA). This latter project was designed to collect community-derived spatial information about ocean use activities occurring in marine waters offshore of Oregon, Washington and the Hawaiian Islands (D'Iorio et al., 2015).

BOEM's Renewable Energy Program is responsible for granting leases, easements and rights-of-way for orderly, safe and environmentally responsible renewable energy development activities (www.boem.gov/Renewable -Energy/). To support the leasing process, and offer a forum for stakeholder engagement, the PROUA team convened a series of facilitated workshops. These were planned to allow coastal and ocean users and experts to map spatial patterns, and provided the team with community perspectives and historical knowledge for over 30 different types of ocean use activities.

The PROUA organizers sought ocean use experts and community representatives through a cascading (sometimes referred to as 'snowball') referral process. Workshop candidates who were identified through initial local consultations were invited to attend the workshops and asked to refer other qualified potential invitees (NOAA, 2014). This referral invitation process continued until a diverse and representative group of ocean use experts were confirmed. Nearly 400 stakeholders participated in the 15 workshops held throughout the study area. Through facilitated mapping exercises, workshop participants documented use areas on an interactive map (Figure 7.6). They noted areas of general and high use (dominant use areas), as well as offering anecdotal and historical information to supplement spatial pattern data. Various groups mapped the same set of uses providing for redundancy, which could be used to identify outliers or pattern inconsistencies during the data processing and compilation.

Primary products derived from the process were a series of maps and spatial datasets indicating patterns and overlap of ocean use activities (see Figure 7.7). These products will help BOEM to better understand the use of

(a)

(b)

FIGURE 7.6
Ocean use experts work in groups to document spatial patterns on digital maps using interactive whiteboards (a) and digital tablets (b). (From D'Iorio, M. et al., *The Pacific Regional Ocean Uses Atlas, Data and Tools for Understanding Ocean Space Use in Washington, Oregon and Hawaii.* U.S. Department of the Interior, Bureau of Ocean Energy Management, Pacific OCS Region, Camarillo, CA, 2015.)

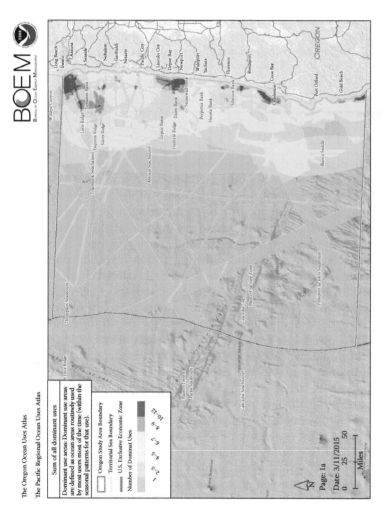

FIGURE 7.7
VGI gathered through PROUA participatory mapping workshops reveal patterns of overlapping ocean uses in the marine waters offshore Washington and Oregon. (From D'Iorio, M., Selbie, H., Gass, J., and Wahle, C. 2015. *The Pacific Regional Ocean Uses Atlas, Data and Tools for Understanding Ocean Space Use in Washington, Oregon and Hawaii.* U.S. Department of the Interior, Bureau of Ocean Energy Management, Pacific OCS Region, Camarillo, CA. OCS Study BOEM 2015–014.)

the ocean, and the proposed lease area, by the affected communities. This information can also assist to identify and evaluate potential stakeholder conflicts with renewable energy developments.

While spatial data and maps were the primary outputs from the PROUA, the participatory mapping process also provided BOEM with a forum to engage with ocean stakeholders. This forum could be used to discuss community concerns, and identify key issues related to offshore renewable energy development from the communities' perspective. These insights can help guide BOEM's future leasing decision making processes, and offer insight to other potential applications of VGI and related social data collection for marine spatial planning.

7.4 Methods and Scale

Goodchild (2008) discusses the 'rise of local expertise to replace centralization' in geography and map-making through the establishment of VGI. This new role of citizens in contributing to geographic information that cannot be sensed remotely, and in identifying and correcting information errors, is important for understanding and planning the use of the ocean space and informing marine resource management. Even when maps of human activities in the ocean may be produced through other methods, participatory information gathering may be preferred due to its perceived inclusivity. Nuanced social data are increasingly recognized as an important component of MSP (Kittinger et al., 2014), and participatory mapping techniques can frequently offer the necessary mechanism for capturing this information. For example, while it is possible to map fishing effort without surveying fishers themselves, for example, by use of aerial surveys, VMS transponder data and dockside landings data, these technologies do not include information about perceptions and values critical to management plans supported by stakeholders. Fishing grounds that are 'special' or 'valuable' to fishers, sometimes for cultural reasons, may not be objectively assessed by empirical evaluation methods alone yet, by not capturing these subjective values, managers run the risk of alienating important stakeholders.

VGI obtained by harvesting and analysing visitor photos posted on social networking sites like Flickr has emerged in recent times as an important way of gaining insights regarding the cultural values provided at various sites of perceived heritage, landscape or other importance (Richards and Friess, 2015). Once integrated with empirical datasets relating to visitor numbers, and the key biophysical properties of sites of interest, scientists and managers are able to better understand trade-offs and comparative values placed on these resources (Griffin et al., 2015).

Using harvested geotagged web content for informing decision making and policy development can provide a grand perspective that is not easily achieved using participatory mapping methods. Conversely, participatory methods can gather more detailed and nuanced information even though the understanding of the perceptions of value may require substantive investment in time. It requires effort to extract views and opinions on the often privileged or personal nature of how people value space. The development of novel methods to optimize and expand the added value of geo-tagged web-content and for it to be more representative of communities is an ongoing challenge for scientists and practitioners.

Knowing stakeholder values and community perceptions is useful for decision making processes. Important value information must often come from a specific user group, rather than the general public, as in the case of characterizing ocean use and values by discrete industries (Ehler and Douvere, 2009). In the context of a public policy process, it would not be appropriate to replace participatory mapping with the compilation of geo-tagged web content since stakeholder engagement forms a key part of the policy process. However, web content may provide insight into the opinion of the general public and short-term visitors, and may be an appropriate proxy for engagement for these large amorphous groups.

7.5 Trust Issues

The use of VGI to inform decision making and policy development is fraught with trust issues. For decision makers, questions of data quality are of key importance, and have been well described in the literature (Comber et al., 2013; Foster-Smith and Evans, 2003; Goodchild and Li, 2012; Heipke, 2010; Hyder et al., 2015; Thiel et al., 2014). Brown et al. (2015) have reviewed these issues within the context of the planning process, and suggest some best-practices guidelines for ensuring adequate data quality.

Equally important is the contributor's trust in the processes of data acquisition, and the motives driving the research. This trust is of key importance to promote volunteer participation and encourage active and honest contributions (Glenn et al., 2012; Klain and Chan, 2012). It may also be determined or influenced by the methods used to collect the data. Gathering ocean use data from fishers, for example, may require meeting fishers at the docks where they are working (Johnson and McClintock, 2013; Scholz et al., 2006). If web-based tools for data collection are used, these may need to be integrated with websites or applications with which contributors are already familiar. Another approach to build trust can be to pair stakeholders with similar interest and to interview them together as part of the data collection process.

In some cases, data collected through VGI methods may include personal and privileged information, such as the spatial extent of practices that are considered trade secrets (e.g. favourite or productive fishing grounds), or that constitute information that might be of competitive advantage to the individual, or that might even amount to an admission of illegal or socially undesirable practices that could place the informant in jeopardy. In such situations, contributors will need to be assured of the integrity and privacy of the information they provide, and how it will be used. On a practical note, there are processes and technical steps that can help overcome these constraints by, for example, allowing data to be aggregated and anonymized so that productive information may be obtained while at the same time masking the individual details (Kittinger et al., 2014).

A common thread shared among many MSP stakeholder engagement approaches is the challenge of ensuring that all relevant voices will be heard, accurately translated and respected throughout the planning process (Ehler and Douvere, 2009; Gopnik et al., 2012; Ritchie and Ellis, 2010). Some stakeholders may want to be actively engaged and consulted throughout a process but may also fear that the information they share cannot be accurately represented with the mapping tools provided. They also may have concerns for the integrity and confidentiality of shared information. There are also concerns for how data will be shared and whether contributions can be traced to them as a source, which could lead to potential retribution, stigmatization or prosecution (Hartter et al., 2013). Other fears that may prevent stakeholders from contributing knowledge include concerns that the process is being conducted for improper motives (e.g. to appease the public rather than address genuinely held worries); suspicions that contributed knowledge may not be used to inform the process fairly or objectively or, worse, that information provided may be used to develop policies that are contrary to the contributor's interests (Klain and Chan, 2012).

In-person meetings and opportunities to communicate face-to-face in a shared physical setting may assist in building the required social capital (Glenn et al., 2012). In most cases, this promotes more meaningful engagement that results in higher quality information. In some instances, a personal engagement prevents the submission of misinformation, or omission of key information by stakeholders or groups.

For in-person participatory mapping methods, transparency of motives and purpose, and time are important ingredients to establish a trust relationship between data contributors and collectors. Transparency results from a clear articulation of the purpose for which information is collected, and how that information will be stored, synthesized and used. Before asking for knowledge contributions, it is important for communities to be afforded an opportunity to learn about, discuss and ask questions regarding the motives and purpose of the data collection effort (Glenn et al., 2012). Equally important to scientists and managers is the investment in time to understanding the nature and people's perceptions of the information. Understanding the context of the information is important to inform its appropriate representation and its use, which, in turn, demonstrates to the community that their concerns were taken into account and that the data they contributed have value (Levine and Feinholz, 2015).

Conversations initiated with appropriate questions and with time to 'talk story' will also provide opportunities for clarification and verification of data.

7.6 Conclusions

The emergence of web-based tools continues to broaden the opportunity to engage larger audiences and gather more representative information and perspectives. Mobile technologies like location-aware (GPS enabled) smartphones offer attractive prospects for the collection of precise, place-based information by a multitude of potential contributors. Additionally, remote participation tools allow stakeholders to be engaged throughout a public process. This maintains open channels of communication, rather than engaging at stakeholder meetings that may be few and far apart.

Participatory mapping and social science methods offer stakeholders an opportunity to submit more nuanced information on how they use and value marine space. Access to digital tools is a prerequisite when determining the most effective way to promote and collect VGI, and enhance stakeholder participation (Carver, 2001; Elwood, 2002; Geertman and Stillwell, 2009). Resource managers and researchers alike must focus on these goals when selecting the tools to meet their data and information requirements (Cravens, 2015). This is particularly relevant as increasingly sophisticated technology becomes more readily accessible to wider user groups, and VGI is recognized as an important component of planning resource management.

References

Berg, B.L. and Lune, H. 2012. *Qualitative Research Methods for the Social Sciences,* 8th ed. Upper Saddle River, NJ: Pearson Education, Inc.

Black, B.D., Adams, A.J. and Berg, C. 2015. Mapping of stakeholder activities and habitats to inform conservation planning for a national marine sanctuary. *Environmental Biology of Fishes,* 98, 2213–2221.

Brown, G., Weber, D. and de Bie, K. 2015. Is PPGIS good enough? An empirical evaluation of the quality of PPGIS crowdsourced spatial data for conservation planning. *Land Use Policy,* 43, 228–238.

Buganza, T., Dell'Era, C., Pellizzoni, E., Trabucchi, D. and Verganti, R. 2015. Unveiling the potentialities provided by new technologies: A process to pursue technology epiphanies in the smartphone app industry. *Creativity and Innovation Management,* 24(3), 391–414.

California Academy of Sciences. 2014. iNaturalist (version 2.6.4). [Mobile application software]. Retrieved from http://www.itunes.apple.com.

Carver, S., Evans, A., Kingston, R. and Turton, I. 2001. Public participation, GIS, and cyberdemocracy: Evaluating on-line spatial decision support systems. *Environment and Planning B,* 28(6), 907–922.

CITT. 1992. Information technology – Digital compression and coding of continuous-tone still images – Requirements and guidelines. Recommendation T.81. Geneva, Switzerland: International Telegraph and Telephone Consultative Committee of the International Telecommunications Union. Available online at https://www.w3.org/Graphics/JPEG/itu-t81.pdf, accessed 4 April 2016.

Comber, A., See, L., Fritz, S., Van der Velde, M., Perger, C. and Giles, F. 2013. Using control data to determine the reliability of volunteered geographic information about land cover. *International Journal of Applied Earth Observation and Geoinformation*, 23, 37–48.

Corbett, J.M. and Rambaldi, G. 2009. Geographic information technologies, local knowledge and change. In M. Cope and S. Elwood (Eds.), *Qualitative GIS: A Mixed Methods Approach*. Thousand Oaks, CA: Sage Publications, pp. 75–91.

Cravens, A.E. 2014. Needs before tools: Using technology in environmental conflict resolution. *Conflict Resolution Quarterly*, 32.1, 3–32.

D'Iorio, M., Selbie, H., Gass, J. and Wahle, C. 2015. *The Pacific Regional Ocean Uses Atlas, Data and Tools for Understanding Ocean Space Use in Washington, Oregon and Hawaii*. U.S. Department of the Interior, Bureau of Ocean Energy Management, Pacific OCS Region, Camarillo, CA. OCS Study BOEM 2015-014.

Davies, T.K., Stevens, G., Meekan, M.G., Struve, J. and Rowcliffe, J.M. 2012. Can citizen science monitor whale-shark aggregations? Investigating bias in mark-recapture modeling using identification photographs sourced from the public. *Wildlife Research*, 39, 696–704.

DeGraff, A.K. and Ramlal, B. 2015. Participatory mapping: Caribbean small island developing states. Forum on the Future of the Caribbean. St. Augustine, Trinidad, University of the West Indies, Trinidad & Tobago Ministry of Foreign Affairs, and the United Nations. Dordrecht: Springer Science & Business Media.

Douvere, F. 2008. The importance of marine spatial planning in advancing ecosystem-based sea use management. *Marine policy*, 32(5), 762–771.

Dunn, C.E. 2007. Participatory GIS – A people's GIS? *Progress in Human Geography*, 31(5), 616–637.

Ehler, C. and Douvere, F. 2009. *Marine Spatial Planning: A Step-By-Step Approach Toward Ecosystem-Based Management*. Intergovernmental Oceanographic Commission and Man and the Biosphere Programme. IOC Manual and Guides No. 53, ICAM Dossier No. 6. Paris: UNESCO.

Elwood, S. 2008. Volunteered geographic information: Key questions, concepts and methods to guide emerging research and practice. *GeoJournal*, 72(3/4), 133–135.

Elwood, S.A. 2002. GIS use in community planning: A multidimensional analysis of empowerment. *Environment and Planning*, 34(5), 905–922.

Elwood, S., Goodchild, M.F. and Sui, D.Z. 2012. Researching volunteered geographic information: Spatial data, geographic research, and new social practice. *Annals of the Association of American Geographers*, 102(3), 571–590.

Foster-Smith, J. and Evans, S.M. 2003. The value of marine ecological data collected by volunteers. *Biological Conservation*, 113(2), 199–213.

Garau, C. 2012. Focus on citizens: Public engagement with online and face-to-face participation – A case study. *Future Internet*, 4(2), 592–606.

Geertman, S. and Stillwell, J. 2009. *Planning Support Systems Best Practice and New Methods,* London: Springer, p. 429.

Glenn, H., Tingley, D., Sánchez Maroño, S., Holm, D. et al. 2012. Trust in the fisheries scientific community. *Marine Policy,* 36(1), 54–72.

Goldberg, E.G. and McClintock, W.J. 2015. Integrated participation tools facilitate science-based spatial planning with a web-based GIS. *CoastGIS2015 – 12th International Symposium for GIS and Computer Cartography for Coastal Zone Management.* Cape Town, South Africa, pp. 154–157.

Goodchild, M.F. 2007. Citizens as sensors: The world of volunteered geography. *GeoJournal,* 69(4), 211–221.

Goodchild, M.F. 2008. Commentary: Whither VGI? *GeoJournal,* 72(3), 239–244.

Goodchild, M.F. 2010. Towards geodesign: Repurposing cartography and GIS? *Cartographic Perspectives,* 66, 7–22.

Goodchild, M.F. and Li, L. 2012. Assuring the quality of volunteered geographic information. *Spatial Statistics,* 1, 110–120.

Gopnik, M., Fieseler, C., Cantral, L., McClellan, K., Pendleton, L. and Crowder, L. 2012. Coming to the table: Early stakeholder engagement in marine spatial planning. *Marine Policy,* 36(5), 1139–1149.

Great Barrier Reef Marine Park Authority, GBRMPA. 2014. Eye on the reef program. http://www.gbrmpa.gov.au/managing-the-reef/how-the-reefs-managed/eye -on-the-reef, accessed 19 Feb. 2016.

Griffin, R., Chaumont, N., Denu, D., Guerry, A., Kim, C.-K. and Ruckelshaus, M. 2015. Incorporating the visibility of coastal energy infrastructure into multi-criteria siting decisions. *Marine Policy,* 62, 218–223.

Guan, X. and Chen, C. 2014. Using social media data to understand and assess disasters. *Natural Hazards,* 74(2), 837–850.

Haklay, M. 2013. Citizen science and volunteered geographic information: Overview and typology of participation. *Crowdsourcing Geographic Knowledge.* The Netherlands: Springer, pp. 105–122.

Hall-Arber, M., Pomeroy, C. and Conway, F. 2009. Figuring out the human dimensions of fisheries: Illuminating models. *Marine and Coastal Fisheries: Dynamics, Management, and Ecosystem Science,* 1(1), 300–314.

Hartter, J., Ryan, S.J., MacKenzie, C.A., Parker, J.N. and Strasser, C.A. 2013. Spatially explicit data: Stewardship and ethical challenges in science. *PLoS Biol,* 11(9), e1001634.

Heipke, C. 2010. Crowdsourcing geospatial data. *ISPRS Journal of Photogrammetry and Remote Sensing,* ISPRS Centenary Celebration Issue, 65(6), 550–557.

Hyder, K., Townhill, B., Anderson, L.G., Delany, J. and Pinnegar, J.K. 2015. Can citizen science contribute to the evidence-base that underpins marine policy? *Marine Policy,* 59, 112–120.

Jarvis, R., Breen, B.B., Krägeloh, C.U. and Billington, D.R. 2015. Citizen science and the power of public participation in marine spatial planning. *Marine Policy,* 57, 21–26.

Jarvis, R.M., Breen, B.B., Krägelo, C.U. and Billington, D.R. 2016. Identifying diverse conservation values for place-based spatial planning using crowdsourced voluntary geographic information. *Society & Natural Resources* (2016), 1–14.

Jeffrey, A. 2015. iSeahorse (version 1.1.1). [Mobile application software] Retrieved from http://itunes.apple.com.

Johnson, A. and McClintock, W. 2013. 3 steps to community-driven ocean zoning. http://voices.nationalgeographic.com/2013/12/16/3-steps-to-oceanzoning/.

Kaplowitz, M.D. and Hoehn, J.P. 2001. Do focus groups and personal interviews reveal the same information for natural resource valuation? *Ecological Economy*, 36, 237–247.

Katsanevakis, S., Stelzenmüller, V., South, A., Sørensen, T.K., Jones, P.J.S., Kerr, S. Badalamenti, F. et al. 2011. Ecosystem-based marine spatial management: Review of concepts, policies, tools, and critical issues. *Ocean & Coastal Management*, 54(11), 807–820.

Kittinger, J.N., Koehn, J.Z., Le Cornu, E., Ban, N.C., Gopnik, M. et al. 2014. A practical approach for putting people in ecosystem-based ocean planning. *Frontiers in Ecology and the Environment*, 12(8), 448–456.

Klain, S.C. and Chan, K.M.A. 2012. Navigating coastal values: Participatory mapping of ecosystem services for spatial planning. *Ecological Economics*, 82, 104–113. http://doi.org/10.1016/j.ecolecon.2012.07.008.

Koehn, J.Z., Reineman, D.R. and Kittinger, J.N. 2013. Progress and promise in spatial human dimensions research for ecosystem-based ocean planning. *Marine Policy*, 42, 31–38.

Kordi, M.N., Collins, L.B., O'Leary, M. and Stevens, A. 2016. ReefKIM: An integrated geodatabase for sustainable management of the Kimberley Reefs, North West Australia. *Ocean & Coastal Management*, 119, 234–43. doi:10.1016/j.ocecoaman.2015.11.004.

Lauriault, T.P. and Mooney, P. 2014. Crowdsourcing: A geographic approach to public engagement (November 2, 2014). Available at SSRN: http://ssrn.com/abstract=2518233.

Levine, A. and Feinholz, C.L. 2015. Participatory GIS to inform coral reef ecosystem management: Mapping human coastal and ocean uses in Hawaii. *Applied Geography*, 59, 60–69.

Marshall, N.J., Kleine, D.A. and Dean, A.J. 2012. CoralWatch: Education, monitoring, and sustainability through citizen science. *Frontiers in Ecology and the Environment*, 10, 332–334.

McClintock, W. J. 2013. GeoDesign: Optimizing stakeholder-driven marine spatial planning. *Coast Guard Journal of Safety & Securty of Sea, Proceedings of the Marine Safety & Security Council*, 70(3), 63–67.

Mehdipoor, H., Zurita-Milla, R., Rosemartin, A., Gerst, K.L. and Weltzin, J.F. 2015. Developing a workflow to identify inconsistencies in volunteered geographic information: A phenological case study. *PLoS ONE*, 10(10), 1–14.

Merrifield, M.S., McClintock, W., Burt, C., Fox, E., Serpa, P., Steinback, C. and Gleason, M. 2013. MarineMap: A web-based platform for collaborative marine protected area planning. *Ocean & Coastal Management*, 74, 67–76.

NOAA Office of Coastal Management. 2014. *Guidebook to Participatory Mapping of Ocean Uses*. Charleston, SC.

Obama, B. 2010. Executive order 13547: Stewardship of the ocean, our coasts, and the great lakes. Washington, DC, July 19, 2010.

Pettit, C.J. and Nelson, A. 2004. Developing an interactive web based public participatory planning support system for natural resource management. *Journal of Spatial Science*, 49(1), 61–70.

Pimm, S.L., Alibhai, S., Bergl, R., Dehgan, A., Giri, C., Jewell, Z., Joppa, L. et al. 2015. Emerging technologies to conserve biodiversity. *Trends in Ecology & Evolution*, 30(11), 685–696.

Pomeroy, R. and Douvere, F. 2008. The engagement of stakeholders in the marine spatial planning process. *Marine Policy*, 32(5), 816–822.

Pomeroy, R.S., Baldwin, K. and McConney, P. 2014. Marine spatial planning in Asia and the Caribbean: Application and implications for fisheries and marine resource management. *Desenvolvimento e Meio Ambiente*, 32, 151–164.

Richards, D.R. and Friess, D.A. 2015. A rapid indicator of cultural ecosystem service usage at a fine spatial scale: Content analysis of social media photographs. *Ecological Indicators*, 53, 187–195.

Ritchie, H. and Ellis, G. 2010. 'A system that works for the sea'? Exploring stakeholder engagement in marine spatial planning. *Journal of Environmental Planning and Management*, 53(6), 701–723.

Scholz, A., Steinback, C. and Mertens, M. 2006. Commercial fishing grounds and their relative importance off the Central Coast of California. Report submitted to the California Marine Life Protection Act Initiative (2006). Ecotrust.

Silvertown, J. 2009. A new dawn for citizen science. *Trends in Ecology and Evolution*, 24, 467–471.

Sui, D., Elwood, S. and Goodchild, M. (Eds.). 2013. *Crowdsourcing Geographic Knowledge: Volunteered Geographic Information (VGI) in Theory and Practice*. Springer.

Sullivan, B.L., Wood, C.L., Iliff, M.J., Bonney, R.E., Fink, D. et al. 2009. eBird: A citizen-based bird observation network in the biological sciences. *Biological Conservation* 142(10), 2282–2292.

Sullivan, C.M., Conway, F.D.L., Pomeroy, C., Hall-Arber, M. and Wright, D.J. 2015. Combining geographic information systems and ethnography to better understand and plan ocean space use. *Applied Geography*, 59, 70–77.

Taylor, D.F. and Lauriault, T. 2014. *Developments in the Theory and Practice of Cyber-cartography, Applications and Indigenous Mapping*, 2nd ed. Elsevier Science.

Thiel, M., Penna-Díaz, M.A., Luna-Jorquera, G. and Stotz, W.B. 2014. Citizen scientists and marine research: Volunteer participants, their contributions, and projection for the future. *Oceanography and Marine Biology: An Annual Review*, 52, 257–314.

Town, C., Marshall, A. and Sethasathien, N. 2013. Manta matcher: Automated photographic identification of manta rays using keypoint features. *Ecology and Evolution*, 3(7), 1902–1914.

Tulloch, A.I.T., Possingham, H.P., Joseph, L.N., Szabo, J. and Martin, T.G. 2013. Realising the full potential of citizen science monitoring programs. *Biological Conservation*, 165, 128–138.

Waitt Foundation. 2013. Barbuda Blue Halo Initiative: Description and Workplan. Waitt Foundation. Available at: http://barbuda.waittinstitute.org/wp-content/uploads/2013/10/Detailed-Workplan.pdf.

Wood, J.S., Moretzsohn, F. and Gibeaut, J. 2015. Extending marine species distribution maps using non-traditional sources. *Biodiversity Data Journal*, (3), e4900.

Wood, S.A., Guerry, A.D., Silver, J.M. and Lacayo, M. 2013. Using social media to quantify nature-based tourism and recreation. *Scientific Reports*, 3, 2976.

8

Geoinformatic Applications in Marine Management

Lauren McWhinnie and Kate Gormley

CONTENTS

8.1 Introduction

Particularly in recent years, geographical information systems (GIS) and other geoinformatics technologies have come to play an increasing role in managing marine and coastal areas, especially where environmental and socioeconomic factors need to be balanced and integrated within a comprehensive framework. The objective of these geospatial applications is to allow coastal managers, policy analysts, natural and social scientists and planners to make better informed, evidence-based decisions, as well as helping identify and address problems and recognize opportunities that might exist within the marine environment. Example application areas include assessment of vulnerability of coastal communities to flooding hazards (Bagdanavičiūtė et al., 2015; Seenath et al., 2016), planning restoration of species such as eelgrass (de Jonge et al., 2000), understanding non-point source pollution (Shen et al., 2013), managing marine protected areas (MPAs) (Gormley et al., 2014; Levine and Feinholz, 2015; Habtemariam and Fang, 2016) and tracking potential conflicts within an environment, for example, marine mammals and industrial developments and activities (Kuletz et al., 2015; Cavazzi and Dutton, 2016) and marine spatial planning (MSP) (Gimpel et al., 2015). Although comprehensive spatial management schemes have long existed within terrestrial environments (Harris and Elmes, 1993), the application of planning and management concepts and instruments to the marine sector is still relatively new and evolving rapidly, and the use of geoinformatics to support these endeavours is likewise still in its comparative infancy.

Many coastal dwellings and infrastructure, activities and developments rely on their surrounding environment to provide them with natural resources, often referred to as ecosystem goods and/or services (Beaumont et al., 2007). Commercial and industrial activities such as shipping, port expansions, renewable energy, oil and gas exploration and decommissioning could all result in negative impacts on marine systems and a concomitant reduction in their ability to provide essential goods and services (Villasante et al., 2016). A challenge facing planners, managers and decision makers will therefore be managing any impacts associated with expanding marine activities, so as to allow social and economic development while at the same time mitigating any adverse environmental effects on both human and ecosystem health.

This chapter draws on a range of studies from around the world, to demonstrate and review the expanding role of geoinformatics, and especially GIS, in various aspects of marine management. This includes coastal and ocean scale applications, industrial expansion, achieving conservation objectives and implementing approaches such as ecosystem-based management (EBM) within planning frameworks. The global increase in marine conservation directives and legislation (Boyes and Elliott, 2014), brought about partly by an expansion in marine industries, and an increased demand for marine goods and services (Visbeck et al., 2014), has resulted in the production of a significant volume of spatial and environmental data. Many other data sources have arisen directly as an outcome of monitoring initiatives set up in response to these new developments and accompanying legislation (e.g. those related to the marine energy sector). The chapter also details the application of GIS, that is, technology and approaches, for managing such data, as well as its use in analyzing, modelling and decision making related to aspects of marine management. The tools and solutions presented include: species distribution modelling, predictive conservation management, oil spill sensitivity mapping, stakeholder engagement and fisheries sensitivity mapping.

8.2 Marine Management

Due to the increasing demand for resources, and the growing potential for user conflicts within a dynamic and three-dimensional multi-use marine space, environmental protection and conservation measures are priorities for most coastal nations. Managers, planners and policy makers require comprehensive knowledge of resources, uses and stakeholders, in order to minimize conflicts or environmental concerns while resolving or mitigating those that do arise. At the same time, these decision makers may also be required to facilitate and support the expansion of economic activities linked to the sea. In order to reconcile these objectives, accurate and reliable information must reach decision makers in appropriate and easy to use formats, so that they can make informed decisions on behalf of stakeholders. Historically, where such information has existed at all, it has often been difficult to access or use due to a variety of institutional, political and technical reasons.

In many cases, important data and information required for sound decision making have been acquired and held by individuals working in federal, provincial/state, government departments and research institutes, rather than being treated as an enterprise-wide corporate resource. While considerable efforts have been made in recent years to rectify this situation, through collaborative projects (Ricketts, 1992), and moves towards greater sharing of information through the use of spatial data infrastructures and geoportals, management of marine and coastal data holdings remains a significant challenge.

In the majority of instances, GIS is utilized to resolve a very specific set of problems and usually incorporates a custom database specific to that project. In contrast, marine managers and planners more often require access to larger, varied and ultimately more complex geo-databases (spatial data). The extensive geo-databases used for environmental and resource management also require the capabilities of GIS (hardware and software), to manage the multiple layers of data and query information about the location of a wide range of spatial objects or properties. With advancements in technology and software capabilities, easier access to training and reduction in data storage costs, the utilization of large-scale GIS within marine planning and management are now far more commonplace than they were a decade ago (Nath et al., 2000).

In addition, habitat modelling (Gonzalez-Mirelis and Buhl-Mortensen, 2015), sensitivity analysis (Wood and Dragicevic, 2007), pollution monitoring/control (Martens and Huntington, 2012; Lu et al., 2014) and contingency planning for natural and human disasters (Nelson et al., 2015) are useful approaches that scientists and managers can use to make appropriate and timely decisions (e.g. in the event of an oil spill). In many instances, decision makers no longer lack data (Mascia, 2003) but it has been shown that abundant data and information alone does not necessarily aid improved decision making, and can, in fact, cause confusion and inaction (Cossarini et al., 2014; Cvitanovic et al., 2015). A current challenge when developing GIS-based tools to aid decision making and marine management in general is therefore to provide simple tools and/or datasets that are easy to use and understand, without a high-level of technical training. These tools must also still be able to cope with large, complex and dynamic data.

8.2.1 Ecosystem-Based Management and Marine Spatial Planning

Ecosystem-based management (EBM) is an integrated approach to managing natural resources and biodiversity by maintaining ecosystem processes, functions and services (Crowder and Norse, 2008; Cavanagh et al., 2016; Nishida et al., this volume). The goal of EBM is to maintain healthy ecosystems capable of providing a range of benefits (McLeod et al., 2005; Lester et al., 2010). These benefits include food, energy, recreational opportunities and shoreline protection, many of which are declining or are seriously compromised in coastal and ocean ecosystems around the world (UNEP, 2006; Foley et al., 2010; Lester et al., 2010).

Marine spatial planning (MSP), a place-based management approach, has now become a much advocated tool to improve the previously fragmented, single-sector approach to EBM (Lester et al., 2010; Buhl-Mortensen et al., 2016). UNESCO defines MSP as 'a public process of analysing and allocating the spatial and temporal distribution of human activities in marine areas to achieve ecological, economic and social objectives that usually have been specified through a political process' (Ehler and Douvere, 2009). A requirement for the MSP process is the ability to map both resources and the environment within a common framework (see Case Study 8.1). This is where geoinformatics generally, and GIS in particular, can play a key role.

The use of GIS within EBM and the MSP process can be as simple as the collation of layered spatially explicit data (e.g. Douven et al. [2003] used a simple overlay analysis to show the largest and most dense fields of seagrass beds in Banten Bay, Indonesia). However, it is important to understand that such a simple representation may lack information about functional relationships within ecosystems (Altman et al., 2014). Expert knowledge is often required to aid the interpretation of the spatial data in relation to the issue (see Case Study 8.2).

In Europe, the Marine Strategy Framework Directive* (MSFD) and MSP are combined to facilitate management actions for sustainable use of coastal and marine resources. In the United Kingdom, MSP is one of the functions of Marine Scotland and the Marine Management Organisation (MMO) under the Marine (Scotland) Act 2010 and the Marine and Coastal Access Act 2009, respectively, and is a tool for implementing EBM. Environmental policies such as the MSFD focus on preventing and reducing adverse and undesirable changes to natural systems as the result of human activities (Harrald and Davies, 2010), and if required, mitigation of undesirable changes (Borja et al., 2010), which

---

* Directive 2008/56/EC of the European Parliament and of the Council of 17 June 2008 establishing a framework for community action in the field of marine environmental policy (Marine Strategy Framework Directive).

can be visualized using a GIS. Marine management as part of the political agenda has migrated towards integrated and ecosystem-based management systems; quantitative cumulative impact assessments and analytical mapping; and integrated monitoring (see Case Study 8.3).

The successful implementation of MSP using an EBM approach requires the use of the best available approaches, tools and techniques. Some of the emerging geoinformatics tools include geospatial analysis (Li et al., 2011), remote sensing (Uitz et al., 2015), geovisualization analysis (Portman, 2014), telemetry (Weber et al., 2013), modelling and quantitative analysis (Critchell et al., 2015). These and others are useful for understanding the spatial and

CASE STUDY 8.1   SHETLAND'S MARINE PLAN

Shucksmith et al. (2014) report on the use of GIS mapping in regional marine spatial planning (MSP) as part the Shetland Islands' Marine Spatial Plan (SIMSP). The SIMSP was initiated by the Scottish Government in 2006 as a pilot in preparation for the development of MSP (a legal requirement in the United Kingdom) across Scotland through the Scottish Sustainable Marine Environment Initiative (SSMEI). The overarching aim was to develop and test the effectiveness of differing management approaches to deliver sustainable development in Scotland's coastal and marine environment (Shucksmith et al., 2014). The SIMSP incorporates spatial data on existing marine and coastal environmental, socioeconomic and cultural features (see Figure 8.1) and activities into decision making, through the use of GIS.

A total of 127 datasets from a range of data sources was collated, with local knowledge used to verify the evidence through extensive stakeholder engagement and participatory mapping. The data and mapping tool developed within the SIMSP has been used by a range of users, such as developers and decision makers, in planning and assessing areas for development, allowing for potential mitigation of conflicts early in the development process (Shucksmith et al., 2014). The Shetland Islands Council has adopted the fourth edition of the SIMSP as 'Supplementary Guidance' (SG) to the Shetland Local Development Plan (LDP). The SIMSP provides an overarching policy framework to guide marine development and activity out to 12 nautical miles; and based on the ecosystem approach to marine planning, ensures spatial planning of the marine environment where practical that facilitates climate change mitigation. The Shetland LDP and SG (e.g. data and maps outlined in the SIMSP) sets out the policies and criteria against which all proposals for developments and infrastructure (e.g. planning applications and work licences) in the coastal zone will be considered and it is hoped it will be a resource used by all users of the marine environment.

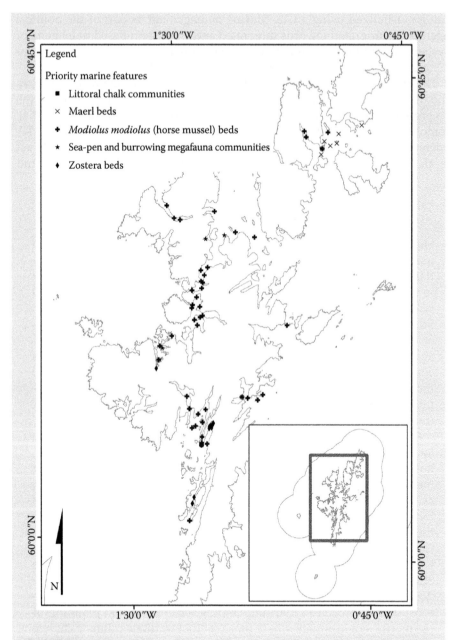

FIGURE 8.1
An example of one of the datasets featured in the Shetland Marine Plan, a map of priority marine features (PMFs).

CASE STUDY 8.2   SHETLAND OIL TERMINAL ENVIRONMENTAL ADVISORY GROUP

The Shetland Isles are the most northerly archipelago of the British Isles, lying approximately 400 miles (640 km) south of the Arctic Circle, which creates a unique and diverse marine ecosystem. Mapping of this important environment is challenging, but critical. This has been achieved through a number of management initiatives, such as the Shetland Oil Terminal Environmental Advisory Group (SOTEAG). Established in 1977, SOTEAG provides environmental advice and continual monitoring of the sensitive environment in the vicinity of the Sullom Voe Terminal; the largest oil and liquefied gas terminal in Europe, occupying more than 400 hectares (receiving crude oil and gas from offshore fields in the North Sea and Atlantic Margin).

The current monitoring programmes include biodiversity analysis of intertidal rocky shores and subtidal sediment, ornithological monitoring and pollution monitoring. SOTEAG is in the process of archiving the datasets with the archive for marine species and habitats data (DASSH) making the data more readily available, with a view to mapping these datasets in the future.

temporal dynamics of marine ecosystems in relation to environmental variation (Katsanevakis et al., 2011). A study recently released by Gavin et al. (2015) outlines many of the newly developed bio-cultural approaches to conservation management and governance. In due course, all of these tools and techniques will need to be combined and made interoperable, in order to successfully coordinate marine uses and activities by implementing MSP and EBM (Curtin and Prellezo, 2010).

8.2.2 Coastal Management

The aim of integrated coastal zone management (ICZM) is to achieve sustainable use and development of coastal resources by balancing environmental, economic, social, cultural and recreational objectives (Khakzad et al., 2015). The implementation of ICZM requires an understanding and knowledge of the physical, social and biological environment, as well as the relationships between these agents (Rodríguez et al., 2009). GIS is increasingly being used to support this process, by enabling the homogenization and integration of social, environmental and economic information within a single geodatabase. This enables easier access to data, generation of thematic cartography and increased use of spatial and geostatistical analysis for evidence-based decision making (Laitinen and Neuvonen, 2001).

CASE STUDY 8.3   SHETLAND OIL SPILL CONTINGENCY PLANNING

Shetland's marine and terrestrial environments are still close to pristine, and oil spill contingency planning is an important function of the Shetland Oil Terminal Environmental Advisory Group (SOTEAG); and as such, is a globally recognized authority. In 1989, SOTEAG was consulted on best practice for oil spill response following the *Exxon Valdez* oil spill in Prince William Sound, Alaska, based on its excellent long-term environmental management record. They were also responders to the *Braer* oil spill. The *Braer* tanker ran aground on the south coast of Shetland, in extreme weather in 1993, losing its entire cargo of crude and fuel oil (87,000 tonnes).

In order to protect the marine environment around the terminal, SOTEAG has created environmental oil spill sensitivity maps as part of the Shetland Marine Pollution Contingency Plan and the Sullom Voe Harbour Oil Spill Plan. The maps are used to identify areas that are environmentally sensitive in the event of an oil spill and how the response can therefore be coordinated. The data was collected from a variety of sources, both from publicly available national and local data, and through the collection of bespoke datasets, including the ongoing monitoring programme data. The local community and stakeholders were consulted throughout the process through an e-working group, stakeholder engagement meetings and consultation with data contributors. This consultation process and local knowledge was the most important aspect of the data collection and mapping, with some mapping methodology reverting to simple drawing on maps (e.g. for bird sightings and feeding grounds) to collate the data, with them subsequently being converted into GIS format. These maps are available online (http://www.soteag.org.uk/) and updated annually.

GIS lends itself to dealing with the complex task of managing and integrating a diversity of spatial data, and is also useful to understand the spatial and temporal evolution of dynamic processes as well as the factors that control their behaviour. The use of GIS to support ICZM allows for the analysis of management scenarios in order to evaluate the impacts of various actions and activities. Local coastal management is contextual and requires unique management strategies (Rodríguez et al., 2009).

GIS can also support engagement by stakeholders and managers, in planning and management of marine resources at all government levels. Mahboubi et al. (2015), for example, developed a GIS-based tool to identify important social-ecological hotspots in the marine environment. This was used to capture local biodiversity, cultural and economic knowledge and values. Spatial analyses were then used to assign and quantify relative social-ecological importance of marine spaces, which could then be visualized with a high degree of spatial significance. Mahboubi et al. (2015) also analyzed and determined the relative importance of how people value their environment. GIS tools such as this can serve as a useful 'first filter' for the spatial distribution of important marine locations.

Another example of the use of GIS for ICZM is the Brazilian National Sea Turtle Conservation Program (TAMAR) that has developed a geospatial tool to identify key areas for sea turtle nesting (Lopez et al., 2015). The sensitivity and detail of the map was specifically designed to support coastal management policies, and recommendations for management of human activities at nesting beaches.

8.2.3 Ocean Management

The need for management of ocean ecosystems at a larger seascape-scale is becoming increasingly recognized (Paxinos et al., 2008; Blau and Green, 2015; Bigagli, 2016). Traditional planning and management initiatives are usually confined to national boundaries. However, many sea uses, such as shipping and fisheries, span these boundaries and require management interventions at larger scales (Katsanevakis et al., 2011). In the open ocean, GIS has application in fisheries management (Stelzenmüller et al., 2008; Nishida et al., this volume), MPA network design (Cabral et al., 2015), predictive modelling (Burkhard et al., 2011), environmental assessment and regulatory requirements (Yates et al., 2015), scientific research (Murphy and Wheeler, this volume) and search and rescue operations (Mathew and Power, this volume), amongst others.

Ocean mapping using GIS tends to focus on generically useful methods for understanding transboundary resource-use patterns, as well as the integrated physio-ecological understanding of marine and coastal environments (Queffelec et al., 2009). Transboundary integration is particularly challenging due to the sectoral and inconsistent nature of government administrations (Cooper, 2011). However, the development of initiatives such as the European Integrated Maritime Policy demonstrate the desire for a more comprehensive approach to maintaining the benefits of and managing ecosystem resources and services in the offshore marine environment (Queffelec et al., 2009; Perry et al., 2010).

Ocean management will require spatial information on the distribution of species assemblages, habitats and other ecological features (Schmiing et al., 2014). While biophysical data of the ocean environments is often available, information relating to human uses of and connections to specific marine places is often lacking. GIS provides a platform for aggregating, displaying and analyzing marine spaces to inform management strategies.

Two contrasting initiatives, from Canada and China, respectively, illustrate this well. In Canada, the Eastern Scotian Shelf Integrated Management (ESSIM)

Initiative was a collaborative management and planning process led by the Oceans and Coastal Management Division (OCMD), Fisheries and Oceans Canada (DFO; see DFO, 2006), Maritimes Region. It was designed as an intergovernmental and multi-stakeholder management and planning process, in order to develop and implement an integrated ocean management plan for the large biogeographic area involved (Rutherford et al., 2005), and has four overarching objectives:

- Integration of management of all measures and activities
- Management for conservation, sustainability and responsible use of the ocean
- Restoration and maintenance of biological diversity and productivity
- Provision of opportunities for economic diversification and sustainable wealth generation

Meanwhile, China has similarly established legislation and management schemes to manage its large marine area. Instead of having specific objectives, this latter initiative is based on three driving principles (Li, 2006; Douvere, 2008): (1) the right to the sea-use authorization scheme (according to Chinese law, the seas around China are the property of the state); (2) a marine functional zoning scheme; and (3) a user-fee system.

In essence, both of these different approaches deal with the multiple-use of the sea area, the Chinese scheme through the adoption of a zoning plan and the Canadian scheme by integrating their various management measures. It could be argued that the Canadian scheme followed a more ecosystem-based holistic approach than that of the Chinese system, with greater emphasis being placed on maintaining the integrity of the wider ecosystem. In contrast, the Chinese scheme appears to perhaps focus more on the management of its resources and improving coordination of its activities.

8.2.4 Ocean Zoning

Ocean zoning seeks to partition marine space, and allocates activities to specifically designated zones in order to reduce conflict within a multi-use and multi-user space (Curtin and Prellezo, 2010; Katsanevakis et al., 2011), and the cumulative impacts of these uses. It can be based on a number of parameters taken either separately or in combination, including ocean conditions, the distribution of resources, social development needs and national security agreements (Chien et al., 2012). Zoning can be seen as one or more regulatory measures that implement and translate the objectives of planning frameworks such as the ESSIM and the Chinese management scheme into practice, and is also a mechanism for implementing the objectives of MSPs.

There are several software tools for calculating and rationalising zoning schemes. For example, Marxan is a set of freely available tools for conservation planning (www.uq.edu.au/marxan). Within this comprehensive package, 'Marxan with Zones' (Watts et al., 2009) is an extension to the core product that is specifically designed to provide land-use zoning options in geographical regions, on-land or in the marine environment, for biodiversity conservation purposes. It uses simulated annealing approaches to create alternative zoning configurations that can maximize defined social, economic and ecological objectives while minimising the overall social, economic and ecological costs (Katsanevakis et al., 2011).

However, despite their proven utility, techno-centric approaches are often perceived to lack an understanding of underlying social issues. Other tools for developing zoning schemes use analytical methods that focus on mapping the cumulative impacts of different sectors of human activities. The total impact of all human activities on the oceans can be assessed and specific activities can be included or excluded from consideration in order to determine which 'suite' of activities can best meet the objectives of a given zone (Douvere, 2008; Katsanevakis et al., 2011; McWhinnie et al., 2014). For example, in the Great Barrier Reef Marine Park (GBRMP) of Australia, various human activities (e.g. fishing and tourism) are permitted within certain zones while more stringent protection applies to other areas (Douvere, 2008). The GBRMP is currently the largest zoned area on the planet, and has demonstrated that a simple zoning classification is important for public acceptance (Katsanevakis et al., 2011). Similar management initiatives are being implemented across Europe and North America, for example: the Coastal Zone Management Program in Massachusetts (MOCZM, 2011; EEA, 2015), the Florida Keys Marine Sanctuary Management Plan (NOAA, 2007) and for the Baltic Sea (Zaucha, 2014).

Many management agencies have now committed to developing ocean zoning and marine spatial plans (Douvere, 2008; Agardy, 2010). These emphasize the regulation of ocean spaces (such as marine protected areas) as opposed to open-ocean uses (e.g. catch or gear). Zoning maps produced by these schemes present information so that planners, policy makers and managers can easily process the information (Chien et al., 2012). The assumption is that managers have access to spatial information, and while biophysical information about the open ocean is often available, human use and associated data are often limited or lacking at this scale. Given the need for improved spatial information and better understanding, the field of geography can provide critical insights towards documenting and theorizing vast marine spaces in order to better inform open-ocean management strategies (Levine et al., 2015).

8.3 Conservation Management

New governance structures based on decentralization and stakeholder involvement are commonly introduced as solutions to the ecological and social complexities associated with conservation management (Sandström et al., 2015). In many areas this re-evaluation and structuring of traditional management approaches has resulted in a renewed commitment to better management of the marine environment. One such area is Canada's Pacific coast, where GIS and the decision support tool Marxan have been used to develop two main products, an atlas of known ecological values and human uses; and an analysis of conservation and human use high-value areas (Ban et al., 2013).

Identifying spatial patterns in the distribution of species, and understanding the processes that create these patterns, often underpin many management and conservation decisions, and are becoming increasingly important in the face of an ever-changing climate, pollution, invasive species and habitat destruction. The species distribution model (SDM) is one approach that can provide practical information on the spatial distribution of a species. Conservation planning has been for the most part the primary objective of marine SDM research over the past 20 years (Marshall et al., 2014). There are several SDM models including: correlative (Guisan and Zimmermann, 2000; Pearson and Dawson, 2003), coupled correlative and process-based (Smolik et al., 2010) and mechanistic approaches (Kearney and Porter, 2009). Each of these has its advantages and disadvantages, with correlative models seemingly the most popular choice. This approach correlates species occurrence (presence or absence) records with environmental data in a geographic place, and seeks to explain the distribution of species. The application of SDMs in the marine environment, although varied, is still relatively limited. Robinson et al. (2011) considered the progress in SDM applications and highlighted some key considerations that are rarely incorporated into SDM. These included species interactions and dispersal patterns, and issues of data quality and model evaluation.

One criticism of SDM within conservation is that it has generally been seen as 'passive', and the desired uses of the model output have rarely been defined beyond general claims that they are available to support marine conservation management. Therefore, a stage has been reached in the development and application of SDMs where it is now crucial to consider how their outputs might be used more than as just a map, whether metadata associated with the models should have a standardized format, and how to integrate their output layers with other conservation software such as Marxan. If SDM are to be applied within a real-world marine conservation management context, then these and other questions urgently need to be addressed.

Marine protected areas (MPAs) are another prominent conservation management tool that have received increasing levels of international support in recent decades, with the UN Convention on Biodiversity setting the target of 10% of the world's oceans being designated as MPAs by 2020 (Levine et al., 2015). Predictive modelling, such as the SDM models just described, can support the identification of priority sites for conservation and spatial assessments of MPA networks, while predictive habitat mapping is an important tool in spatial planning and management (Schmiing et al., 2014; see also Nishida et al., this volume). Given the ongoing interest to designate and network additional national and transboundary MPAs coupled with rapidly evolving marine planning initiatives, GIS models using SDM have many applications and benefits for marine conservation management, one of which is described in Case Study 8.4.

CASE STUDY 8.4   PREDICTING CONSERVATION MANAGEMENT EFFORTS

A study carried out by Gormley et al. (2014) used SDM and GIS to examine the level of management complexity (simultaneously occurring licensed human activities) within MPAs in Scottish waters. The aim of this study was to determine whether it is possible to develop a suitable indicator to predict how challenging or time consuming the management of MPAs in Scottish waters might be, using GIS mapping and available data. The study utilized data for priority marine features (PMFs), modelled 'most suitable' PMF distribution (MAXENT) and environmental data consigned to GIS grids. Layers were overlaid in GIS and scores representing 'most suitable habitat', 'PMF' and 'management complexity' were created (Figure 8.2).

Two prediction indicator methods were developed. The first was based on ranking the MPAs against the statistically significant correlations between the different variables and the second was developed to provide a formula by which to calculate an approximate number of 'casework events' (defined as any work or statutory consultation associated with an MPA, such as planning applications, discharges or new fisheries) required for each MPA based on the linear regression relationship between variables.

Results showed that human activities within 5 km of an MPA had no impact on the number of casework events. However, the number of casework events was influenced by the location, number of features, the type of features and most suitable habitat of an MPA. This work demonstrates possible cost-effective evaluation options for an MPA network and that some MPAs are likely to be more efficient than others in terms of management time. Furthermore, this work has produced a predictive calculation method that could be applied to future MPAs to provide an assessment of the level of management that might be required.

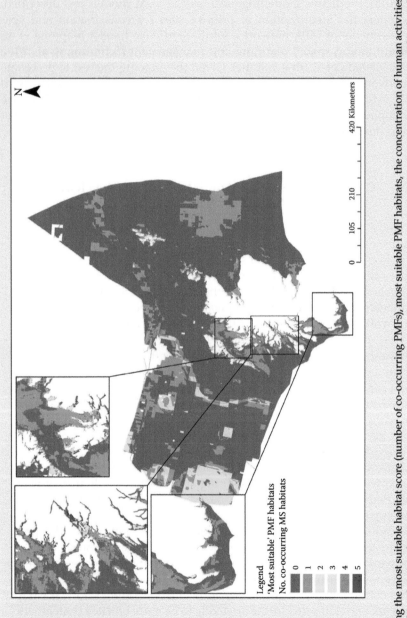

FIGURE 8.2
Maps showing the most suitable habitat score (number of co-occurring PMFs), most suitable PMF habitats, the concentration of human activities and the number of casework events. *(Continued)*

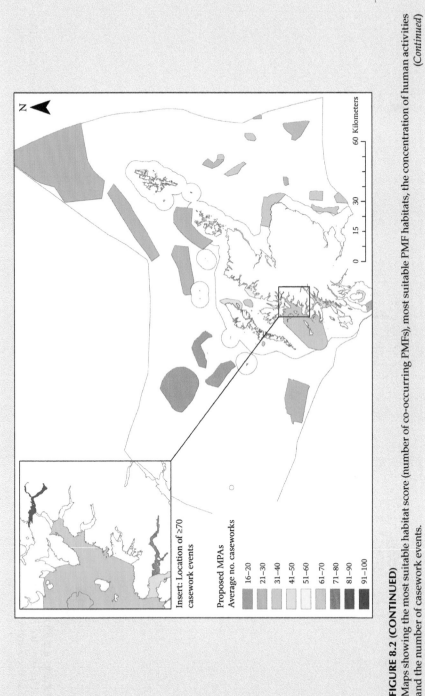

FIGURE 8.2 (CONTINUED)

Maps showing the most suitable habitat score (number of co-occurring PMFs), most suitable PMF habitats, the concentration of human activities and the number of casework events.

(Continued)

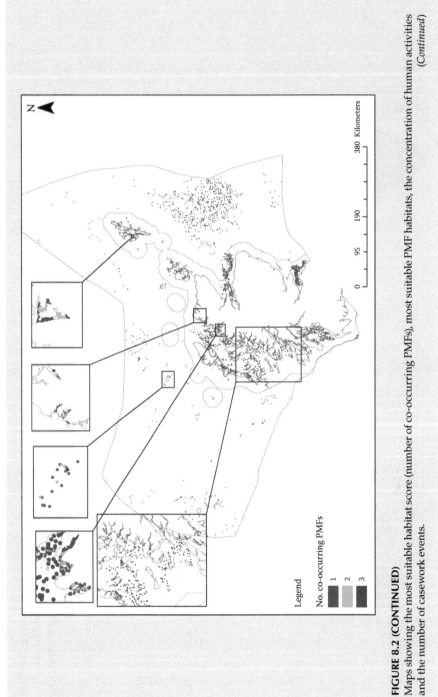

FIGURE 8.2 (CONTINUED)

Maps showing the most suitable habitat score (number of co-occurring PMFs), most suitable PMF habitats, the concentration of human activities and the number of casework events.

(Continued)

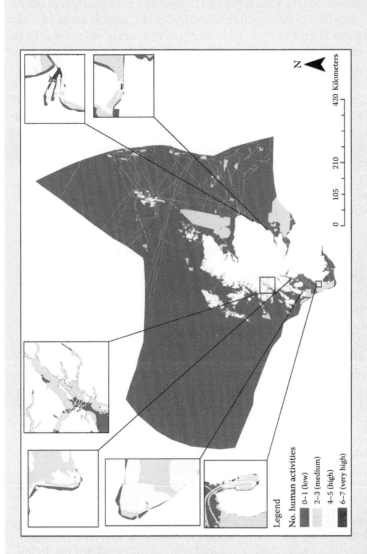

FIGURE 8.2 (CONTINUED)

Maps showing the most suitable habitat score (number of co-occurring PMFs), most suitable PMF habitats, the concentration of human activities and the number of casework events. PMF, priority marine feature. (Reprinted from *Marine Policy*, 47, Gormley, K.S.G., McWhinnie, L.H., Porter, J.S., Hull, A.D., Fernandes, T.F. and Sanderson, W.G. Can management effort be predicted for marine protected areas? New considerations for network design, 138–146. Copyright (2014). With permission from Elsevier.)

8.4 Resources Management

Managing marine resource use is important to reduce user conflicts and to mitigate environmental impacts. A precautionary and preventative management strategy requires a better understanding and quantification of risks to the environment from the extraction and use of marine resources. To be effective, this requires innovative spatially explicit information systems to inform stakeholder engagement, and decision and policy making.

Global coastal development pressure creates an urgent need for decision makers to employ tools to manage marine spaces, assess their diversity of social, commercial or ecological values and to quantify the relative importance of these. In particular, there is widespread and growing recognition of the importance of considering and including people and human behaviour when managing development in ocean and coastal regions (Levine and Feinholz, 2015).

It can be assumed, for example, that many countries will increasingly seek alternatives to coal-based energy generation, including those based on

CASE STUDY 8.5   BP SENSITIVITY MAPPING

As part of oil spill preparedness activities, BP identified the need for sensitivity mapping of the coastline near to where they operate and, in the United Kingdom, have gathered a variety of GIS data on biological, social and economic sensitivities on the Scottish coastline that will facilitate this mapping.

The sensitivity maps were created based on the National Oceanic and Atmospheric Administration's (NOAA) Environmental Sensitivity Index (ESI) guidelines (http://response.restoration.noaa.gov/esi_guide lines). These maps provide a concise summary of coastal resources that may be at risk, should an oil spill occur, and enable BP to overlay oil spill modelling outputs to assess which types of oil spills, and at what locations, would have the highest impact on the coastline (Figure 8.3).

The maps are based on publicly available data held by a number of organizations and agencies. Once the data are collected and mapped, a sensitivity ranking (measured from low to very high) is applied. BP aims to enhance these map resources further, through the development of seasonal sensitivity mapping, to ensure the long-term relevance of their resource. BP engaged with Oil and Gas UK throughout the process to ensure the datasets were accessible industry wide as an example of best available datasets for oil spill preparedness and response. As a result, the maps are being made available on Marine Scotland's National Marine Plan Interactive (NMPi; https://marinescotland.atkins geospatial.com/nmpi/).

FIGURE 8.3
Oil spill sensitivity data for part of the Shetland Islands, Scotland, overlaid on a Google Earth base map.

conventional hydrocarbons, nuclear power and a range of marine renewables (see Chapter 12 by Green, this volume, for a discussion on the role of geoinformatics in developing and managing these last). Nuclear power stations are usually situated on the coast for easy access to sea water for cooling the reactors, but this can have resulting social and environmental implications. GIS-based multicriteria decision making are often used for siting nuclear power plants (Abudeif et al., 2015) to help identify key questions, and to collate, standardize and weight appropriate data before final spatial suitability is analyzed.

The offshore renewable and hydrocarbon industries are subject to strict environmental legislation and regulations and as part of environmental impact assessments (EIA), other assessment processes and potential pollution management (e.g. oil spill contingency planning; see Case Study 8.5). Mapping of constraints and sensitive environmental features is important for designing mitigation strategies for energy installations, and the availability of environmental data, both spatial and temporal, is a key part of developing management options. In the United Kingdom, for example, oil and gas operators have for many years been submitting seabed survey data to the industry trade associated (Oil and Gas UK) and the regulators (Marine Scotland, CEFAS and DECC) on a voluntary basis. This has resulted in a database of environmental data spanning five decades that is also available for research and industry. The database has recently been converted into spatial formats and layers, along with sediment and oceanography data provided by the British Geological Survey and the National Oceanographic Center through the Natural Environment Research Council (NERC) North Sea Interactive project (http://www.northseainteractive.hw.ac.uk/) and the project has also been expanded to cover data produced for the West of Shetland, Atlantic Margin area. As a next phase, it is intended that academics will work with the industry to create an online mapping portal to assist marine managers with these and other planning and management tasks in the future.

8.5 Summary, Conclusions and Future Plans

Marine management policies and practices are increasingly becoming area-based and dependent on the use of spatially explicit data and supporting technologies. The drivers for this change in emphasis include requirements and obligations to comply with MSP policy and legislation; recognition of the need to promote sustainable stewardship of ecosystem goods and services for the benefit of coastal communities; and increasing awareness of human and natural disasters, many of which can be linked to changing climate and industrial activities. Nonetheless, while recognition of the need is increasing, to date the range and number of geoinformatics applications developed to support marine management still lags well behind those used for management of terrestrial environments.

One clear trend that may be discerned is the evolution of GIS applications from addressing the management needs of single sectors of the marine economy to now a much more holistic and integrative and multisectorial area-based management of transboundary marine spaces. The ability to analyze data to understand temporal cumulative impacts, combined with the inherent ability of GIS and related technologies to support communication and understanding of complex, often contested use of three-dimensional and dynamic marine spaces, across a range of spatial and temporal scales, makes it a particularly useful tool to inform, and negotiate planning scenarios with stakeholders and managers alike.

Recent progress in the application of geoinformatics to improve governance of the world's oceans is partly due to a significant improvement in data availability and quality. Much of this improved access to data has come about through the application of remote sensing (Chapter 4 of this volume) and autonomous or semi-autonomous sensors (Chapter 5), as well as expanded use of GIS and other analytical tools to convert these data into useful information products. Sharing of data between individuals, organizations, industry and governments has also been facilitated and encouraged by a change in the understanding of the value of shared data, the willingness of institutions to collaborate more widely, the development of ontologies (Chapter 6 of this volume) to facilitate semantic and technical interoperability between datasets, and the growing availability of data and mapping portals and other online tools that enable greater worldwide data discovery and mobility.

These same developments have also resulted in the wider general public and non-specialists being able to view, contribute to, download and use data, especially in conjunction with free and open source GIS software. Support for enhanced stakeholder involvement and community or end-user participation in decision making and planning is one of the exciting current developments in the field of geoinformatics (see Chapters 7 and 10 in this volume). However, as GIS technology becomes more widely used, and the interfaces become ever more user-friendly, non-specialist users need to be aware that the attractive maps are only as good as the underlying data that they draw on, and that often the information presented is highly interpreted. This is important when presenting information to aid decision making.

Despite, or perhaps because of, these caveats, it is reasonable to predict that GIS applications will increasingly support and contribute to marine management. As data storage, software and hardware become more affordable, 'Big Data' ventures will become more commonplace, as will remote/mobile data storage and handling capabilities. Access to more and better quality data will make it more feasible to seek and achieve ambitious management and conservation objectives, and the scale of management interventions will increasingly focus on transboundary management of marine ecosystems.

Acknowledgements

Case study contributors:

- Melanie Netherway, BP Exploration Operating Company Ltd., Aberdeen
- Dr. Lorraine Gray, Marine Atlas Consultants Ltd., Aberdeenshire
- Meriem Kayoueche-Reeve and Dr. Rebecca Kinnear, SOTEAG, University of St. Andrews
- Robert Watret, Marine Scotland Science, Aberdeen

References

Abudeif, A.M., Abdel Moneim, A.A., and Farrag, A.F. (2015). Multicriteria decision analysis based on analytical hierarchy process in GIS environment for siting nuclear power plants in Eygpt. *Annals of Nuclear Energy*, 75, 682–692.

Agardy, T., Notarbartolo di Sciara, G., and Christie, P. (2011). Mind the gap: Addressing the shortcomings of marine protected areas through large scale marine spatial planning. *Marine Policy*, 35(2), 226–232.

Aires, C., González-Irusta, J.M., and Watret, R. (2014). Updating fisheries sensitivity maps in British waters. *Scottish Marine and Freshwater Science*, Edinburgh: Scottish Government, 5(10), 88.

Altman, I., Boumans, R., Gopal, S., Roman, J., and Kaufman, L. (2014). An ecosystem accounting framework for marine ecosystem-based management in the sea. In: *The Sea: Volume 16: Marine Ecosystem-Based Management*, M. Fogarty and J. McCarthy (Eds.). Cambridge, MA: Harvard University Press, pp. 245–276.

Bagdanavičiūtė, I., Kelpšaitė, L., and Soomere, T. (2015). Multi-criteria evaluation approach to coastal vulnerability index development in micro-tidal low-lying areas. *Ocean and Coastal Management*, 104, 124–135.

Ban, N.C., Bodtker, K.M., Nicolson, D., Robb, C.K., Royle, K., and Short, C. (2013). Setting the stage for marine spatial planning: Ecological and social data collection and analyses in Canada's Pacific waters. *Marine Policy*, 39, 11–20.

Beaumont, N.J., Austen, M.C., Atkins, J.P., Burdon, D., Degraer, S., Dentinho, T.P., Derous, S. et al. (2007). Identification, definition and quantification of goods and services provided by marine biodiversity: Implications for the ecosystem approach. *Marine Pollution Bulletin*, 54, 253–265.

Bigagli, E. (2016). The international legal framework for the management of the global oceans social-ecological system. *Marine Policy*, 68, 155–164.

Blau, J. and Green, L. (2015). Assessing the impacts of a new approach to ocean management: Evidence to date from five ocean plans. *Marine Policy*, 56, 1–8.

Borja, A., Elliott, M., Carstensen, J., Heiskanen, A.S., and van de Bund, W. (2010). Marine management – Towards an integrated implementation of the European Marine Strategy Framework and the Water Framework Directives. *Marine Pollution Bulletin*, 60, 2175–2186.

Boyes, S., and Elliott, M. (2014). Marine legislation – The ultimate 'horrendogram': International law, European directives and national implementation. *Marine Pollution Bulletin*, 86(1–2), 39–47.

Buhl-Mortensen, L., Galparsoro, I., Fernández, T.V., Johnson, K., D'Anna, G., Badalamenti, F., Garofalo, G. et al. (2016). Maritime ecosystem-based management in practice: Lessons learned from the application of a generic spatial planning framework in Europe. *Marine Policy* (Article in Press), doi:10.1016/j.marpol.2016.01.024.

Burkhard, B., Opitz, S., Lenhart, H., Ahrendt, K., Garthe, S., Mendel, B., and Windhorst, W. (2011). Ecosystem based modelling and indication of ecological integrity in the German North Sea – Case study offshore wind parks. *Ecological Indicators*, 11, 168–174.

Cabral, R.B., Mamanag, S.S., and Aliño, P.M. (2015). Designing a marine protected areas network in a data-limited situation. *Marine Policy*, 59, 64–76.

Cavanagh, R.D., Hill, S.L., Knowland, C.A., and Grant, S.M. (2016). Stakeholder perspectives on ecosystem-based management of the Antarctic krill fishery. *Marine Policy*, 68, 205–211.

Cavazzi, S. and Duton, A.G. (2016). An offshore wind energy geographic information system (OWE-GIS) for assessment of the UKs offshore wind energy potential. *Renewable Energy*, 87, 212–228.

Chien, L., Tseng, W., Chang, C., and Hsu, C. (2012). A study of ocean zoning and sustainable management by GIS in Taiwan. *Ocean and Coastal Management*, 69, 35–49.

Cooper, J.A.G. (2011). Progress in integrated coastal zone management (ICZM) in Northern Ireland. *Marine Policy*, 35, 794–799.

Cossarini, D.M., MacDonald, B.H., and Wells, P.G. (2014). Communicating marine environmental information to decision makers: Enablers and barriers to use of publications (grey literature) of the Gulf of Marine Council on the Marine Environment. *Ocean and Coastal Management*, 96, 163–172.

Coull, K.A., Johnstone, R., and Rogers, S.I. (1998). Fisheries Sensitivity Maps in British Waters. Published and Distributed by UKOOA Ltd.

Critchell, K., Grech, A., Schlaefer, J., Andutta, F.P., Lambrechts, J., Wolanski, E., and Hamann, M. (2015). Modelling the fate of marine debris along a complex shoreline: Lessons from the Great Barrier Reef. *Estuarine, Coastal and Shelf Science*, 167, 414–426.

Crowder, L. and Norse, E. (2008). Essential ecological insights for marine ecosystem-based management and marine spatial planning. *Marine Policy*, 32(5), 772–778.

Curtin, R. and Prellezo, R. (2010). Understanding marine ecosystem based management: A literature review. *Marine Policy*, 34, 821–830.

Cvitanovic, C., Hobday, A.J., van Kerkhoff, L., Wilson, S.K., Dobbs, K., and Marshall, N.A. (2015). Improving knowledge exchange among scientists and decision makers to facilitate the adaptive governance of marine resources: A review of knowledge and research needs. *Ocean and Coastal Management*, 112, 25–35.

De Jonge, V.N., de Jong, D.J., and van Katwijk, M.M. (2000). Policy plans and management measures to restore eel grass (*Zostera marina* L.) in the Dutch Wadden Sea. *Helgoland Marine Research*, 54, 151–158.

Department of Fisheries and Oceans, Canada (DFO). (2005). Eastern Scotian Shelf Integrated Ocean Management Plan. ESSIM Planning Office, Oceans and Coastal Management Division, Fisheries and Oceans Canada, 72 pp. Downloaded (19/04/2016): http://www.inter.dfo-mpo.gc.ca/Maritimes/Oceans/OCMD/ESSIM/Strategic-Plan/ESSIM-Plan.

Douven, W.J.A.M., Buurman, J.J.G., and Kiswara, W. (2003). Spatial information for coastal zone management: The example of the Banten Bay seagrass ecosystem, Indonesia. *Ocean and Coastal Management*, 46(6–7), 615–634.

Douvere, F. (2008). The importance of marine spatial planning in advancing ecosystem-based sea use management. *Marine Policy*, 32, 762–771.

Ehler, C. and Douvere, F. (2009). Marine spatial planning: A step-by-step approach towards ecosystem-based management. Intergovernmental Oceanographic Commission and Man and the Biosphere Programme, IOC Manual and Guides No. 53, ICAM Dossier No. 6. Paris: UNESCO (English).

Ellis, J.R., Milligan, S.P., Readdy, L., Taylor, N., and Brown, M.J. (2012). Spawning and nursery grounds of selected fish species in UK waters. *Sci. Ser. Tech. Rep.*, CEFAS Lowestoft, 147.

Executive Office of Energy and Environmental Affairs (EEA). (2015). Commonwealth's first-ever Ocean Management Plan 2015 Massachusetts Ocean Management Plan. Commonwealth of Massachusetts. Downloaded (19/04/2016): http://www.mass.gov/eea/.

Foley, M.M., Halpern, B.S., Micheli, F., Armsby, M.H., Caldwell, M.R., Crain, C.M., Prahler, E. et al. (2010). Guiding ecological principles for marine spatial planning. *Marine Policy*, 34, 955–966.

Gavin, M.C., McCarter, J., Mead, A., Berkes, F., Stepp, J.R., Peterson, D., and Tang, R. (2015). Defining biocultural approaches to conservation. *Trends in Ecology and Evolution*, 30(3), 140–145.

Gimpel, A., Stelzenmüller, V., Grote, B., Buck, B.H., Floeter, J., Núñez-Riboni, I., Pogoda, B., and Temming, A. (2015). A GIS modelling framework to evaluate marine spatial planning scenarios: Co-location of offshore wind farms and aquaculture in the German EEZ. *Marine Policy*, 55, 102–115.

Gonzalez-Mirelis, G. and Buhl-Mortensen, P. (2015). Modelling benthic habitats and biotopes off the coast of Norway to support spatial management. *Ecological Informatics*, 30, 284–292.

Gormley, K.S.G., McWhinnie, L.H., Porter, J.S., Hull, A.D., Fernandes, T.F., and Sanderson, W.G. (2014). Can management effort be predicted for marine protected areas? New considerations for network design. *Marine Policy*, 47, 138–146.

Guisan, A. and Zimmermann, N.E. (2000). Predictive habitat distribution models in ecology. *Ecological Modelling*, 135, 147–186.

Habtemariam, B.T. and Fang, Q. (2016). Zoning for a multiple-use marine protected area using spatial multi-criteria analysis: The case of Sheik Seid Marine National Park in Eritrea. *Marine Policy*, 63, 135–143.

Harrald, M. and Davies, I. (2010). The Saltire Prize Programme: Further Scottish leasing round (Saltire Prize projects) scoping study. *Marine Scotland*. March.

Harris, T.M. and Elmes, G.A. (1993). The application of GIS in urban and regional planning: A review of the North American experience. *Applied Geography*, 13, 9–27.

Katsanevakis, S., Stelzenmüller, V., South, A., Sørenson, T.K., Jones, P.J.S., Kerr, S., Badalamenti, F. et al. (2011). Ecosystem-based marine spatial management: Review of concepts, policies, tools and critical issues. *Ocean and Coastal Management*, 54, 807–820.

Kearney, M. and Porter, W. (2009). Mechanistic niche modelling: Combining physiological and spatial data to predict species' ranges. *Ecology Letters*, 12, 334–350.

Khakzad, S., Pieters, M., and Van Balen, K. (2015). Coastal cultural heritage: A resource to be included in integrated coastal zone management. *Ocean and Coastal Management*, 118, 110–128.

Kuletz, K.J., Ferguson, M.C., Hurley, B., Gall, A.E., Labunski, E.A., and Morgan, T.C. (2015). Seasonal spatial patterns in seabirds and marine mammal distribution in the eastern Chukchi and western Beaufort Seas: Identifying biologically important pelagic areas. *Progress in Oceanography*, 136, 175–200.

Laitinen, S. and Neuvonen, A. (2001). BALTICSEAWEB: An information system about the Baltic Sea environment. *Advances in Environment Research*, 5, 377–383.

Lee, A.J. and Ramster, J.W. (1981). *Atlas of the Seas around the British Isles*. Lowestoft: MAFF.

Lester, S.E., McLeod, K.L., Tallis, H., Ruckelshaus, M., Halpern, B.S., Levin, P.S., Chavez, F.P. et al. (2010). Science in support of ecosystem-based management for the U.S. West Coast and beyond. *Biological Conservation*, 143, 576–587.

Levine, A.S. and Feinholz, C.L. (2015). Participatory GIS to inform coral reef ecosystem management: Mapping human coastal and ocean uses in Hawaii. *Applied Geography*, 59, 60–69.

Levine, A.S., Richmond, L., and Lopez-Carr, D. (2015). Marine resource management: Culture, livelihoods and governance. *Applied Geography*, 59, 56–59.

Li, H. (2006). The impacts and implications of the legal framework for sea use planning and management in China. *Ocean and Coastal Management*, 49, 717–726.

Li, W., Chen, G., Kong, Q., Wang, Z., and Qian, C. (2011). A VR-Ocean system for interactive geospatial analysis and 4D visualisation of the marine environment around Antarctica. *Computers and Geoscience*, 37, 1743–1751.

Longdill, P.C., Healy, T.R., and Black, K.P. (2008). An integrated GIS approach for sustainable aquaculture management area site selection. *Ocean and Coastal Management*, 51, 612–624.

Lopez, G.G., Saliés, E.C., Lara, P.H., Tognin, F., Marcovaldi, M.A., and Serafini, T.Z. (2015). Coastal development at sea turtles nesting grounds: Efforts to establish a tool for supporting conservation and coastal management in Northern Brazil. *Ocean and Coastal Management*, 116, 270–276.

Lu, F., Chen, Z., and Liu, W. (2014). A GIS-based system for assessing marine water quality around offshore oil platforms. *Ocean and Coastal Management*, 102(A), 294–306.

Mahboubi, P., Parkes, M., Stephen, C., and Chan, H.M. (2015). Using expert informed GIS to locate important marine social-ecological hotspots. *Journal of Environmental Management*, 160, 342–352.

Marshall, C.E., Glegg, G.A., and Howell, K.L. (2014). Species distribution modelling to support marine conservation planning: The next steps. *Marine Policy*, 45, 330–332.

Martens, J. and Huntington, B.E. (2012). Creating a GIS-based model of marine debris 'hot-spots' to improve efficiency of lobster trap debris removal program. *Marine Pollution Bulletin*, 64(5), 949–955.

Mascia, M.B. (2003). The human dimension of coral reef marine protected areas: Recent social science research and its policy implications. *Conservation Biology*, 17(2), 630–632.

Massachusetts Office of Coastal Zone Management (MOCZM). (2011). Policy Guide. Commonwealth of Massachusetts. Downloaded (19/04/2016): http://www.mass.gov/eea/agencies/czm/about-czm/czm-policy-guide/.

McLeod, K.L., Lubchenco, J., Palumbi, S.R., and Rosenberg, A.A. (2005). Scientific Consensus Statement on Marine Ecosystem-Based Management. Communication Partnership for Science and the Sea.

McWhinnie, L., Briers, R., and Fernandes, T.F. (2014). The development and testing of a multiple-use zoning scheme for Scottish waters. *Ocean and Coastal Management,* 103, 34–41.

Nath, S.S., Bolte, J.P., Ross, L.G., and Aguilar-Manjarrez, J. (2000). Application of geographical information systems (GIS) for spatial decision support in aquaculture. *Aquacultural Engineering,* 23(1–3), 233–278.

National Oceanic and Atmospheric Administration (NOAA). (2007). Florida Keys National Marine Sanctuary Revised Management Plan. National Oceanic and Atmospheric Administration. Downloaded (19/04/2016): http://floridakeys .noaa.gov/.

Nelson, J.R., Grubesic, T.H., Sim, L., Rose, K., and Graham, J. (2015). Approach for assessing coastal vulnerability to oil spills for prevention and readiness using GIS and the Blowout and Spill Occurrence Model. *Ocean and Coastal Management,* 112, 1–11.

Paxinos, R., Wright, A., Day, V., Emmett, J., Frankiewicz, D., and Goecker, M. (2008). Marine spatial planning: Ecosystem-based zoning methodology for marine management in South Australia. *Journal of Conservation Planning,* 4, 37–59.

Pearson, R.G. and Dawson, T.P. (2003). Predicting the impacts of climate change on the distribution of species: Are bioclimate envelope models useful? *Global Ecology and Biogeography,* 12, 361–371.

Perry, R.I., Ommer, R.E., Barange, M., and Werner, F. (2010). The challenge of adapting marine social-ecological systems to the additional stress of climate change. *Current Opinion in Environmental Sustainability,* 2, 356–363.

Portman, M.E. (2014). Visualisation for planning and management of oceans and coasts. *Ocean and Coastal Management,* 98, 176–185.

Queffelec, B., Cummins, V., and Bailly, D. (2009). Integrated management of marine biodiversity in Europe: Perspectives from ICZM and the evolving EU Maritime Policy framework. *Marine Policy,* 33, 871–877.

Ricketts, P.J. (1992). Current approaches in geographic information systems for coastal management. *Marine Pollution Bulletin,* 25(1–4), 82–87.

Robinson, L.M., Elith, J., Hobday, A.J., Pearson, R.G., Kendall, B.E., Possingham, H.P., and Richardson, A.J. (2011). Pushing the limits in marine species distribution modelling: Lessons from the land present challenges and opportunities. *Global Ecology and Biogeography,* 20, 789–802.

Rodríguez, I., Montoya, I., Sánchez, M.J., and Carreño, F. (2009). Geographic information systems applied to integrated coastal zone management. *Geomorphology,* 107, 100–105.

Rutherford, R.J., Herbert, G.J., and Coffen-Smout, S.S. (2005). Integrated ocean management and the collaborative planning process: The Eastern Scotian Shelf Integrated Management (ESSIM) Initiative. *Marine Policy,* 29, 75–83.

Sanström, A., Bodin, Ö., and Crona, B. (2015). Network governance from the top – The case of ecosystem-based coastal and marine management. *Marine Policy,* 55, 57–63.

Schmiing, M., Diogo, H., Serrao Santos, R., and Afonso, P. (2014). Assessing hotspots within hotspots to conserve biodiversity and support fisheries management. *Marine Ecology Progress Series,* 513, 187–199.

Seenath, A., Wilson, M., and Miller, K. (2016). Hydrodynamic versus GIS modelling for coastal flood vulnerability assessment: Which is better for guiding coastal management? *Ocean and Coastal Management*, 120, 99–109.

Shen, Z.Y., Chen, L., Liao, Q., Liu, R.M., and Huang, Q. (2013). A comprehensive study of the effects of GIS data on hydrology and on-point source pollution modelling. *Agricultural Water Management*, 118, 93–102.

Shucksmith, R., Gray, L., Kelly, C., and Tweddle, J. (2014). Regional marine spatial planning – The data collection and mapping process. *Marine Policy*, 50, 1–9.

Smolik, M.G., Dullinger, S., Essl, F., Kleinbauer, I., Leitner, M., Peterseil, J., Stadlr, L.M., and Vogl, G. (2010). Integrating species distribution models and interacting particle systems to predict the spread of an invasive alien plant. *Journal of Biogeography*, 37, 411–422.

Stelzenmüller, V., Maynou, F., Bernard, G., Cadiou, G., Camilleri, M., Crećhriou, R., Criquet, G. et al. (2008). Spatial assessment of fishing effort around European marine reserves: Implications for successful fisheries management. *Marine Pollution Bulletin*, 56, 64–76.

Uitz, J., Stramski, D., Reynolds, R.A., and Dubranna, J. (2015). Assessing phytoplankton community from hyperspectral measurements of phytoplankton absorption coefficient and remote-sensing reflectance in open-ocean environments. *Remote Sensing of the Environment*, 171, 58–74.

UNEP. (2006). Marine and Coastal Ecosystems and Human Wellbeing: A Synthesis Report Based on the Findings of the Millennium Ecosystem Assessment. [UNEP], United Nations Environment Programme.

Verfaillie, E., Degraer, S., Schelfaut, K., Willems, W., and Lancker, V.V. (2009). A protocol for classifying ecologically relevant marine zones, a practical approach. *Estuarine, Coastal and Shelf Science*, 83, 175–185.

Villasante, S., Lopes, P.F.M., and Coll, M. (2016). The role of marine ecosystem services for human well-being: Disentangling synergies and trade-offs at multiple scales. *Ecosystem Services*, 17, 1–4.

Visbeck, M., Kronfeld-Goharani, U., Neumann, B., Rickels, W., Schmidt, J., van Doorn, E., Matz-Lück, N. et al. (2014). Securing blue wealth: The need for a special sustainable development goal for the oceans and coasts. *Marine Policy*, 48, 184–191.

Watts, M.E., Ball, I.R., Stewart, R.S., Klein, C.J., Wilson, K., Steinback, C., Lourival, R. et al. (2009). Marxan with Zones: Software for optimal conservation based land- and sea-use zoning. *Environmental Modelling and Software*, 24(12), 1513–1521.

Weber, N., Weber, S.B., Godley, B.J., Ellick, J., Witt, M., and Broderick, A.C. (2013). Telemetry as a tool for improving estimates of marine turtle abundance. *Biological Conservation*, 167, 90–96.

Wood, L. and Dragicevic, S. (2007). GIS-based multicriteria evaluation and fuzzy sets to identify priority sites for marine protection. *Biodiversity and Conservation*, 16(9), 2539–2558.

Yates, R.L., Schoeman, D.S., and Klein, C.J. (2015). Ocean zoning for conservation, fisheries and marine renewables energy: Assessing trade-offs and co-location opportunities. *Journal of Environmental Management*, 152, 201–209.

Zaucha, J. (2014). Sea basin maritime spatial planning: A case study of the Baltic Sea region and Poland. *Marine Policy*, 50(Part A), 34–45.

. . . .

. . . .

. . . .

. . . .

. . . .

. . . .

9

Navigating a Sea of Data: Geoinformatics for Law Enforcement at Sea

Phillip Saunders

CONTENTS

9.1 Introduction

The management of coastal and marine areas includes the protection of the coastal state's security interests, including law enforcement and crime prevention. This extends to the identification of and enforcement against criminal activities which occur fully within maritime areas (e.g. piracy, crimes on board vessels, attacks on maritime installations), and those that originate in maritime areas but which are targeted at the commission of offences on land or in the internal waters of the state (e.g. drug or arms smuggling, human trafficking, terrorist intrusions). In the post-9/11 era, the importance of the seaborne threat to security interests acquired a new significance, and events such as the 2008 Mumbai attack (originating from the sea with the hijacking of the trawler *Kuber*), as well as the continuing threat of piracy in a number of regions, and mass migrations across the Mediterranean (see Chapter 15 by Mathews and Power of this volume), have only served to heighten the concerns. At the same time, the development and mandatory application of new technologies for vessel identification and tracking have increased the potential sources of relevant data by orders of magnitude, challenging the ability of authorities to not only monitor activity, but also to analyze and identify what presents an actual threat to be addressed.

The availability of large volumes of geospatial data, coupled with the need to filter and analyze that information for decision making purposes, is naturally suited to the application of geoinformatics. These techniques are familiar, and permit the analysis and representation of information from diverse data sources in an understandable (and actionable) manner, but it must be remembered that in the maritime domain one is not always dealing with the sovereign territory of the state, but rather with varying zones of jurisdiction, as well as areas outside national jurisdiction, which will affect what is and is not a 'crime' at a given location, and will likewise limit the capability of a government authority to enforce against foreign ships and nationals, regardless of the fact that illicit activity may be taking place.

This chapter considers, primarily from a legal perspective, the application of geoinformatics approaches, broadly considered, in the context of crime prevention and enforcement in the maritime areas under the jurisdiction of a coastal state, under the following general topics:

- The primary sources of threats which must be considered in terms of law enforcement in coastal and marine areas; that is, the criminal activities which either occur at sea, or originate at sea and threaten the territory of the state.
- The jurisdictional entitlements provided under international law for the coastal state (and other states), with particular reference to jurisdiction over criminal activities in maritime zones, including the high seas.
- The concept of Maritime Domain Awareness (MDA) is considered, and the availability of relevant data sources (some now mandated under international law) which contribute to the monitoring, control and surveillance of the maritime domain.
- The potential contribution of geoinformatics in allowing national authorities to utilize existing data sources in a timely and accurate manner in the prevention of and enforcement against criminal activities, including the challenges facing achievement of this objective.

Somewhat arbitrary dividing lines have been drawn in this chapter, to exclude the more clearly military aspects of national security (as with, e.g. efforts to prevent the non-proliferation of nuclear weapons), as well as the enforcement of 'quasi-crimes' in the fisheries and environmental contexts. While these are important and closely related problems, often engaging the same monitoring and enforcement capabilities as for criminal enforcement, they raise particular issues which are beyond the scope of this review. It is, however, clear that activities such as terrorism and piracy blur the lines between law enforcement and military responsibilities, while certain fisheries and environmental offences are treated as crimes in a number of jurisdictions. The role of geoinformatics in fisheries management, including the enforcement of regulations, is considered in more detail by Nishida et al. in Chapter 13 of the present volume.

9.2 Criminal Activity at Sea

The scope of criminal activity at sea is potentially as unlimited as on land, with the added complexity of varying jurisdictional entitlements and, in some cases, occurrences beyond the geographical jurisdiction of any state. Categories of particular interest include the following:

- Criminal acts onboard vessels, not involving any other vessel – these could include, for example, 'simple' cases of robbery, assault, sexual assault and murder, crimes involving the cargo or actual mutiny against the lawful authority of the captain.
- Criminal acts on board offshore installations such as oil and gas drilling and production rigs.
- Acts of armed robbery or piracy against merchant vessels or private vessels.*
- Acts of smuggling of contraband substances into a state (including, e.g. contraband, drugs and weapons).
- The trafficking of migrants by sea, and the associated entry into a state's jurisdiction for that purpose (UNDOC, 2013).
- Human trafficking, which involves elements of force or coercion and ultimate exploitation of the trafficked persons (UNDOC, 2013).

---

* The definition of piracy is dealt with below, and is not considered to apply in internal waters, the territorial sea and archipelagic waters. In these zones similar acts are simply referred to as armed robbery against ships.

- Acts of violence and/or terrorism conducted against offshore installations.
- Illegal entry into a state by sea, which could include entry for the purpose of other criminal actions, such as terrorism within the state or illegal transport of migrants.

One significant characteristic of some of these activities is their *transnational* character, while others can also occur entirely within one state's jurisdiction – smuggling, human trafficking, trafficking of migrants and illegal entry are all inherently transnational in nature, and are also often engaged in by transnational organized crime. Piracy, though it will most often have a transnational element, if only because of the interference with foreign vessels, can still occur entirely between vessels of one state, while other forms of criminal activity may never go beyond the confines of one state's jurisdiction.

Finally, it is at least worth noting the obvious point that in the vast majority of cases a vessel (however small) is required for the commission of these crimes, even where carried out fully on board one ship.* Any monitoring and surveillance efforts, and any analysis of the resulting data, will focus first and foremost on the vessel from which they are carried out; this is not to say that this is the only type of data required, as discussed below, but the acquisition and analysis of accurate information about the vessel and its actions must be a starting point.

9.3 The 'Legal Seascape'

In the context of law *enforcement* at sea, a fundamental starting point is the jurisdiction of a coastal state to exert enforcement powers over the relevant vessel and/or persons involved in the target activity, in that particular location. The structure of the various zones of maritime jurisdiction (Figure 9.1), as set out in the 1982 *United Nations Convention on the Law of the Sea* (UNCLOS or LOS 1982) and customary international law provide the essential spatial definitions of the scope of powers to *legislate* and, as here, to *enforce* against foreign persons and vessels.† As such, it functions as a 'legal overlay' on the geographical areas in question, critical to understanding the status of some acts as lawful or not by virtue of location, and to the ability of a state to act against perpetrators. At the outset, two essential points about jurisdiction in this context should be

---

* Similar actions can be carried out by air over maritime spaces, but these are subject to a different regime, and will not be considered here.
† In international law there is a fundamental distinction between a state's jurisdiction to legislate (or prescribe), and its jurisdiction to enforce, whether in particular places or against particular parties.

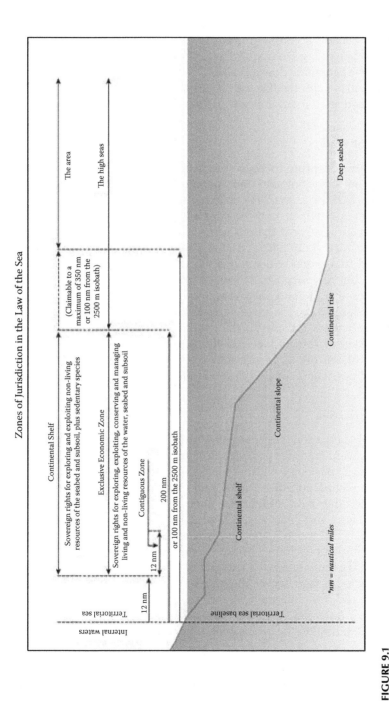

FIGURE 9.1

Maritime zones. (From Fisheries and Oceans Canada, Defining Canada's Maritime Zones, http://www.dfo-mpo.gc.ca/science/publications/article/2011/10-11-11-eng.html).

noted: first, states have jurisdiction over their own nationals and the vessels flying their flag (i.e. duly registered in the state), wherever they go (although they may not assert that jurisdiction in all cases); and, second, the concept of 'flag state jurisdiction' accords significant rights and responsibilities to the flag state (i.e. the state of registration), which is critically important in zones such as the Exclusive Economic Zone (EEZ) and continental shelf, where the default jurisdiction will be with the flag state. Coastal state jurisdiction in these zones must be justified as an interference with flag state jurisdiction.*

9.3.1 Internal Waters

Internal waters constitute those marine areas 'on the landward side of the baseline of the territorial sea' and can include bays, harbours and other indentations (LOS 1982, Art. 8(1)). Internal waters may also be enclosed by 'straight baselines', which may be applied where the coastline is 'deeply indented and cut into, or if there is a fringe of islands along the coast in its immediate vicinity' (LOS 1982, Arts. 5–7, 9–14). In practice, straight baselines are employed in a much wider range of geographic situations (Roach and Smith, 2000), but for purposes of this chapter the validity of the baselines is not considered, and generally the waters on the landward side of a baseline are considered to be the sovereign territory of the state for all purposes, including the seabed and subsoil, the water column and the airspace above, and the state may exert its law-making and law enforcement functions in the same manner as on land, with the exception of areas newly enclosed by straight baselines, where a right of innocent passage survives (LOS 1982 8(2)).

9.3.2 Territorial Sea

The territorial sea is a zone of coastal state jurisdiction extending up to 12 nautical miles seaward from the coastal baselines (whether the low water line or a defined straight baseline), within which the coastal state exerts sovereignty, including the seabed, subsoil, water column and airspace, but 'subject to' the LOS 1982 and 'other rules of international law' (LOS 1982, Arts. 2, 3 and 4, Klein, 2011; Kraska and Pedrozo, 2013). In practical terms, this means that the coastal state has full sovereignty to prescribe and enforce its laws, subject to any positive limitations found in the convention and general international law. The primary limitation is the recognition of the right of innocent passage, which applies to passage *through* the territorial sea, whether from one zone to another, or to and from internal waters of the state, in a continuous and expeditious manner (LOS 1982, Art. 18). Passage will be considered non-innocent where it is 'prejudicial to the peace, good order or security of the coastal State'

---

* The following descriptions are necessarily brief; for fuller descriptions of the various jurisdictional zones with a focus on security and law enforcement, see Klein (2011), Chapters 3 and 5; Kraska and Pedrozo (2013), Chapter 8 and Kaye (2007).

(LOS 1982, Art. 19). The mere fact of a criminal act being committed in passage, if confined to the vessel, may not be enough to permit intervention and enforcement (LOS 1982, Art. 27; Klein, 2011), but if the consequences extend to the territory of the coastal state, or if the crime is of a defined category such as smuggling or narcotics trafficking, enforcement is clearly permitted.*

9.3.3 Contiguous Zone

The contiguous zone comprises a zone, extending up to 12 nautical miles beyond the outer limits of the territorial sea, within which the coastal state may act to 'prevent infringement of its customs, fiscal, immigration or sanitary laws and regulations within its territory or territorial Sea' or to 'punish infringement' of the same laws or regulations, where a violation has occurred within the territory or territorial sea (LOS 1982, Art. 33). The jurisdiction of the coastal state within this zone, then, is focussed on the *prevention* of acts which are about to occur within the territorial sea or territory, or *enforcement* action against a vessel which has already been 'inside' and committed the relevant offence (Klein, 2011; Kraska and Pedrozo, 2013).

9.3.4 Exclusive Economic Zone (EEZ)

The EEZ is a zone of *functional* jurisdiction, not sovereign territory, extending from the outer limits of the territorial sea to a maximum of 200 nautical miles from the state's baselines, and requiring actual proclamation by the coastal state; it includes the seabed, subsoil and water column, but does not extend to the superjacent airspace (LOS 1982, Arts. 55, 57). In the EEZ the coastal state exercises 'sovereign rights' (not sovereignty) for the 'purpose of exploring and exploiting, conserving and managing the natural resources, whether living or non-living' of the zone, and 'with regard to other activities for the economic exploitation and exploration of the zone, such as the production of energy from the water, currents and winds' (LOS 1982, Art. 56(1)(a)). Additionally, the coastal state has 'jurisdiction as provided in the Convention over offshore islands and installations, marine scientific research and marine pollution' (LOS 1982, Art. 56(1)(b)). Other states maintain rights of navigation and overflight, and with respect to the laying and maintenance of submarine cables and pipelines (LOS 1982, Art. 58).

The law enforcement powers of the coastal state in the EEZ are restricted in important ways. First, the coastal state's enforcement powers are generally limited to those laws and regulations related to the aforementioned subject matters of jurisdiction (resource exploitation, economic activities, marine scientific

---

* A special regime of passage – 'transit passage' – applies in territorial sea areas within 'straits used for international navigation', as set out in LOS 1982, Part III. Apart from some limitations on the status of environmental and navigational rules which may be applied (i.e. to adopt international rules), the main impact in comparison to innocent passage is on military uses of the straits.

research and marine environmental protection in particular) (Rothwell and Stephens, 2010; Klein, 2011). This was confirmed in 1997 by the International Tribunal for the Law of the Sea (ITLOS), in the case of the *MV Saiga*, in which it was found that the enforcement of customs laws in the EEZ was beyond the scope of coastal state jurisdiction (ITLOS, *M/V Saiga Case*, 1997). Common criminal offences connected to the functional areas of jurisdiction, such as an assault on a fisheries officer, could, however, still be enforced.*

Second, even within the subject areas of recognized jurisdiction, UNCLOS goes to great lengths to define and constrain the scope of a state's *enforcement* powers, as with the limits on punishment for fishing offences (no corporal punishment or imprisonment), and the system of graduated levels of enforcement against foreign vessels for pollution offences, only escalating to full enforcement powers depending on the strength of the evidentiary grounds for believing an offence has occurred, and the seriousness and wilfulness of the incident (LOS 1982, Art. 220; Klein, 2011). These limits, as well as the requirement to apply international standards with respect to pollution offences in the EEZ, reflect the concerns of the states party to UNCLOS with the maintenance of freedom of navigation in the vast ocean areas encompassed by EEZs.

9.3.5 Continental Shelf

A coastal state has jurisdiction (whether formally asserted or not) over the non-living resources and sedentary species of the continental shelf, defined as the continental margin (i.e. the shelf, the slope and the rise), and to at least 200 nautical miles where the geological shelf does not extend that far[†] (LOS 1982, Arts. 76(1), 77 and 78). Ireland is one example of a state where jurisdiction applies to the continental shelf extending beyond the 200-mile limit (see Chapter 2 by Scott et al. of this volume). As with the EEZ, the jurisdiction is functional in nature, in that the coastal state only has 'sovereign rights for the purpose of exploring it and exploiting its natural resources' (LOS 1982, Art. 77). Coastal state entitlements include enforcement with regard to activities related to the exploration and exploitation of natural resources, 'exclusive jurisdiction' over artificial islands, installations and structures on the shelf (as well as in the EEZ) (LOS 1982, Arts. 60, 80) and the right to establish 500 m safety zones around such installations (LOS 1982, Art. 60(4)–(7), Art. 80). Further rights are set out in the 1988 *Protocol for the*

---

* Other powers over platforms and installations are, by the interaction of Arts. 60 and 82, LOS 1982, the same in the EEZ as on the continental shelf, and will be dealt with under the shelf regime. The high seas freedoms, including navigation, overflight and laying of cables and freedom of navigation and pipelines, apply in the EEZ except to the extent they are limited in the Convention; LOS 1982, Arts. 58, 87.

† The outer limits of the continental shelf are determined by a complex set of criteria set out in Art. 76, based on a combination of distance, geological and geomorphological tests. Prescribed limits, to be acceptable as final and binding, must be consistent with recommendations of the Commission on the Limits of the Continental Shelf, established pursuant to UNCLOS, Annex II.

Suppression of Unlawful Acts against the Safety of Fixed Platforms Located on the Continental Shelf (SUA Protocol 1988), a protocol to the 1988 SUA Convention adopted at the same time, which established jurisdiction related to acts of violence against 'fixed platforms' on the continental shelf. The 2005 *Protocol for the Suppression of Unlawful Acts against the Safety of Fixed Platforms Located on the Continental Shelf* (SUA Protocol 2005) extended applicability to various terrorist acts.*

9.3.6 Archipelagic Waters

Archipelagic waters are waters found within the archipelagic baselines which may be established only by 'archipelagic states', states comprised wholly of islands and meeting various criteria related to land-to-water ratios and baseline segment lengths (including, e.g. the Bahamas and Indonesia). In these waters, an archipelagic state exercises full sovereignty, but is subject to the rights of innocent passage and 'archipelagic sealane passage' (LOS 1982, Part IV).

9.3.7 High Seas

On the high seas, or waters beyond any state's national jurisdiction (including waters above the continental shelf beyond 200 nautical miles) (LOS 1982, Art. 86), the default position is 'flag state jurisdiction'; that is, ships generally fall under the exclusive jurisdiction of their state of registry, 'save in exceptional cases expressly provided for' in UNCLOS or other agreements (LOS 1982, Art. 92(1)). There are, however, significant exceptions to this position in the Convention itself (Klein, 2011), examples of which are summarized below:

- Hot pursuit – A coastal state that has 'good reason to believe that the ship has violated the laws and regulations of that State' (including applicable laws in the EEZ and Continental Shelf) while *within* the relevant zone, may continue a pursuit initiated inside the zone, subject to various requirements set out in UNCLOS (LOS 1982, Art. 111).

- Right of visit – UNCLOS (and customary international law) provides for warships operating on the high seas to approach vessels and request documentation 'to verify the ship's right to fly its flag'. It may only proceed to boarding where there are 'reasonable grounds to believe, *inter alia* that the ship is engaged in piracy or the slave trade; the ship is without nationality (i.e. no flag) or the ship is actually of the same nationality as the warship' (LOS 1982, Art. 110).

---

* For detailed reviews of the impact of these protocols, see Klein (2011) (Chapter 3) and Kraska and Pedrozo (2013) (Chapter 21).

- Piracy – Beyond a right of visit, states exercise universal jurisdiction over pirates (though the exercise may be dependent on the existence of applicable *national* laws permitting enforcement) (LOS 1982, Arts. 101, 105). Under UNCLOS, piracy is defined as 'illegal acts of violence or detention, or any act of depredation, committed for private ends by the crew or the passengers of a private ship or a private aircraft', and directed against ships and aircraft on the high seas, or otherwise beyond the jurisdiction of any state (LOS 1982, Art. 101(a)).

In addition to these specific exceptions, UNCLOS also allows for enforcement with consent of a flag state, and subject to treaty-based provisions (binding only among parties to the relevant agreement) which may provide for some higher level of jurisdiction for particular offences. These can be global, regional and bilateral, and a full survey of the many agreements in force is beyond the scope of this brief review.* However, three such instruments are of particular interest. With respect to narcotics trafficking, although UNCLOS requires states to cooperate in the suppression of narcotics trafficking (LOS 1982, Art. 108), it makes no specific provision for boarding or enforcement. However, Article 17 of the 1988 *United Nations Convention against Illicit Traffic in Narcotic Drugs and Psychotropic Substances* (1988 Narcotics Convention), provides as follows:

> A Party which has reasonable grounds to suspect that a vessel … flying the flag or displaying marks of registry of another Party is engaged in illicit traffic may so notify the flag State, request confirmation of registry and, if confirmed, request authorization from the flag State to take appropriate measures in regard to that vessel.

The requirement for flag state authorization limits the effectiveness of this provision, but regional and bilateral arrangements have been more successful in giving this restricted obligation more scope (Kraska and Pedrozo, 2011; UNDOC, 2013).

The related problems of migrant smuggling and human trafficking have been the subject of two protocols to the 2000 *United Nations Convention against Transnational Organized Crime* (UNTOC): the 2000 *Protocol against the Smuggling of Migrants by Land, Sea and Air* (Migrant Protocol)† and the 2000 *Protocol to Prevent, Suppress and Punish Trafficking in Persons, especially Women and Children* (Trafficking Protocol).

Migrant smuggling is defined in Art. 3(a) of the Migrant Protocol as 'the procurement, in order to obtain, directly or indirectly, a financial or other

---

* For a full review of relevant regional agreement, see Kraska and Pedrozo (2013) and UNDOC (2013).
† Not considered here, but of great importance to affected states, are the obligations to preserve human life at sea and to accord appropriate protection to persons considered refugees under international law.

material benefit, of the illegal entry of a person into a State Party of which the person is not a national or a permanent resident'. Human trafficking, by contrast, is defined in Art. 3(a) of the Trafficking Protocol as requiring elements of force or coercion, and a purpose of exploitation such as 'sexual exploitation, forced labour or services, slavery or practices similar to slavery, servitude or the removal of organs'. Only the Migrant Protocol, however, deals specifically with enforcement against foreign flagged vessels in areas outside national jurisdiction. Art. 17 provides for a right of visit where there are 'reasonable grounds' to believe that a vessel is engaged in migrant smuggling, and if it is a foreign flagged vessel (as opposed to a stateless vessel, as will often be the case), the enforcing ship is only entitled to seek authorization from the flag state for further action.

9.3.8 Summary

The jurisdictional regime described above undoubtedly limits the ability of states to exert control over the 'global commons' (Allen, 2007), both to prescribe rules and engage in enforcement action. However, the regime leaves largely untouched the ability of a state to carry out surveillance and data-gathering on activities in maritime spaces beyond national jurisdiction, so long as the existing rules on freedom of navigation for foreign states are observed. For states with global or extensive regional security interests, this may broaden the scope of the data acquisition effort considerably, while other states may be primarily focussed on coastal waters and approaches. It should also be remembered that while international law sets out jurisdictional *entitlements* of states, domestic enforcement actions will depend upon those entitlements having been enacted in national law, including the specification of the outer limits of the various zones.

9.4 Maritime Domain Awareness and Data Sources

The concept of Maritime Domain Awareness (MDA) (also referred to as maritime situational awareness) reflects the modern mandate for the authorities responsible for control of maritime activities and the identification of threats arising from the maritime domain. The following 'key definitions' from the 2005 *National Plan to Achieve Maritime Domain Awareness* encapsulates the U.S. understanding of the breadth of the doctrine:

Maritime Domain is all areas and things of, on, under, relating to, adjacent to, or bordering on a sea, ocean, or other navigable waterway, including all maritime related activities, infrastructure, people, cargo, and vessels and other conveyances.

> **Maritime Domain Awareness** is the effective understanding of any-
> thing associated with the maritime domain that could impact the
> security, safety, economy, or environment of the United States
> (Homeland Security, 2005).*

In Canada, a similar definition was employed in stating the mandate of
the Marine Security Operations Centers (MSOC) in 2013, and the potential
threats were further defined to include 'foreign trans-national organized
crime – drug trafficking, piracy, migrant smuggling – emerging terrorist
activity, over-fishing, and polluters' (Lavigne, 2014). Maritime Domain Awareness is discussed further, in the specific context of marine search and rescue,
in Chapter 15, by Mathews and Power, of the present volume.

9.4.1 Potential Data Sources

The breadth of the mandate envisioned under MDA is extraordinary (and, as
a result, likely only aspirational), and is clearly dependent on the collection,
fusion and analysis of significant volumes of data, from multiple sources.
The adoption of an MDA approach, and similar initiatives, 'have been critical drivers and goals for intelligence gathering at sea, as well as information sharing' (Klein, 2011). In addition to governing the scope of prescription
and enforcement jurisdiction in the various maritime zones, international
legal developments have been responsible for the proliferation of information sources – including geospatial information – on the movement of ships,
including cargos and personnel. Two important sources of data are the automatic identification system (AIS) and the long-range identification and tracking (LRIT) system, both mandated under the 1974 *International Convention for
the Safety of Life at Sea* (SOLAS).

AIS is required under SOLAS, Regulation 19 of Chapter V, for all vessels of 300 GRT and over on international voyages, all passenger ships and
all vessels of 500 GRT and over whether or not on an international voyage
(Boraz, 2009; Klein, 2011; Kraska and Pedrozo, 2013). The system operates
through a VHF link with navigational systems, and transmits data on the
'ship's identity, type, position, course, speed, navigational status and other
safety-related information' (IMO–SOLAS). Originally intended for safety
and collision-avoidance, it allows communication with other AIS-equipped
ships, and shore stations, with a range of approximately 40 nautical miles
for communication to shore stations (IMO–SOLAS, Boraz, 2009; Kraska and
Pedrozo, 2013). One shortcoming of AIS as a source of security information is
the limited range, unless utilized via a satellite link, which is not mandated
under SOLAS (Eriksen et al., 2006) and unreliability of data (Boraz, 2009). In
2006, IMO adopted a further amendment to SOLAS, requiring implementation of the LRIT (SOLAS, Chapter V, Reg. 19-1), a satellite-based tracking

---

* See also the updated 2013 iteration of this Plan (U.S. Government, 2013).

system, for vessels over 300 GRT, all passenger vessels and mobile drilling units (MDUs). The LRIT system is built around national, regional and global data centers that collect information and transmit it via the 'International LRIT Data Exchange', designated as the International Mobile Satellite Organization (IMSO), to parties eligible to receive it (SOLAS, Klein, 2011). The information (including identification, location and time) can be used for safety, search and rescue (SAR), security and environmental protection purposes, and is available to port states where a foreign vessel 'has indicated its intention to enter that port', and to coastal states when a vessel is within 1000 nautical miles of its coast (SOLAS, Klein, 2011).

In addition to these global arrangements, it should also be noted that there are multiple regional measures that have mandated data collection and/or the sharing of information (Kraska and Pedrozo, 2013; Bueger, 2015). Additionally, some regional organizations (including Regional Fisheries Management Organizations and the European Union) provide for the use of vessel monitoring systems (VMS) for fishing vessels which will not be caught by AIS requirements (Flewwelling et al., 2002).

In addition to the masses of data generated via AIS and LRIT, surveillance operations rely on 'coastal radars, space-based imagery and other sensors, to form a picture in which the operator can recognize complex patterns and make decisions' (Snidaro et al., 2013; see also Boraz, 2009). Moreover, the effective identification of security threats and possible criminal activity requires the acquisition and analysis of data from numerous other sources, including, potentially, vessel owners, cargo, previous port visits, passengers and crew and other intelligence information (Boraz, 2009; Homeland Security, 2005; U.S. Government, 2013). For the category of cargo alone, the complexity can be daunting:

> The control of cargo transition is essential to the security of a country. Cargo ships can constitute a threat with weapons of mass destruction, other weapons, drugs or other illegal items on board. Having access to the ship's manifest is a start, but manifests can be falsified. Therefore it is important to know who owns the vessels, so that links between maritime transport and criminal organizations can be identified. In addition, certain countries are known for their lax control of goods, and criminal organizations and terrorists can exploit these vulnerabilities. (Martineau and Roy, 2011)

9.5 Applications of Geoinformatics in Monitoring and Surveillance Operations

The discussion above presents only a partial picture of the complex array of data available to those charged with ensuring effective levels of MDA, as well as the ability to act in a timely manner on the awareness which is established. At one level, the collection of geographically referenced information on past locations of criminal activity (e.g. through known smuggling routes), vulnerable locations and known targets (e.g. port facilities and offshore installations) and shipping patterns could allow for the mapping of priority areas for the location of limited enforcement resources. This has been done in other contexts, for example, with respect to the placement of SAR assets in relation to known incident patterns (Malik et al., 2014; Mathews and Power, Chapter 15 this volume). Similarly, the tracking of fishing patterns via a regional registry and database of vessels in the South Pacific Forum Fisheries Agency, coupled with reporting requirements including location of fishing effort (later reinforced by VMS data), fed into more effectively targeted use of surveillance and enforcement assets through the 1980s and 1990s (Slade, 1990; Flewwelling et al., 2002, Annex D).

The more important requirement, however, is a real-time ability to respond to developing threats in order to permit an effective enforcement response – to move from the acquisition of data through to the full understanding of the situation, assessment of threats and the possibility of timely action. A fundamental problem facing the operators of maritime surveillance systems is the absorption and analysis of the large amounts of data that are now available, coming from diverse sources (some geospatial in nature and others not), and fusing them into an awareness of the situation which permits more precise determination of threats:

> ...MDA is not just about the blips [*on a screen*]; it's about whether the blips matter. Aggregating disparate datasets to generate a useful operational picture is an increasingly complex task ... Fusing and analyzing those data may find anomalies that point to threat activity of interest to decision makers. (Boraz, 2009)

Martineau and Roy (2011) have identified seven conceptual steps in the development of MDA and the detection of potential threats:

1. Data and information acquisition – As noted, this function has progressed in recent decades to the point that the volume and diversity of the data is itself a major challenge (Bueger, 2015).

2. Data and information fusion – Fusion of data is required across different types of sensor information, but also in dealing with what Martineau and Roy refer to as 'the higher-level aspect of data and information'. Fusion of spatial and non-spatial data (e.g. ownership, history of a vessel, cargo records, intelligence reports) in an analytical approach is both a necessity and a widely recognized challenge (Riviero, 2011; Homeland Security, 2005; Boraz, 2009). The creation and use of appropriate ontologies (see Lassoued and Leadbetter, Chapter 6 of this volume) may be beneficial in overcoming this difficulty.

3. Situational awareness – The acquisition and fusion of data 'of sufficient quality and quantity can provide a fair representation of the current state of reality' (Martineau and Roy, 2011), such as the presentation of identity, position and speed of ships, linked as necessary to information such as ownership, cargo and other relevant data to establish situational awareness for an operator (Riviero, 2011; Snidaro et al., 2013).

4. Anomaly detection – Beyond situational awareness, the possibility of automated detection of anomalies – departures from expected patterns indicative of a threat – has been the focus of significant work (Laxhammer et al., 2009; Riviero and Falkman, 2014), and remains an important challenge in the support of decision making (see below).

5. Putting anomalies in context – The mere existence of an anomaly does not constitute a threat, but may be explained by contextual information (e.g. a stopped vessel in response to an environmental circumstance). The provision of contextual information, as part of situational awareness, is essential (Radon et al., 2015).

6. Threat assessment – While an advanced system may be able to provide some broad classification of events – for example, as a 'vessel of interest', 'benign', 'not explained' or 'threat' (Martineau and Roy, 2011) – the ultimate assessment must be that of a human operator.

7. Dissemination and presentation of results – Once threats have been identified, the results of the analysis will be disseminated (Martineau and Roy, 2011); in the case of law enforcement, this will presumably include the authorities responsible for taking that action – and dissemination is likely to be multi-agency and, in some cases, multinational, as in regional monitoring programmes (Bueger, 2015).

Significant progress has been made in a number of national and regional authorities in the acquisition of data, the fusion of multiple data sources and their representation for operators in graphical form to present situational awareness, and in some cases the automated provision of 'alerts' or identification of vessels of interest, as in the following examples:

• In Canada, Maritime Security Operations Centers (MSOCs) were established in a project running from November 2005 to January 2016 as interdepartmental centers including Canada Border Services Agency, Fisheries and Oceans Canada and Canadian Coast Guard, Department of National Defence/Canadian Armed Forces, Royal Canadian Mounted Police and Transport Canada. In collaboration with a private sector partner, the project developed systems 'that access, gather and share Maritime Domain Awareness information efficiently in an automated, secure manner', as well as 'a situational awareness tool that displays positional and contextual information about vessels' on GIS maps, and a vessel selection system that allows 'detailed analysis of vessels-of-interest' (Government of Canada, 2016).

- The European Maritime Safety Agency (EMSA) offers an integrated service to member states, whereby it 'collects, processes and exchanges' data and provides vessel tracking analytical tools including visual situational awareness representations, in areas related to fisheries, pollution, piracy and maritime border control. Data sources accessed include AIS, Satellite AIS, LRIT, Coastal Radar, Synthetic Aperture Radar Satellite Images, Ship and Voyage Information and meteorological and oceanographic information. 'Vessel behaviour monitoring' is described as being under development (EMSA, 2014).
- The Maritime Safety and Security Information System (MSSIS), a U.S.-supported network involving over 70 state participants, provides 'near real-time' access to global AIS data (obtained from participating states), as well as a software package (Transview/TV32) that allows access to MSSIS and visual representation of the data, and that can also support connections to national MDA systems for fusion with broader data sources (U.S. Department of Transportation, 2015).

While existing systems have successfully integrated multiple sources of data in graphical representations of situational awareness, which support identification of threats and decision making by operators, anomaly detection remains more difficult. The objective is to provide 'automated ways to search through mundane surveillance tracks and contacts and extract a smaller number of anomalous events that are worthy of an operator's attention' (Roy and Davenport, 2009). Advances have been made in this regard as well,* but significant challenges remain. Problems include: the definition of 'normal' and 'anomalous' behaviour of a vessel (Riviero and Falkman, 2014; Lane et al., 2010; Vandecasteele and Napoli, 2012),† which is highly specific to the consumer of the information (Riviero, 2011); a high number of 'false alerts' (Roy and Davenport, 2009); the need to integrate contextual information (e.g. weather, navigational threats) which explain vessel movements (Radon et al., 2015) and categorization of threats. An important element in the development of more effective anomaly detection is the engagement of subject-matter experts (SMEs) in the development of a 'taxonomy of maritime

---

* See above, developments in the Canadian and European systems, and note the availability of commercially developed software which claims to extend to both data fusion and anomaly detection and alerts (see, e.g. Larus Technologies, 2013).
† Lane et al. (2010) describe a number of data-driven aspects of anomalous behaviour, including deviation from expected routes, unexpected entry to a port, entry into a protected or otherwise designated zone and close approach to certain targets.

anomalies' (Roy and Davenport, 2009; Riviero, 2011), which can include 'static' and 'dynamic' anomalies.*

9.6 Issues Arising in the Application of Geoinformatics to Law Enforcement

The application of the technological advances described above in the specific context of law enforcement in maritime areas give rise to some particular concerns that should be taken into account by those responsible for ongoing development of these systems. First, any systems that support decision making on enforcement actions must be built with a full understanding of where different levels of enforcement jurisdiction may be applied, under the law of the sea.

Second, as noted earlier, the nature of the maritime sector means that the data on which the surveillance and monitoring systems are based will necessarily be acquired, shared and analysed through joint operations centers, incorporating vessel traffic control, national security, search and rescue, fisheries and environmental enforcement and criminal law enforcement.[†] This is unavoidable, but it does mean that for the purpose of law enforcement activities (especially if prosecution is an eventual outcome), close attention must be paid to the reliability, preservation and probative value of the evidence generated by the joint system. This can be as simple as ensuring the preservation of data that prove locations and vessel actions, or an awareness of the varying authorities for boarding and inspection of vessels in different locations, as determined by the applicable jurisdictional zone and domestic law (e.g. requirements for existence of some reasonable grounds for intervention). Third, the collection and sharing of information on vessels and persons may give rise to privacy concerns under national law (Homeland Security, 2012). This concern is exacerbated by the increasingly internationalized approach to MDA, whether through regional or global data-sharing (Bueger, 2015), incorporating both personal information and commercially sensitive data that are transferred across multiple jurisdictions (European Commission, 2008).

Fourth, as subject-matter experts are drawn into the design and development of automated anomaly detection systems, and the identification of potential enforcement actions, the range of expertise accessed by system developers should go beyond surveillance operators to include legal experts,

---

* Static anomalies would include issues related to the name, radio call sign and licensing of a vessel. Dynamic anomalies would include both 'kinematic' facts such as course, location and movements, and 'non-kinematic' anomalies, for example, crew and cargo (Roy and Davenport, 2009).

† Government of Canada (2016), EMSA (2014) and UK Secretary of Defense (2014).

both to ensure compliance with international law and to assist in the categorization of potentially criminal activities.

9.7 Conclusion

Law enforcement in maritime areas, including prevention and long-range detection of threats, cannot be dependent on technological solutions alone. Important gaps remain, including the use of small vessels which are not tracked by any of the existing systems (as with the recent migrant crisis in the Mediterranean – see Chapter 15 of the present volume), or simple noncompliance by criminals with the monitoring system requirements. Other sources of criminal intelligence and surveillance will continue to be essential, but automated systems do offer the promise of fully exploiting the vastly expanded data sources which now exist.

It is clear that international and national legal developments have been partly responsible for accelerating the development of monitoring and control systems. The assertion of jurisdiction over larger areas of marine space, as codified in UNCLOS, has expanded the scope for coastal state legislative and enforcement action from what was possible in an era when the high seas began at the limits of the territorial sea and contiguous zone. More directly, the creation of new legal reporting requirements for vessels, including technical specifications, has enabled the acquisition of the large volumes of geospatial data, often in real or near-real time. These data are now available to feed the sophisticated systems that fuse and analyze the material, and will increasingly assist in decision making through automation of threat detection. The experience to date suggests that close cooperation between the specialists developing the analytical systems and legal experts will continue to be required. The creation of decision making support systems should incorporate legal knowledge among the subject-matter expertise, as noted above, to ensure more effective anomaly detection. Moreover, as new data sources are introduced and incorporated into existing systems, data acquisition and processing should be structured with a view to the requirements for eventual utilization in enforcement proceedings, and with the necessary protections of privacy interests.

References

Allen, C., 2007. Command of the commons boasts: An invitation to lawfare. *International Law Studies* 83: 21–50.

Boraz, S.C., 2009. Maritime domain awareness: Myths and realities. *Naval War College Review* 62/3: 137–146.

Bueger, C., 2015. From dusk to dawn? Maritime domain awareness in Southeast Asia. *Contemporary Southeast Asia* 37/2: 157–182.

European Maritime Safety Agency (EMSA), 2014. Study to Assess the Future Evolution of SSN to Support CISE and Other Communities. Final Report. EMSA ITT No. EMSA/OP/07/09/Lot2/RFP. Version 1.8, September 23, 2014. Lisbon: GMVIS Skysoft S.A. for EMSA.

Eriksen, T., Høye, G., Narheim, B., and Meland, B.J., 2006. Maritime traffic monitoring using a space-based AIS receiver. *Acta Astronautica* 58: 537–549.

European Commission, Directorate-General for Maritime Affairs and Fisheries, 2008. Legal aspects of maritime monitoring & surveillance data. Brussels: European Commission.

Flewwelling, P., Cullinan, C., Balton, D., Sautter, R.P., and Reynolds, J.E., 2002. Recent trends in monitoring, control and surveillance systems for capture fisheries. FAO Fisheries Technical Paper No. 415. Rome: FAO.

Government of Canada, Secretary of National Defense, 2016. MSOC Project Description. http://www.msoc-cosm.gc.ca/en/description.page (accessed 10 March 2016).

International Maritime Organization (IMO), 2016. SOLAS. http://www.imo.org/en/About/Conventions/ListOfConventions/Pages/International-Convention-for-the-Safety-of-Life-at-Sea-%28SOLAS%29,-1974.aspx (accessed 25 Feb. 2016).

Kaye, S., 2007. Threats from the global commons: Problems of jurisdiction and enforcement. *International Law Studies* 83: 69–82.

Klein, N., 2011. *Maritime Security and the Law of the Sea.* Oxford: Oxford University Press.

Kraska, J. and Pedrozo, R., 2013. *International Maritime Security Law.* Leiden, Boston: Martinus Nijhoff.

Lane, R., Nevell, D., Hayward, S., and Beaney, T., 2010. Maritime anomaly detection and threat assessment. 13th International Conference on Information Fusion, Edinburgh, July 2010. http://www.richardlane.net/lane_2010_maritime_anomaly_threat_assessment.pdf (accessed 10 March 2016).

Larus Technologies, 2013. Total Maritime Domain Awareness. https://www.larus.com/wp-content/uploads/2013/05/Larus_Total_Insight_ds.pdf (accessed 8 April 2016).

Lavigne, V., 2014. Maritime Visual Analytics Prototype – Final Report. Scientific Report Defense R&D Canada-2014-R182. http://publications.gc.ca/site/eng/9.807437/publication.html (accessed 30 Jan. 2016).

Laxhammer, R., Falkman, G., and Sviestins, E., 2009. Anomaly detection in sea traffic – A comparison of the Gaussian mixture model and the kernel density estimator. *12th International Conference on Information Fusion*, Seattle, WA: 756–763.

Malik, A., Maciejewski, R., Jang, Y., Oliveros, S., Yang, Y., Maule, B., White, M., and Ebert, D.S., 2014. A visual analytics process for maritime response, resource allocation and risk assessment. *Information Visualization* 13(2): 93–110.

Martineau, E. and Roy, J., 2011. Maritime anomaly detection: Domain introduction and review of selected literature. Technical Memorandum Defense R&D Canada, Valcartier TM-2010-460. http://pubs.drdc-rddc.gc.ca/BASIS/pcandid/www/engpub/DDW?W%3DSYSNUM=535375 (accessed 25 Feb. 2016).

Radon, A.N., Wang, K., Glässer, U., When, H., and Westwell-Roper, A., 2015. Contextual verification for false alarm reduction in maritime anomaly detection. *IEEE Conference on Big Data*, Santa Clara, CA. https://www.cs.sfu.ca/~wangk/pub/BigData2015CameraReady_Radon.pdf (accessed 10 Feb. 2016).

Riviero, M., 2011. *Visual Analytics for Maritime Anomaly Detection*, Örebro Studies in Technology 46 Örebro, Sweden: Örebro University.

Riviero, M. and Falkman, G., 2014. Detecting anomalous behaviour in sea traffic: A study of analytical strategies and their implications for surveillance systems. *International Journal of Information Technology & Decision Making* 13/2: 317–360.

Roach, J.A. and Smith, R., 2000. Straight baselines: The need for a universally applied norm. *Ocean Development and International Law* 31/1–2: 47–80.

Rothwell, D. and Stephens, T., 2010. *The International Law of the Sea*. Oxford: Hart Publishers.

Roy, J. and Davenport, M., 2009. Categorization of maritime anomalies for notification and alerting purpose. *NATO Workshop on Data Fusion and Anomaly Detection for Maritime Situational Awareness*, La Spezia.

Slade, N., 1990. Forum fisheries agency and the next decade: The legal aspects. In *The Forum Fisheries Agency: Achievements, Challenges and Prospects*, R. Herr, Ed. Fiji: Institute for Pacific Studies, pp. 296–313.

Snidaro, L., Visentini, I., and Bryan, K., 2013. Fusing uncertain knowledge and evidence for maritime situational awareness via markov logic networks. *Information Fusion* 21: 159–172.

United Kingdom, Secretary of Defense, 2014. *UK Strategy for Maritime Security*. London: Govt. of UK.

United Nations Office on Drugs and Crime (UNDOC), 2013. Combating Transnational Organized Crime Committed at Sea, UN, Vienna. https://www.unodc .org/documents/organized-crime/GPTOC/Issue_Paper_-_TOC_at_Sea.pdf (accessed 10 March 2016).

United States Government, 2013. National Maritime Domain Awareness Plan. Washington, DC: U.S. Govt. http://nmio.ise.gov/Portals/16/Docs/National _MDA_Plan_NMDAP-20131216_Final.pdf?ver=2015-12-04-124431-527 (accessed 30 March 2016).

United States Department of Homeland Security, 2012. Privacy Impact Assessment for the Automated Targeting System. Washington, DC: U.S. Govt. https://www.dhs .gov/xlibrary/assets/privacy/privacy_pia_cbp_ats006b.pdf (accessed 28 March 2016).

United States Department of Homeland Security, 2005. The National Plan to Achieve Maritime Domain Awareness. Washington, DC: U.S. Govt. https://www.dhs .gov/sites/default/files/publications/HSPD_MDAPlan_0.pdf (accessed 15 March 2016).

United States Department of Transportation, 2015. Maritime Safety and Security Information System (MSSIS). https://www.volpe.dot.gov/infrastructure-systems -and-technology/situational-awareness-and-logistics/maritime-safety-and (accessed 5 April 2016).

Vandecasteele, A. and Napoli, A., 2012. Spatial ontologies for detecting abnormal maritime behaviour. Oceans 2012, Yeosu, South Korea. https://hal.archives -ouvertes.fr/file/index/docid/741035/filename/vandecasteele-napoli _oceans2012.pdf (accessed 8 April 2016).

Conventions and Case Law

1974 Convention for the Safety of Life at Sea (SOLAS) 1184 UNTS 3.
1982 United Nations Convention on the Law of the Sea (LOS 1982) 1833 UNTS 3.

1988 United Nations Convention against Illicit Traffic in Narcotic Drugs and Psychotropic Substances 1019 UNTS 175.

1988 Convention for the Suppression of Unlawful Acts Against the Safety of Maritime Navigation (SUA Convention) 1678 UNTS 221.

1988 Protocol for the Suppression of Unlawful Acts against the Safety of Fixed Platforms Located on the Continental Shelf (Protocol to SUA) 1678 UNTS I-29004.

2005 Protocol of 2005 to the Protocol for the Suppression of Unlawful Acts against the Safety of Fixed Platforms Located on the Continental Shelf IMO DOC LEG/CONF.15/22 (Nov. 1, 2005).

2000 United Nations Convention against Transnational Organized Crime 2225 UNTS 209.

2000 Protocol against the Smuggling of Migrants by Land, Sea and Air 2241 UNTS 507.

2000 Protocol to Prevent, Suppress and Punish Trafficking in Persons, especially Women and Children 2237 UNTS 319.

The M/V Saiga Case (St. Vincent and the Grenadines v. Guinea), International Tribunal for the Law of the Sea, Case #1, 1997.

10

Geospatial Technologies and Indigenous Knowledge Systems

Shankar Aswani

CONTENTS

10.1 Introduction

During the twentieth and twenty-first centuries, pressure on coastal ecosystems has amplified and resulted in the widespread degradation of adjacent marine and terrestrial habitats globally (Burke et al., 2001). The ecosystem services provided by coastal habitats, including coastal protection and food procurement, have been heavily compromised by anthropogenic disturbance such as overfishing, pollution, sedimentation and alteration of coastal vegetation (Costanza et al., 1997; Agardy et al., 2009). In the context of small islands, this continued degradation in tandem with the ongoing effects of climate change is putting the livelihoods of coastal peoples at risk (e.g. Bell et al., 2009). While international efforts at curtailing these negative trends are ongoing, many researchers are working directly with coastal local/indigenous communities to seek more effective management of coastal terrestrial and marine resources. Among various approaches, researchers are increasingly incorporating local knowledge systems for designing resource management and conservation plans (e.g. Gadgil et al., 1993). In this respect, of particular significance is the incorporation of indigenous ecological knowledge (IEK) into management schemes for the conservation of biological diversity (e.g. Berkes et al., 2000).

There is a history of advocacy for the inclusion of local and indigenous ecological knowledge in fisheries management (Johannes, 1978; Ruddle, 1998). More recently, the inclusion of local understandings and institutions for ameliorating resource degradation (through such measures as marine spatial planning and marine protected areas) and for building resilience to human-generated environmental and climate change have gained importance (Mercer et al., 2010; Alexander et al., 2011; Cornu et al., 2014; Weeks et al., 2014). But it is important to realize that local knowledge systems are increasingly being transformed, syncretized or simply being lost altogether. Research around the world is showing that there is an increasing inter-generational loss in indigenous people's capacity to classify their environment, manage their terrestrial and marine resources and understand spatio-temporal changes locally (e.g. Turvey et al., 2010; Le Guen et al., 2013; Padilla and Kofinas, 2014). Simultaneously, while there is an inter-generational transformation of ecological knowledge, new knowledge is also produced across people's lifetime (Godoy et al., 2006).

Facing the increasing transformation of IEK, a number of researchers have begun using geospatial technologies for archiving and spatially mapping local knowledge systems. This knowledge is used for understanding epistemological diversity and for resource management (e.g. Rundstrom, 1995; Ulluwishewa et al., 2008). For instance, Harmsworth et al. (2002) have shown that geospatial technologies are advantageous for Maori indigenous peoples to store and transfer knowledge and information, while sustaining the verbal and oratory nature of their culture. Chapin et al. (2005) similarly have shown that the inclusion of participatory mapping and participatory GIS/remote sensing can be a powerful means for indigenous people's empowerment around the world. Participatory geospatial mapping allows communities to spatially represent their local knowledge systems, map their territorial estates (for land and sea claims) and manage their own natural resources in a way that acknowledges their own worldview and which is culturally appropriate.

This chapter draws on geospatial and human ecology research by the author in the Solomon Islands (Roviana and Vonavona Lagoons, or referred to as 'Roviana' henceforth; Figure 10.1). It is cross-referenced with similar research to illustrate conceptual incorporation of local/indigenous ecological knowledge and coastal resident behaviour into a participatory GIS database. This chapter examines: (1) spatial settlement dynamics across time, (2) human foraging strategies and spatio-temporal variation, (3) marine local/indigenous ecological knowledge systems, (4) people's detection of rapid ecological change, (5) local understandings of protracted environmental and climatic changes and (6) indigenous views of spatial conflicts and their relation to development.

Documenting and incorporating local knowledge systems and concomitant behavioural patterns into a geospatial database, for subsequent visualization and analysis, can furnish a broader understanding of localized

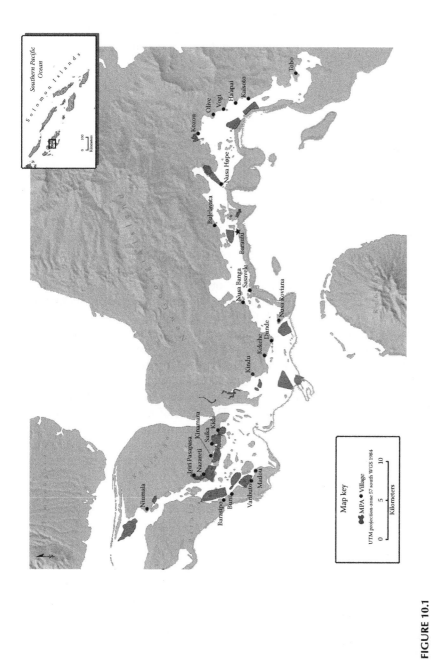

FIGURE 10.1

The Roviana and Vonavona Lagoons, New Georgia, Solomon Islands. (MPA sites established under our research and conservation programme in collaboration with local communities shown in dark grey.)

human environmental interactions and environmental impacts. It can also assist in designing and implementing locally contextualized management and conservation strategies that are cost-effective, participatory and inclusive. For instance, participatory GIS and interdisciplinary research for designing community-based marine protected areas (MPAs) suggests that integrated work can better inform artisanal fisheries management and biodiversity conservation in Oceania and beyond (e.g. Anuchiracheeva et al., 2003; Aswani and Lauer, 2006a).

10.2 Regional Context

The ecological health of the Solomon Islands has been sustained by low human populations and relatively small development pressure, although this is currently changing. In the absence of a modern economy (e.g. available waged labour and a welfare system), indigenous people's subsistence needs and developmental aspirations can only be met through the capital extraction of their natural resources. Land-based activities such as logging, in tandem with over-fishing, disease and water quality degradation and climate change have accelerated the degradation of marine environments (Albert et al., 2012). The degradation of coastal habitats presents a great challenge for local communities who rely on fish for their livelihoods and subsistence (Bell et al., 2009). In the past, people managed their marine environment through a complex system of sea tenure combined with various forms of harvest restrictions. In Roviana today, these local systems of customary management have been greatly affected by economic development and westernization, and in some places these developmental pressures have led to the breakdown of local marine management practices. This lack of management is aggravated further by ineffective regional and national government management strategies. Though a number of communities in the region have begun implementing marine protected areas under their sea tenure regimes, the current environmental degradation trends pose unparalleled challenges to community-based management. Thus, this ongoing degradation needs to be addressed with a combination of different forms of knowledge and management actions, which incorporate the various uses of geospatial technologies.

10.3 Human Spatio-Temporal Settlement Dynamics

Grasping how and why humans move across land- and seascapes through space and time is important for a number of reasons, and geospatial research can assist in this respect. For instance, Turner (2003) analyzed how spatial patterns of grazing pressures across agro-pastoral landscapes were determined by different local ideas and external socio-political and economic processes.

Similarly, Moran et al. (1994) have clearly showed how social class differentiation among various stakeholders in Amazonia influences deforestation patterns. In regard to governance of marine space, Mohamed and Ventura (2000) used geospatial analysis to show how indigenous land and sea tenure systems are spatially distributed in particular regions, and how they change across time. In the Roviana case, geospatial analysis has been very important for identifying population movement trends over space and time, and for extrapolating the possible effects of population distributional changes on the spatial characteristics of governance systems.

Spatial settlement patterns across time were mapped by initially determining the spatial distribution of current stakeholders to the principal land and sea estates of the region. Using compiled census data of more than 15 hamlets, members (both parents) of sampled households (at least 20% of randomly selected households per village) were asked to identify their main tribal affiliations and concomitant territorial rights. This was done to discern the proportion of community members with secure tenure over their occupying and adjacent land and sea estates, as well as those who do not have tenure. Also, key elder informants (selected through a snowball sample) were asked to map on aerial photographs all new settlements established over the informant's life (across their youth, maturity and old age) to plot historical changes in the spatial distribution of various sea and land estate stakeholders across the lagoons. Finally, informants were asked to rank settlements according to a rough estimate of population size (e.g. single family, extended family, village, etc.). The average distance between settlements at the three temporal points was measured and plotted in a set of aerial photographs according to estimated year of establishment, and all of the above information was georeferenced to understand spatial patterns of settlement and associated land/sea use changes. The goal, from a management perspective, was to extrapolate from these data the possible impacts of people's movements across the seascape on regional tenure systems across time.

The spatial analysis of the distribution of households with at least one member belonging to the major territories showed that the observed association between contemporary villages/settlements and existing tribal affiliations is significantly non-random. Over the past century (and centuries before), various tribal groups have competed for natural resources, leading to periods of political and territorial volatility (through migration and intermarriage). This resulted in the scattering of entitlement holders away from their ancestral territories, or conversely in some cases, to their neighbouring nucleation. These settlement patterns have resulted in differential territorial rights, affording some claimants several possibilities to access and control various estates while denying others the same opportunities. Underlying this dynamism is the local kinship system, which affords landowners the right to land and sea estates either by matrilineal and patrilineal descent, or both (cognatic descent), and to accrue these rights across generations (cumulative filiation).

This kinship system can blur boundaries between social groups and rights to their concomitant ancestral estates, if people were evenly distributed across the lagoons. But as mentioned, people's distribution across the seascape is not randomly distributed, thus giving some groups more power and accessibility to sea territories than others, and their potential capacity for rearranging territorial limits, particularly under the pressures of commercial exploitation. Uneven settlement histories, then, have determined the current choices and responses of fishers as they use and access marine resources within and across coastal territories, and has resulted in various systems of sea tenure governance in the region. Visualizing the mosaic of tenure rights and associated governance capabilities has implications for the management of marine resources (e.g. establishing marine protected areas), as it is important to establish and envision an estate's ownership (by both researchers and local stakeholders), and the concomitant capacity of inclusive stakeholders to curtail interloping and free-riding for managing resources successfully.

10.4 Human Foraging Strategies

Recognizing (and visualizing) human foraging strategies in small scale artisanal and/or subsistence fisheries can entail studying various dimensions including: (1) spatio-temporal human resource exploitation patterns (e.g. seasonal changes in targeted habitats and species), (2) human responses to variability in inter- and intra-habitat relative productivity (as determined by catch rates) and the influence of this variability on fishing strategies and (3) gauging human pressures on certain marine environments. At a fine spatial scale, optimal foraging theory models drawn from human behavioral ecology (HBE) can be a useful framework for analyzing the seasonal movements of fishers, the decisions that fishers make in the types and abundance of fish that they target, the use frequency of certain marine habitats and the rise or decrease of fishing effort resulting from seasonal environmental variation. This information along with biophysical information can then be incorporated into a GIS database for a visual representation of local foraging patterns, which, in turn, can better assist in designing marine management programmes that consider human foraging strategies at finer spatial scales. Similar geospatial approaches to human foraging strategies have been used by archaeologists to understand the relationship between the seasonality of particular available foods and site location selection by prehistoric foragers (e.g. Allen et al., 1990; Smith, 2012). In a fisheries context, De Freitas and Tagliani (2009) have used geospatial technologies to map target species, harvest areas and catch per unit of fishing effort (CPUE) for integrating local and scientific knowledge for more comprehensive management of artisanal fisheries in southern Brazil.

In the Solomon Islands study, optimal foraging models (the patch–choice model and the marginal value theorem [MVT]) (Stephens and Krebs, 1986) were used as a theoretical and methodological framework to assess how fishers selected patches for foraging and used their time across resource spatiotemporal variability. Drawing from the patch–choice model, for instance, it was possible to test the hypotheses that overall time allocation to a habitat type (set of patches) increases when seasonal productivity for that habitat increases and is higher than that of other habitats, as well as the inverse. Focal follows and self–reporting diaries were used to test the hypothesis. Focal follows is a method of keeping time–motion records of fishers while foraging (to calculate the caloric expenditure of their effort) and measuring their catches for each visited patch (e.g. each fishing ground). On the other hand, the diary method entails recruiting randomly selected subjects to individually keep a diary of their fishing activities (e.g. measuring time and catch in each patch visited). Through these two methods, it was possible to collect time-series foraging data for a period of 10 years (1994–2004) encompassing more than 10,000 fishing events and 15,000 hours of foraging activities. The data were used to examine the effects of village and habitat type on 'mean net return rates' and fishing event duration (time-allocation to patches). The mean net return rate measurement is equivalent to the energy gained during fishing (the kcal value of the edible catch) minus the labour input (labour costs incurred during foraging, including travel, search and handling times) divided by the total residence time at a fishing spot. Broadly, results indicated that overall effort was directed to the habitats with the highest yields and that fishermen shifted between habitat types, fishing grounds and species assemblages across different seasons to maximize their mean net return rate.

For a visual representation of these processes, the non-spatial attribute data, or the foraging dataset, were imported into a GIS and connected to the cartographic spatial dataset of indigenously defined resource patches (i.e. the actual fishing grounds), which were composed of 615 fishing sites collected with GPS receivers (Figure 10.2). This permitted the visualization (presented in maps for local communities to see) of the spatio-temporal relationships between particular marine patches (and related habitats) and changes in their relative productivity and associated temporal increases or decreases in foraging effort by fishermen of various hamlets in the region. For instance, a query was run with the GIS that extracted the fishing events associated with each of the three locally recognized tidal seasons in the Roviana Lagoon. The visual representation of the foraging analysis made details more apparent and thus provided a deeper understanding of inter- and intra–habitat relative productivity variability and human responses to this tidal season changeability (Figure 10.3) (see Aswani and Lauer, 2006a for further discussion).

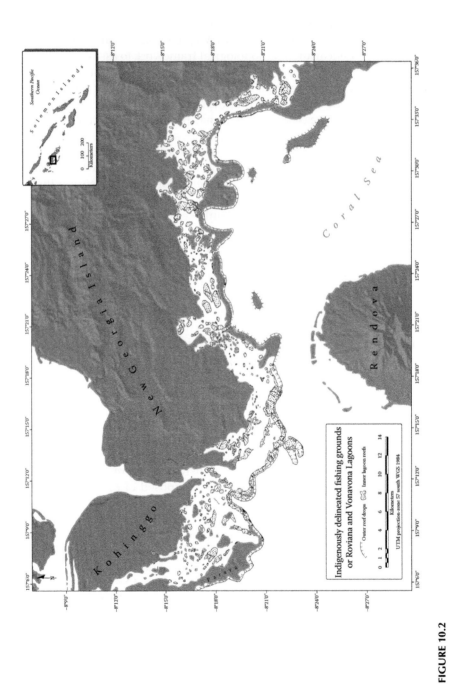

FIGURE 10.2
Locally delineated fishing grounds and associated marine habitats. (From Aswani, S. and Lauer, M. (2006a). *Human Organization*, 65, 80–102.)

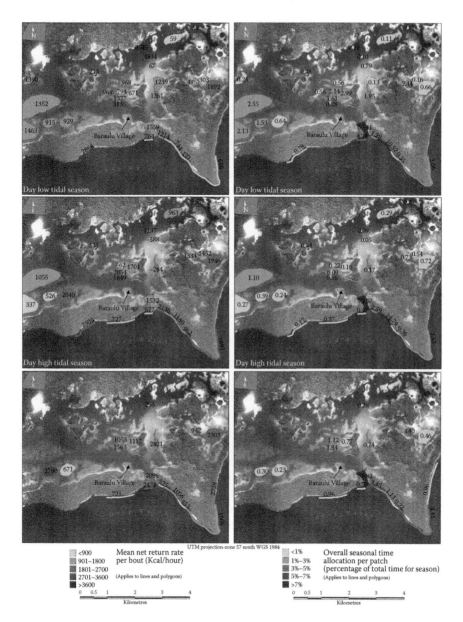

FIGURE 10.3
Visual representation of spatio-temporal characteristics of fishing grounds around Baraulu Village. (Seasonal mean net rate of return for fishing grounds of different habitat types are shown on the left side of the figure, and the percentage of total seasonal foraging time in those same fishing grounds is shown on the right side.) (From Aswani, S. and Lauer, M. (2006a). *Human Organization*, 65, 80–102.)

Visualization of these processes provided a better understanding of (1) the distribution of fishing methods and the geographical disparities in yield and effort, (2) different relative habitat productivities across seasonal and spatial variations and (3) the changes in time use as a response to resource abundance or scarcity. This information can help in the design of permanent and seasonal closures modelled in accordance with human seasonal foraging patterns. In fact, this data assisted in the designation of a local marine protected area that was designed to manage fish stocks and protect biodiversity while simultaneously being less disruptive to people's fishing strategies (e.g. not closing major foraging areas permanently). In sum, integrating local fishing behaviour and strategies into a GIS-driven management programme can enhance people's compliance with management in coastal artisanal fisheries.

10.5 Local/Indigenous Ecological Knowledge Systems

Increasing numbers of researchers are incorporating local/indigenous ecological knowledge into GIS databases to document and better understand human environmental cognition and spatio-temporal resource-use behaviour, and to build partnerships with local communities for marine resource conservation. For instance, Baldwin and Oxenford (2014) employed participatory GIS with Caribbean Grenadine Islands fishing communities to produce habitat maps that were informative to resource managers (ecologically valuable) and which represented local perceptions of habitat use and distribution. Other researchers have used participatory GIS for demarcating and cataloguing Miskito Indians reefs in Nicaragua (Nietschmann, 1995), for mapping fishing spots in southeastern Brazil and assisting local fishermen to use this knowledge for territorial defense (Begossi, 2001; De Freitas and Tagliani, 2009) and for documenting and organizing indigenous ecological knowledge to guide fishery management in Bang Saphan Bay, Thailand (Anuchiracheeva et al., 2003). While such efforts can empower local communities (St. Martin, 2009), there are challenging epistemological issues in reconciling local and scientific knowledge systems (Agrawal, 1995). In this context, however, geospatial technologies can serve as a medium to integrate and represent indigenous and western forms of knowledge for marine resource management successfully and equitably (Close and Hall, 2006), particularly as geospatial tools are increasingly accessible and now include some free and open source GIS packages (e.g. QGIS) (even if many indigenous and local communities don't have good Internet access as compared to modern societies; the gap is narrowing).

In the Roviana context, documenting local knowledge and changes to ethnobiological knowledge over the past 24 years (since 1992) was achieved using various ethnographic methods. Open-ended and structured interviews as well as focus groups have been used to interview hundreds of very young (13–19), young (19–35), middle-aged (36–55) and elderly (56>) men and women from across villages in the lagoons to document local knowledge and to gauge differences in taxonomic distinctiveness capacity across different age and gender groups. During these interviews, a number of social and biological factors related to marine foraging were documented including the following: (1) the name and ecological composition of recognized fishing grounds and associated habitats; (2) the attendant species of fish, molluscs and crustaceans found in each habitat category; (3) seasonal variations in the availability of different taxa found within each habitat type; (4) the occurrence (and mapping) of particular seasonal events such as spawning aggregations; (5) varying weather, tidal and lunar conditions and their impacts on habitat types and fauna and (6) local uses for each habitat (e.g. fishing methods) and its associated species.

Emic (indigenously defined categories) environmental types were matched with corresponding Western categories to designate habitat composition and biotic taxonomies. All organisms were identified through photographs and specimen collections, particularly shells, using scientific sources to identify the binomial nomenclatures for organisms. Informants identified over 50 fishing methods, hundreds of vertebrate and invertebrate marine species and around 14 major habitat types and 6 minor ones, as well as the mentioned 615 locally delineated fishing grounds throughout the region. The latter were first mapped onto aerial photographs of the Roviana and Vonavona Lagoons. Ninety-one black-and-white aerial photos were digitized using a high-resolution scanner and then georectified, so that they could be used as base maps. Thereafter, the digitized aerial photos were brought into the GIS and merged to create a mosaic of the lagoons. This established the spatial foundation for the subsequent georeferencing of local data and input into a GIS database for visual representation and analysis (Figure 10.2).

For the participatory mapping exercise, the photo mosaic maps were used for participatory image interpretation exercises in villages across both lagoons to identify fishing grounds, coral reefs and other marine habitat types. During these focus group workshops, fishers were asked to delineate the named patches on the maps. Then, the selected members of the focus group in each community took the researchers with GPS receivers and mapped the indigenously defined biophysical areas, fishing grounds and spots and associated coral reefs and other marine habitats. Fishermen from each community guided a researcher in a small boat around the perimeter of each named area and the locations of spawning, nursery, burrowing and aggregating sites for particular species within each recognized area were recorded and pinpointed with the GPS. The spatial extent of each spot represented as either lines or polygons and the location of particular biological characteristics represented as points were consolidated into a large file and imported into our GIS database as a layer. Eventually, local knowledge was ground-truthed via *in situ* habitat

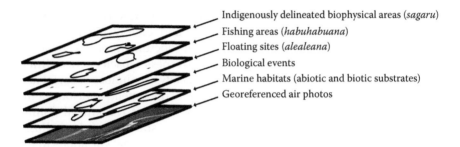

Indigenously delineated biophysical areas (*sagaru*)
Fishing areas (*habuhabuana*)
Floating sites (*alealeana*)
Biological events
Marine habitats (abiotic and biotic substrates)
Georeferenced air photos

FIGURE 10.4
Local cognition of the seascape as represented by layers (or themes) in the GIS. (From Aswani, S. and Lauer, M. (2006a). *Human Organization*, 65, 80–102.)

mapping and underwater visual census (UVC) surveys at various locations (see Aswani and Lauer, 2006b).

During this exercise we also developed participatory maps of the benthos and associated biological communities for designing particular marine protected areas in the region (established under our programmes). Using local knowledge, an emic classification of benthic habitats commonly found in the region was formulated (Figure 10.4). Then, 2 × 4 foot maps of the planned MPAs were printed. Informants (five men and women) – who were chosen as photo interpreters based on their knowledge of the local marine

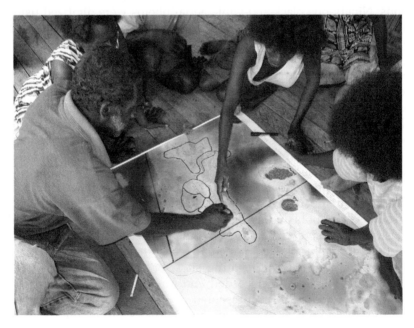

FIGURE 10.5
Local informants participating in a collective marine habitat mapping exercise.

environment – identified the main habitats and predominant benthic characteristics. They then cooperatively drew the boundaries of abiotic and biotic substrates on the image (Figure 10.5). The resulting paper maps with the particular benthic types as locally conceptualized were scanned, and the image files were loaded into the GIS for georeferencing. After, each of the boundaries was traced using on-screen digitizing techniques that created polygons (shape files) of each of the benthic substrates in the MPAs (Figure 10.6).

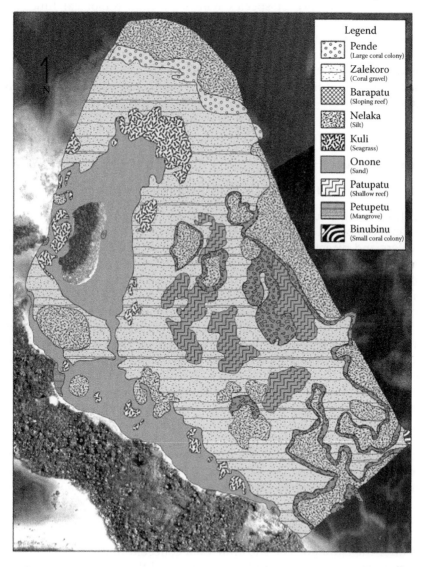

FIGURE 10.6
Informants' demarcation of predominant abiotic and biotic substrates on the aerial photographs of the Olive MPA. (Illustrated by UCSB project staff.)

Eventually, these spatial data were ground-truthed for some sites with reef check underwater visual surveys (see Aswani and Lauer, 2006a,b).

Generally, coupling spatially delineated indigenous knowledge with marine science knowledge allowed us to identify species and associated habitats that most urgently need management. Also, we generally found commensurability between indigenous ecological knowledge and marine science for identifying ecological processes that support biodiversity, including the presence of species with significant ecological functions, vulnerable life stages and interconnectivity among populations of certain species (e.g. Olds et al., 2014). In sum, in our participatory conservation efforts, socio-ecological and spatial analyses were fundamental for the successful implementation of MPAs, especially when delimiting the fragmentation and distribution of locally identified vulnerable habitats across the lagoons.

10.6 People's Detection of Rapid Ecological Change

Local and indigenous knowledge can be documented to understand ongoing rapid ecological changes caused by localized processes such as natural disasters and direct anthropogenic disturbances. Particularly interesting is measuring how coastal people gauge ecological change ensuing a catastrophic event (natural or human made) and the time lag it takes local inhabitants to identify the directionality and scale of ecological change. Recently, Fernandez-Llamazares et al. (2000) have suggested that different age groups within the Tsimané in the Bolivian Amazonia are experiencing 'shifting baselines', and thus indigenous knowledge may not be transforming fast enough for younger people to adapt to current ecological changes. In the Roviana research, geomatics were used to measure the direction and periodicity of experimental learning of people after an earthquake and ensuing tsunami in 2007, using similar participatory mapping methods as outlined in this chapter (Figure 10.6). Because there were pre-existing local and scientific benthic baseline data for an area that was under consideration for an MPA, it was possible to compare the results of marine scientific surveys with local knowledge of the benthos across three impacted villages and over three time episodes before and after the natural disaster.

More specifically, GIS was used to compare the marine survey data with the indigenous photo delineation of benthic characteristics before and after the tsunami. The substrate data collected in the marine science survey were displayed as one layer (points and their attributes), together with the layers (polygons and their attributes) created by the local photo interpreters. Then spatial queries selected all of the points from the marine science survey layer found within each polygon of the indigenously defined dominant benthic attribute(s) for comparison. This comparison permitted the addition of an attribute column to the benthic dataset indicating which indigenously defined benthic types were spatially associated with each survey site. This, in turn, allowed for measuring the correspondence between local knowledge of benthic types and marine survey results across periods before and after the catastrophic event (see Aswani and Lauer, 2014).

The research objective was to establish how people recognize biophysical changes in the environment before and after catastrophic events, and if people have the ability to detect ecological changes over short time scales or require lengthier time scales to identify transformations. The results of our study showed that local people were able to detect changes in the benthos over time. However, detection levels differed between the marine science surveys and local ecological knowledge sources over time, albeit overall direction of detection of change was identified for various marine habitats in the area. In sum, as people need to be capable of detecting changes in their environment in order to exploit and manage their resources, this kind of research has an implication for marine conservation and coastal management policies in coastal areas.

10.7 People's Detection of Protracted Ecological and Climate Change

Local and indigenous people's environmental knowledge can furnish important information regarding protracted climatic and ecological changes (e.g. Sagarin and Micheli, 2001), about how local peoples are coping with these changes (e.g. Ford et al., 2010) and to the role of local knowledge in disaster risk reduction and resource management (e.g. Mercer et al., 2010). To spatially delineate these changes, for instance, Griggs et al. (2013) used a GIS database to combine northern Australian Yorta Yorta aboriginal with scientific knowledge to better understand current climate-driven environmental changes and use this knowledge for designing adaptation plans locally. Similarly, Alexander et al. (2011) employed a GIS to explore the connections between indigenous climate-related observations and science. They overlaid the locations of local observations of temperature changes (between 1970 and 2004) across various continents with scientific studies documenting physical and biological climate-related changes. The study found that local observations not only provided data about climate change that has significance in assessing regional changes in climatic patterns, but also a much more nuanced and contextualized interpretation of current climatic and environmental changes locally.

In the Roviana lagoon, local perceptions and effects of environmental and climatic change on various habitat types were analyzed. Indigenous ecological knowledge of environmental change was recorded through two methods: Interviewing and participatory image interpretation. For interviews, semi-structured interviews and free-listed exercises were used. Respondents were asked to describe and list the changes they had observed in various marine and terrestrial habitats (open sea, inner lagoon, land ecology, agriculture and weather) (n = 266). The responses were 'free-listed' allowing each respondent to list as many responses as they wanted. Most of these interview-derived data of observed changes formed the basis of the geospatial analysis. Next, knowledgeable informants were selected through a snowball sample to interpret remotely sensed data via participatory meetings (e.g. identify reef, garden and plantation types) and delineate changes in domains on large-format image printouts as described in this chapter.

The focus was to elicit the spatial dimensions of local perceptions of change over the past 25 years by juxtaposing 1986 with 2011. Participants were specifically not asked about 'climate change' to avoid bias in the responses and the questions were left at the more generic level of 'environmental change'. The photographs of each marked-up satellite image (as per the participatory mapping) were georeferenced, and each drawing was digitized as a particular point, line or polygon feature representing the location of an impact on the environmental domains. The digital features were assigned attributes corresponding to the ancillary written data collected during the mapping exercise. These attributes describe (a) the village of the participants who created the drawing, (b) the domain associated with the picture and (c) a description of the noticed impact on the environment (Aswani et al., 2015a).

Informants recognized changes in both marine and terrestrial ecosystems between 1986 and 2011 (whether related to climate change or not) in areas that are critical for their subsistence needs. For instance, in the marine sphere, important changes identified for particular locations were a decrease in the frequency of certain important species for both subsistence (tunas, barracudas) and for commercial purposes (deep water/red snappers) caused by harvesting pressure (particularly the open sea and the outer reef habitats) and/or climate change (Figure 10.7). They also documented changes in coral damage via bleaching and disease (which cannot be differentiated locally), and reported that since 1986 there has been an increase in dead coral in the outer lagoon reefs (Figure 10.8) rather than in the inner lagoon ones. In the terrestrial sphere, for example, informants recognized a considerable expansion of gardening and plantation areas in the barrier islands between 1986 and 2011 and the concomitant reduction of natural vegetation caused by logging, overuse of land and the increasing commercialization of certain crops such as cocoa and copra.

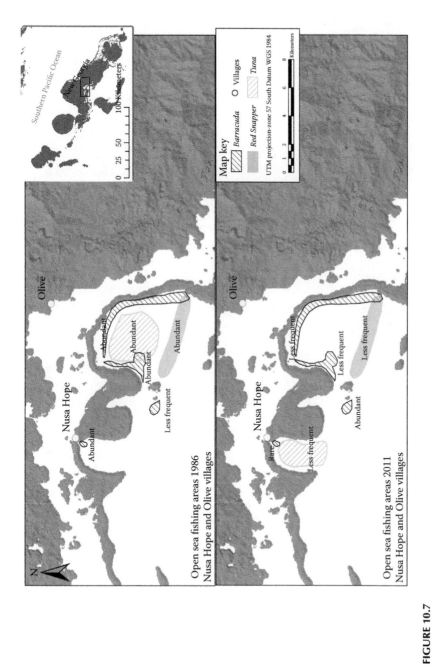

FIGURE 10.7

Locally recognized changes in abundance of key prey species in the Olive and Nusa Hope areas between 1986 and 2011. (From Aswani, S., Vaccaro, L., Abernethy, K., Albert, S. and Fernandez, J. (2015a). *Environmental Management*, 56(6), 1487–1501.)

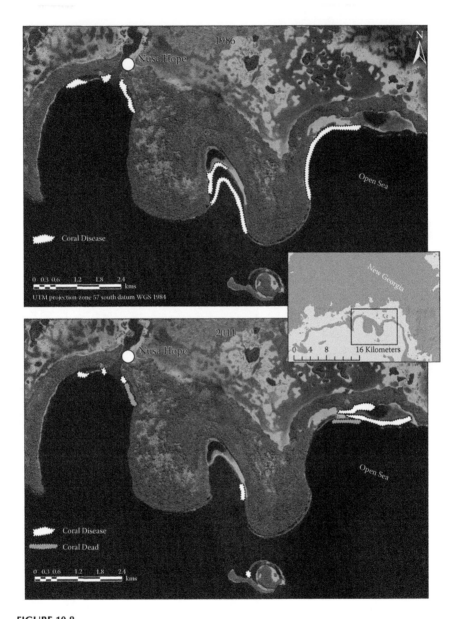

FIGURE 10.8
A comparison of locally recognized distribution of coral reef damage (disease and bleaching) and death in the Olive and Nusa Hope areas for the years 1986 and 2011.

10.8 Development and Spatial Conflicts

In small-scale societies (tribal or otherwise) unplanned development can create havoc, particularly spatial conflicts over terrestrial and marine resources. In new tourism destinations, conflict can grow when local people feel they are losing control and access to their territories and the natural resources that they have traditionally depended on for survival (e.g. Stem et al., 2003). In Oceania, and most notably Melanesia, traditional systems of land/sea ownership have created difficulties for the implementation of economic development such as tourism (Sofield, 1996). Because tourism can be a significant driver of social and ecological change in developing countries, particularly in coastal communities, inclusive participatory methods can be developed to account for local perceptions about future changes. For instance, a participatory mapping approach can be developed to obtain spatially explicit local perceptions of future environmental and social change resulting from tourism development as well as address the different community conflicts that may arise through the introduction of tourism in the community (or any other development such as logging). In the British West Indies, Feick and Hall (2000) used a participatory mapping approach for consensus building and conflict identification in tourism land development planning. Likewise, planners in Kangaroo Island (Australia) developed a public survey method to solicit stakeholder perceived landscape values using participatory mapping, and results showed that positive attitudes were linked with development location, justifying the use of spatial data in tourism planning (Brown, 2006).

In the Roviana project, participatory mapping approach was used (see previous sections) to obtain spatially explicit local perceptions of future environmental and social change resulting from tourism development. The objective was to address the potential community conflicts that may arise through the introduction of tourism in the region in the near future. The geospatial analysis showed that spatial conflicts within a community over territory and associated resources are likely to occur when designing natural resource management and tourism development plans. People's negative projections of potential future impacts included territorial conflicts, overfishing of vulnerable species and habitat damage. However, there were also positive projections including suitable tourists' sites, establishment of conservation areas and intensification of farming and associated increasing household incomes (Figure 10.9). The objective of the spatial analysis was to illustrate how planning for future tourism development in the Solomon Islands and elsewhere can assist in avoiding community conflicts for people who are already experiencing multiple pressures from capital extraction development (e.g. logging and mining). More generally, this kind of information can inform the design of preventive and adaptive management plans locally, as tourism begins to develop in the region. It can also assist policy makers to increase the future

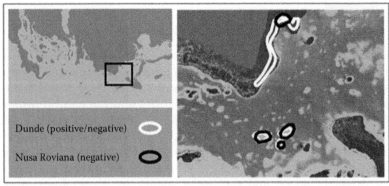

(a)

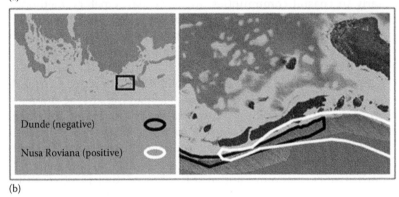

(b)

FIGURE 10.9
Example of conflicting opinions for different mangrove and reef areas resulting from existing land/sea disputes (digitized drawings shown on a Quickbird image). (a) Dunde participants predicted both positive and negative impacts at two inner reefs and one mangrove area, while Nusa Roviana participants predicted negative effects in the same areas. (b) Dunde participants predicted negative impacts to the area along the outer reef while Nusa Roviana participants predicted positive impacts. (From Aswani, S., Diedrich, A. and Currier, K. (2015b). *Society and Natural Resources*, 28, 703–719.)

sustainability of tourism in nascent small island destinations, particularly in vulnerable regions such as Roviana, which have experienced very little tourism development and will likely experience more in the coming years (see Aswani et al., 2015b for further discussion).

10.9 Discussion

This chapter has outlined various ways to combine local/indigenous ecological knowledge and local behaviour into a participatory GIS database to better understand spatio-temporal human-environmental interactions, and build partnerships with local communities for marine resource management and conservation purposes. The geospatial analysis can help researchers authenticate, magnify or reveal site-specific or regional patterns of human demographic, political, economic, socio-cultural and ecological dynamics that may not be visible to researchers on the ground. Thus, from the viewpoint of human ecology theory building, converting peoples' knowledge and socio-ecological behaviour into geospatial representations can assist researchers to: (1) better understand demographic and settlement histories and their effects on governance and resource management regimes; (2) specifically conceptualize human productive activities across land- and seascapes (e.g. spatio-temporal foraging strategies of fishers or hunters) and their concomitant environmental and social impacts; (3) illustrate spatially how local knowledge systems of land- and seascapes transforms into tangible classification of habitats and associated organisms, resource use patterns and resource allocation and governance strategies; (4) comprehend how people identify protracted and rapid ecological and climatic changes and adapt to these (or not) and (5) understand economic development and territorial conflicts among other processes.

Significant here is that the geospatial analysis can enhance the formulation of additional hypotheses (i.e. those not framed at the beginning of the study) regarding ecological and social human responses to a changing environment. It can also bridge the conceptual gap between indigenous and Western cognition of land- and seascapes, and make people on the ground participants of the scientific and resource management process. In fact, there has been a long-standing interest in 'public participation GIS' whereby local concerns, interests and knowledge systems are included in planning a GIS (e.g. Rundstrom, 1995 – see also Chapter 7 by Goldberg, D'Iorio and McClintock in the present volume). This has also been called 'counter-mapping' or an application of technology to map and reclaim rights over resources and represent local interests and values, particularly people who have been dispossessed by the advancement of non-inclusive capitalism (St. Martin, 2009).

So, people's involvement in building a GIS database has the dual advantage of: (1) empowering indigenous peoples to map their land and sea territories and (2) providing a research context for locals to contribute important insights about their environment's processes and functions. Potentially, then, participatory GIS can bridge the conceptual gap between local/indigenous and Western conceptions of landscapes and seascapes (Herlihy and Knapp, 2003) and enhance, if locally sanctioned, local involvement in resource management. For example, many of the outlined approaches can be used for designing and establishing marine protected areas and ecosystem-based management among other management schemes that involve local people in the design, establishing and monitoring process. It is also a cost-effective way for documenting (which could otherwise be lost) and obtaining missing scientific data crucial for selecting biodiversity conservation priority areas – missing data that would otherwise take years to collect (Aswani and Lauer, 2006 a,b).

10.10 Conclusion

This chapter has explored various ways of using geospatial technologies to better understand the social and ecological dynamics of human-environmental relationships. The use of geospatial methods (e.g. participatory GIS/remote sensing) can help in the analysis of various dimensions of human behaviours/ perceptions and concomitant impact on the natural environment. First, this kind of research is also fundamental for the development of hybrid management systems that combine local systems of fishery management with modern fishery management for designing regulatory measures to protect the marine environment (e.g. protect functional groups like parrotfish, minimize user conflicts and to develop marine protected areas) to address environmental challenges in the twenty-first century. Second, spatial research can also contribute towards theory building, as we test on the ground human-ecological interaction (e.g. hypotheses on resilience/vulnerability). Finally, geospatial analysis is a successful platform for integrating natural and social science for problem-solving integrated science. Overall, the illustrated case studies from the Solomon Islands demonstrate the power of geospatial technologies in representing visually site-specific spatio-temporal patterns of human ecological and social dynamics, and their applicability to coastal contexts around the world.

References

Agardy, T., Lau, W. and Hume, A. (2009). *Payments for Ecosystem Services: Getting Started in Marine and Coastal Ecosystems: A Primer*. Forest Trends, Washington DC.

Agrawal, A. (1995). Dismantling the divide between indigenous and western knowledge. *Development and Change*, 26, 413–439.

Albert, S., Grinham, A., Bythell, J., Olds, A., Schwarz, A., Abernethy, K., Aranani, K. et al. (2012). Building social and ecological resilience to climate change in Roviana, Solomon Islands. The University of Queensland, Brisbane.

Alexander, C., Bynum, N., Johnson, E., King, K., Mustonen, T. et al. (2011). Linking indigenous and scientific knowledge of climate change. *BioScience*, 61, 477–484.

Allen, K.M.S., Green, S.W. and Zubrow, E.B.W. (1990). *Interpreting Space: GIS and Archaeology*. London: Taylor & Francis.

Anuchiracheeva, S., Demaine, H., Shivakoti, G.P. and Ruddle, K. (2003). Systematizing local knowledge using GIS: Fisheries management in Bang Saphan Bay, Thailand. *Ocean and Coastal Management*, 46, 1049–1068.

Aswani, S. and Lauer, M. (2006a). Incorporating fishermen's local knowledge and behavior into geographical information systems (GIS) for designing marine protected areas in Oceania. *Human Organization*, 65, 80–102.

Aswani, S. and Lauer, M. (2006b). Benthic mapping using local aerial photo interpretation and resident taxa inventories for designing marine protected areas. *Environmental Conservation*, 33, 263–273.

Aswani, S. and Lauer, M. (2014). Indigenous people's detection of rapid ecological change. *Conservation Biology*, 28, 820–828.

Aswani, S., Vaccaro, I., Abernethy, K., Albert, S. and Fernandez, J. (2015a). Can local perceptions of environmental and climate change in island communities assist in adaptation planning. *Environmental Management*, 56(6), 1487–1501.

Aswani, S., Diedrich, A. and Currier, K. (2015b). Planning for the future: Mapping anticipated environmental and social impacts in a nascent tourism destination. *Society and Natural Resources*, 28, 703–719.

Baldwin, K. and Oxenford, H.A. (2014). A participatory approach to marine habitat mapping in the Grenadine Islands. *Coastal Management*, 42, 36–58.

Begossi, A. (2001). Mapping spots: Fishing areas or territories among islanders of the Atlantic Forest (Brazil). *Regional Environmental Change*, 2, 1–12.

Bell, J., Kronen, M., Vunisea, A., Nash, W., Keeble, G., Demmke, A. et al. (2009). Planning the use of fish for food security in the Pacific. *Marine Policy*, 33, 64–76.

Berkes, F., Colding, J. and Folke, C. (2000). Rediscovery of traditional ecological knowledge as adaptive management. *Ecological Applications*, 10, 1251–1262.

Brown, G. (2006). Mapping landscape values and development preferences: A method for tourism and residential development planning. *International Journal of Tourism Research*, 8, 101–13.

Burke, L., Kura, Y., Kassem, K., Ravenga, C., Spalding, M. and McAllister, D. (2001). *Pilot Assessment of Global Ecosystems: Coastal Ecosystems*. World Re-sources Institute (WRI), Washington, DC.

Chapin, M., Lamb, Z. and Threlkeld, B. (2005). Mapping indigenous lands. *Annual Review of Anthropology*, 34, 619–638.

Close, C.H. and Hall, G.B. (2006). A GIS-based protocol for the collection and use of local knowledge in fisheries management planning. *Journal of Environmental Management*, 78, 341–352.

Cornu, E.L., Kittinger, J.N., Koehn, J.Z., Finkbeiner, E.M. and Crowder, L.B. (2014). Current practice and future prospects for social data in coastal and ocean planning. *Conservation Biology*, 28, 902–911.

Costanza, R., Arge, R., de Groot, R., Farber, S., Grasso, M. et al. (1997). The value of the world's ecosystem services and natural capital. *Nature*, 387, 253–260.

De Freitas, D.M. and Tagliani, R. (2009). The use of GIS for the integration of traditional and scientific knowledge in supporting artisanal fisheries management in southern Brazil. *Journal of Environmental Management*, 90, 2071–2080.

Fernandez-Llamazares, A., Diaz-Reviriego, I., Luz, A., Cabeza, M., Pyhala, A. and Reyes-Feick, R.D. and Hall, G.B. (2000). The application of a spatial decision support system to tourism-based land management in small island states. *Journal of Travel Research*, 39, 163–71.

Feick, R.D. and Hall, G.B., 2000. The application of a spatial decision support system to tourism-based land management in small island states. *Journal of Travel Research*, 39: 163–171.

Ford, J.D., Pearce, T., Duerden, F., Furgal C.H. and Smit, B. (2010). Climate change policy responses for Canada's Inuit population: The importance of and opportunities for adaptation. *Global Environmental Change*, 20, 177–191.

Gadgil, M., Berkes, F. and Folke, C. (1993). Indigenous knowledge for biodiversity conservation. *Ambio*, 22, 151–156.

Godoy, R., Tanner, S., Fitzpatrick, I.C. et al. (2006). Does modernization erode the secular trend of indigenous knowledge? In *Tsimane' Amazonian Panel Study Working Paper*, 29, 1–38.

Griggs, D., Lynch, A., Joachim, L., Zhu, X., Adler, C., Bischoff-Mattson, Z., Wang, P. and Kestin, T. (2013). Indigenous voices in climate change adaptation: Addressing the challenges of diverse knowledge systems in the Barmah-Millewa. National Climate Change Adaptation Research Facility, Gold Coast, Australia.

Harmsworth, G.R., Barclay-Kerr, K. and Reedy, T. (2002). Maori sustainable development in the 21st century: The importance of Maori values, strategic planning and information systems. He Puna Korero. *Journal of Maori and Pacific Development*, 32, 40–68.

Herlihy, P.H. and Knapp, G. (2003). Maps of, by, and for the people in Latin America. *Human Organization*, 62, 303–314.

Johannes, R.E. (1978). Traditional marine conservation methods in Oceania and their demise. *Annual Review of Ecology and Systematics*, 9, 349–364.

Le Guen, O., Iliev, R., Lois, X., Atran, S. and Medin, L.D. (2013). A garden experiment revisited: Inter-generational change in environmental perception and management of the Maya Lowlands, Guatemala. *Journal of the Royal Anthropological Institute*, 19, 771–794.

Mercer, J., Kelman, I., Taranis, L. and Suchet-Pearson, S. (2010). Framework for integrating indigenous and scientific knowledge for disaster risk reduction. *Disasters*, 34, 214–239.

Mohamed, M.A. and Ventura, S.J. (2000). Use of geomatics for mapping and documenting indigenous tenure systems. *Society and Natural Resources*, 13, 223–236.

Moran, E.F., Brondizio, E., Mausel, P. and Wu, Y. (1994) Integrating Amazonian vegetation, land-use and satellite data. *BioScience*, 44, 329–338.

Nietschmann, B. (1995). Defending the Miskito reefs with maps and GPS: Mapping with sail, SCUBA, and satellite. *Cultural Survival Quarterly*, 18, 34–36.

Olds, A., Connolly, R.M. Pitti, K.A., Maxwell, P.S., Aswani, S. and Albert, S. (2014). Incorporating seascape species and connectivity to improve marine conservation outcomes. *Conservation Biology*, 28, 982–991.

Padilla, E. and Kofinas, G.P. (2014). Letting the leaders pass: Barriers to using traditional ecological knowledge in comanagement as the basis of formal hunting regulations. *Ecology and Society*, 19(7). http://dx.doi.org/10.5751/ES-05999-190207.

Ruddle, K. (1998). The context of policy design for existing community-based fisheries management systems in the Pacific islands. *Ocean Coastal Management*, 40, 105–126.

Rundstrom, R. (1995). GIS, indigenous peoples and epistemological diversity. *Cartography and Geographic Information Systems*, 22, 45–57.

Sagarin, R. and Micheli, F. (2001). Climate change in nontraditional data sets. *Science*, 294, 811.

Smith, G. (2012). Highland hunters: Prehistoric resource use in the Yukon Tanana Uplands. Unpublished M.A. thesis, Department of Anthropology, University of Alaska, Fairbanks.

Sofield, T. (1996). Anuha Island Resort, Solomon Islands: A case study of failure. In R. Butler and T. Hinch (Eds.), *Tourism and Indigenous Peoples*, 176–202. Boston: International Thomson Business Press.

St. Martin, K. (2009). Toward a cartography of the commons: Constituting the political and economic possibilities of place. *The Professional Geographer*, 61, 493–507.

Stem, C.J., Lassoie, J.P., Lee, D.R. and Deshler, D.J. (2003). How 'eco' is ecotourism? A comparative case study of ecotourism in Costa Rica. *Journal of Sustainable Tourism*, 11, 322–347.

Stephens, D.W. and Krebs, J.R. (1986). *Foraging Theory*. Princeton, NJ: Princeton University Press.

Turner, M.D. (2003). Methodological reflections on the use of remote sensing and geographical information systems in human ecological research. *Human Ecology*, 31, 255–279.

Turvey, S.T., Barrett, L.A., Yujiang, H., Lei, Z. et al. (2010). Rapidly shifting baselines in Yangtze fishing communities and local memory of extinct species. *Conservation Biology*, 24, 778–787.

Ulluwishewa, R., Roskruge, N., Harmsworth, G. and Antaran, B. (2008). Indigenous knowledge for natural resource management: A comparative study of Māori in New Zealand and Dusun in Brunei Darussalam. *GeoJournal*, 73, 271–284.

Weeks, R., Aliño, P.M., Atkinson, S., Beldia II, P., Binson, A. et al. (2014). Developing marine protected area networks in the Coral Triangle: Good practices for expanding the Coral Triangle marine protected area system. *Coastal Management*, 42, 183–205.

11

Spatial Information and Ecosystem-Based Management in the Marine Arctic

Tom Barry, Tom Christensen, Soffia Guðmundsdóttir,
Kári Fannar Lárusson, Courtney Price and Anders Mosbech

CONTENTS

11.1 Introduction

The Arctic is experiencing a period of intense and accelerating change (CAFF, 2013) and in the past century average temperatures have increased at almost twice the average global rate (IPCC, 2007). The scale and pace of these changes is exerting major influences on Arctic ecosystems and it has become increasingly important to find a way in which these ecosystems can be sustainably managed (CAFF, 2013).

Worldwide there is an increasing paradigm shift in ocean management that comes from an increasing awareness of the cumulative effects of human activities and the need to take a holistic and integrated approach to management to ensure the sustainability of marine ecosystems (Ottersen et al., 2011; O'Boyle and Jamieson, 2006). However while little is yet known on patterns of cumulative effects and the changes these effects may cause, spatial data on the distribution and intensity of human activities in marine areas is essential in establishing a more adaptive and ecosystem-based approach to marine environmental management (Halpern et al., 2008; Ottersen et al., 2010).

BOX 11.1 ECOSYSTEM-BASED MANAGEMENT (EBM)

Ecosystem-based management (EBM) is defined as the comprehensive, integrated management of human activities based on best available scientific and traditional knowledge about the ecosystem and its dynamics, in order to identify and take action on influences that are critical to the health of ecosystems, thereby achieving sustainable use of ecosystem goods and services and maintenance of ecosystem integrity (Arctic Council, 2013).

BOX 11.2 DEFINITION OF THE ARCTIC

There is no agreed-upon definition of the geographical extent of the Arctic and it is left to individual Arctic countries to provide such definition. Some of the Arctic Councils Working Groups (e.g. Conservation of Arctic Flora and Fauna [CAFF]) have defined boundaries for the purpose of monitoring and assessment activities. However, these are defined as much by political as environmental factors and vary among nations.

Geophysically, the Arctic may be defined as the land and sea north of the Arctic Circle. From an ecological point of view, it is more meaningful to use the land north of the tree line, which generally has a mean temperature below 10–12°C for the warmest month, July. With this definition, the Arctic land area comprises about 7.1 million km^2, or 4.8% of the land surface of Earth (CAFF, 2013).

Similarly, Arctic waters are defined by the characteristics of surface water masses. The Arctic Ocean is the smallest of the world's oceans (ca. 10 million km^2) and consists of a deep central basin, the Arctic Basin, surrounded by continental shelves. The circumpolar marine Arctic comprises the Barents Sea, Kara Sea, Laptev Sea, East Siberian Sea, Chukchi Sea, Beaufort Sea, Canadian Arctic Archipelago and Greenland Sea (Figure 11.1).

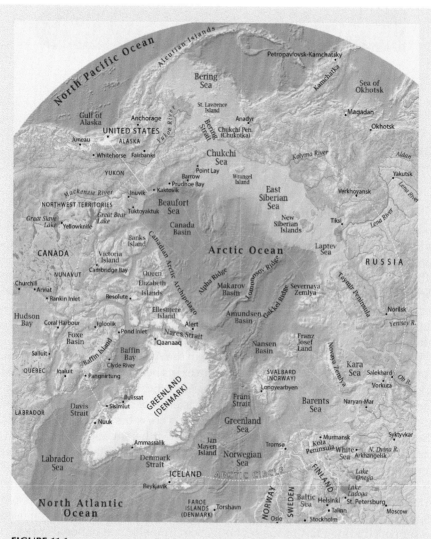

FIGURE 11.1
Topographic map of the Arctic.

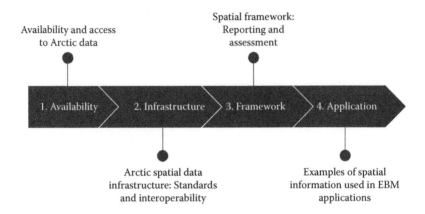

FIGURE 11.2
Components of an EBM framework for the Arctic. (From CAFF, Arctic Biodiversity Trends 2010: Selected Indicators of Change. CAFF Secretariat, Akureyri, 2010.)

Ecosystem-based management has been identified by Arctic states* as key to an adaptive way to sustainably manage Arctic ecosystems. Its interdisciplinary approach takes into account the political, regional and cultural differences found across the region and provides a flexible means to manage the effects of multiple pressures on Arctic ecosystems (Arctic Council, 2013). The Arctic marine ecosystem with its overlapping claims and disputes would benefit from the application of an EBM approach which could help address rights, restrictions and responsibilities, helping find space to reconcile the often conflicting and competing claims of conservation and development.

While a conceptual framework for EBM in the Arctic has been agreed upon, key elements are not yet in place and a gap exists between the concept of EBM and its implementation (Arkema et al., 2006). Critical to the successful implementation of EBM for the Arctic is the existence of a cohesive circumpolar approach (including sharing lessons learnt and cross-border collaboration) to the collection and management of data and the inherent application of compatible frameworks, standards and protocols that this entails. This chapter considers progress towards the development of those components needed to ensure efficient integration, access to and coordination of data across the Arctic's marine environment (Figure 11.2). It focuses on Arctic ecosystems and biodiversity and provides examples where spatial information has been used to support marine EBM processes both at circumpolar and national scales.

---

* The Arctic Council consists of the eight Arctic states: Canada, the Kingdom of Denmark (including Greenland and the Faroe Islands), Finland, Iceland, Norway, Russia, Sweden and the United States. The following organizations are Permanent Participants of the Arctic Council: Aleut International Association (AIA), Arctic Athabaskan Council (AAC), Gwich'in Council International (GCI), Inuit Circumpolar Council (ICC), Russian Association of Indigenous Peoples of the North (RAIPON) and the Saami Council (SC). The Permanent Participants have full consultation rights in connection with the Council's negotiations and decisions.

11.2 Data Sources

In recent years, the Arctic has been subjected to intense scrutiny and as a result a wide array of data has been generated which is spatial in nature. Typically, datasets are compiled for use in circumpolar assessments or monitoring programmes, for example, as conducted via the Arctic Council. Upon release these data tend to remain in formats and publications or locations that do not allow for easy use or access. There also exist a range of online mapping services that focus on various aspects of Arctic marine data, for example, the Arctic Ocean Observing Viewer (AOOV). These seek to integrate diverse Arctic datasets, with a goal of facilitating visualization, assessment and decision making. In addition, global data facilities often contain Arctic data, for example, the Global Biodiversity Information Facility (GBIF), the Oceans oBserving Information System (OBIS) and the International Council for the Exploration of the Sea (ICES) Marine Data Center. Regional and national data centers such as the Polar Data Catalogue (PDC) and the United States National Snow and Ice Data Center (NSIDC) also provide invaluable data services on specific issues. There is also a range of industry services where data are accessible at a cost.

Despite the availability of such multiple data sources, easier access to and more transparent management of data remains a challenge (CBD, 2014). Improved access to these geospatial data would help regional, local and national authorities, indigenous organizations, the larger scientific community, industry and international organizations to better predict, understand and react to changes in the Arctic. The approach to managing these data has been characterized by its fragmented manner with datasets distributed across many organizations and countries. Datasets are often duplicated, not integrated, uncoordinated or incomparable across regions, for example, isolated in national or thematic data sources. There has also been a tendency to focus on provision of online mapping applications as opposed to ensuring interoperability and application of standards. A significant obstacle to combining data from different studies in integrated and trend analysis in support of EBM is that often seemingly comparable datasets may have been obtained using different methods. Therefore, scientifically valid harmonization of datasets is fundamental to facilitate data integration.

In recent years, a number of efforts have focused on trying to standardize and impose more coherence on how these data are collected, stored and shared. These have tended to focus on specific issues as with biodiversity in the case of the CBMP Arctic Marine Biodiversity Monitoring Plan (Barry et al., 2013; Gill et al., 2011) where the focus is on coordination of monitoring and data management. Alternatively initiatives have attempted to address everything concerned with data management across multiple thematic areas and, as in the case of the Sustaining Arctic Observing Network (SAON), face challenges in providing clear added value (SAON, 2014).

BOX 11.3   THE CIRCUMPOLAR BIODIVERSITY MONITORING PROGRAMME (CBMP)

The Circumpolar Biodiversity Monitoring Programme is an international network of scientists, government agencies, indigenous organizations and conservation groups working to harmonize and integrate efforts to monitor the Arctic's living resources. Monitoring is being coordinated and integrated via four ecosystem-based monitoring plans (marine, freshwater, terrestrial and coastal). The CBMP is working to facilitate more rapid detection, communication and response to significant biodiversity-related trends and pressures. Results are channelled into conservation, mitigation and adaptation policies.

These efforts can suffer from lack of focus and a lack of long-term funding to allow for a sustained way in which their objectives might be achieved. They are beneficial in that they facilitate dialogues and help raise awareness of the issues to be tackled and as in the case of the Polar Data Forum (2015), provide a venue where specialists can engage and consider ways forward. Regarding Arctic biodiversity data, the creation of the Arctic Biodiversity Data Service (ABDS) is an important step towards ensuring a data-management framework for managing Arctic biodiversity data. The ABDS serves as the online, interoperable data management system for biodiversity data generated via the Arctic Council. It adopts an interoperable policy and is part of a framework which incorporates Arctic data contained in the PDC and GBIF, and as the Arctic node of OBIS, provides a means to ensure access to Arctic marine data and avoid data duplication. Work is also underway within the Arctic Council to develop comparable data services focusing on Arctic marine shipping activities (PAME, 2016).

11.2.1 Arctic Spatial Data Infrastructure

Key to completing the technical foundation that will facilitate EBM in the marine Arctic is the establishment of an Arctic Spatial Data Infrastructure (Arctic SDI [2013]), which would help harmonize how Arctic data are managed and utilized. The recent signing of a Memorandum of Understanding between the mapping agencies* of the eight Arctic states constitutes an important step towards the provision of reliable and interoperable geospatial reference data for the Arctic.

---

* Arctic SDI participating National Mapping Agencies: Canada Center for Mapping and Earth Observation; Natural Resources Canada, Danish Geodata Agency; National Land Survey of Finland; National Land Survey of Iceland; Norwegian Mapping Authority; Federal Service for State Registration, Cadastre and Mapping of the Russian Federation; Swedish Mapping, Cadastral and Land Registration Authority and the U.S. Geological Survey.

The first building blocks of the Arctic SDI, that is, an Arctic SDI Geoportal and SDI Reference Map are operational (2016) and aligned with global, regional and national geo-data efforts including the United Nations Committee of Experts on Global Geospatial Information Management (UN-GGIM), Global Earth Observation System of Systems (GEOSS), Infrastructure for Spatial Information in the European Community (INSPIRE), U.S. National Spatial Data Infrastructure (NSDI) and Canadian Geospatial Data Infrastructure (CGDI). It adheres to open data principles, including facilitation of open and interoperable data based on the Open Geospatial Consortium (OGC) and International Standards Organization (ISO) standards, specifications, architecture and software.

11.2.2 Spatial Framework

An important first step towards enabling EBM in the marine Arctic was the establishment of a spatial framework through which to coordinate and understand reporting on status and trends in marine ecosystems. This framework consists of a series of agreed upon ecological units to facilitate monitoring, assessment and reporting, whose application will help prepare the way towards future integrated ecological assessments and subsequent management actions in the marine environment (Figure 11.3):

Large Marine Ecosystems (LMEs): Intended to serve as a framework for enabling EBM (PAME, 2013) 18 LMEs have been agreed upon for the marine Arctic and adjacent seas (Figure 11.3a). They are regions of ocean space (200,000 km² or greater) that encompass coastal areas from river basins and estuaries to the outer margins of a continental shelf or the seaward extent of a predominant coastal current. They are defined by ecological criteria, including bathymetry, hydrography, productivity and trophically linked populations. The LME concept for ecosystem-based management focuses on productivity, fish and fisheries, pollution and ecosystem health, socioeconomics and governance (PAME, 2013).

Arctic Marine Areas (AMAs): Eight AMAs were defined which possess similar physical and biogeochemical characteristics (Figure 11.3). These define areas where a suite of common parameters, sampling approaches and indicators are providing a framework aligned with the LME boundaries by which status and trends will be reported upon across the Arctic (Gill et al., 2011) and allow for useful spatial comparisons.

Other initiatives which while not working at a circumpolar scale, impact on areas within the Arctic include work by the OSPAR Commission to conduct status and trends assessments on those areas within OSPARs mandate

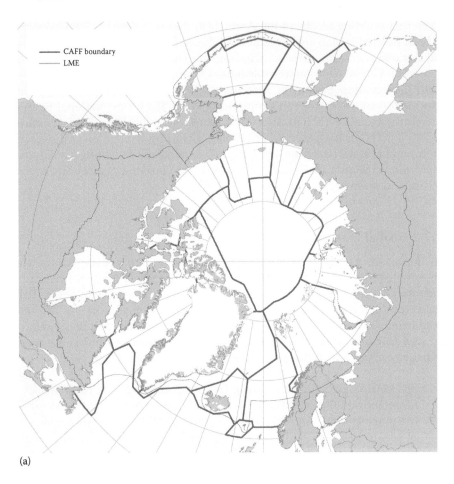

(a)

FIGURE 11.3
Arctic Marine Spatial Framework. (a) Large Marine Ecosystems (LMEs). *(Continued)*

which extend into the Arctic (OSPAR, 2015). Work has also begun on development of a digital elevation model (Arctic Council, 2015) for the circumpolar Arctic which, if it encompasses the seabed, will provide a final component to the Arctic's EBM spatial framework for the marine environment.

While in existence for a number of years, this reporting framework comprising LMEs and AMAs is only now starting to be used to conduct spatial analysis and reporting on Arctic marine ecosystems. Initiatives utilizing this framework include:

Circumpolar Biodiversity Monitoring Programme (CBMP): Is using existing capacity and data amongst Arctic countries to enhance integration and coordination to report in an integrated way and improve the policy and management response to Arctic change.

(b)

FIGURE 11.3 (CONTINUED)
Arctic Marine Spatial Framework. (b) Arctic marine areas.

The CBMPs Arctic Marine Biodiversity Monitoring Plan (CAFF, 2011) represents an agreement across Arctic states on how to generate better results from existing monitoring efforts and identify gaps in knowledge on Arctic marine ecosystems. The State of the Arctic Marine Biodiversity Report (SAMBR), scheduled for completion in 2017, will be the first integrated reporting outcome from the Arctic Marine Biodiversity Monitoring Plan. This will provide a circumpolar assessment of the key biological components of the Arctic environment and represents a first step to better reporting on the Arctic's living marine resources.

International Council for Exploration of the Sea (ICES), in cooperation with Arctic Council Working Groups, is preparing to conduct an integrated ecosystem assessment for the Central Arctic Ocean

(CAO). This initiative would consider potential effects and vulnerability to climate change and human activities such as Arctic shipping and potential future fisheries (ICES, 2015) thereby providing circumpolar information for effective decision making as the CAO becomes the object of increasing interest.

11.3 Applications

A number of recent initiatives involving Arctic states, indigenous organizations, non-government organizations (NGO) and intergovernmental governmental organizations (IGO) have focused on the identification of areas of ecological importance and/or sensitive to pressures from specific activities. The following examples illustrate how such work can help facilitate EBM both at circumpolar and national scales by considering potential impacts from activities in the marine Arctic and subsequent management implications.

11.3.1 Circumpolar

In 2013, the Arctic Council identified ecologically and culturally significant marine areas vulnerable to marine vessel activities in light of changing climate conditions and increasing multiple marine uses (AMAP/CAFF/SDWG, 2013). This process involved the eight Arctic Council member states, indigenous organizations and NGOs. It entailed the compilation of existing information and identification of areas of heightened ecological and cultural significance. These areas were overlapped with existing information on Arctic marine vessel activity to assess their vulnerability. The outcomes of this process then informed the 'Specially Designated Marine Areas in the Arctic high seas report' (Det Norske Veritas, 2014), which explored the need for internationally designated areas in the high seas area of the Arctic Ocean that might warrant protection from risks posed by international shipping activities, and the potential under the International Maritime Organization (IMO) for the application of Special Areas (SA) and the Particularly Sensitive Sea Area (PSSA) measures in the Arctic.

Informed by these efforts, and as part of a global effort to identify ecologically or biologically significant marine areas (EBSAs) (CBD, 2012a), in 2014 the CBD convened a workshop to identify Arctic EBSAs and contribute towards focusing future conservation and management efforts. This process involved experts from governments, indigenous organizations, NGOs and IGOs, who provided a broad range of data in differing formats, scales and detail which were relevant to identifying areas meeting the criteria (CBD, 2012b) to qualify

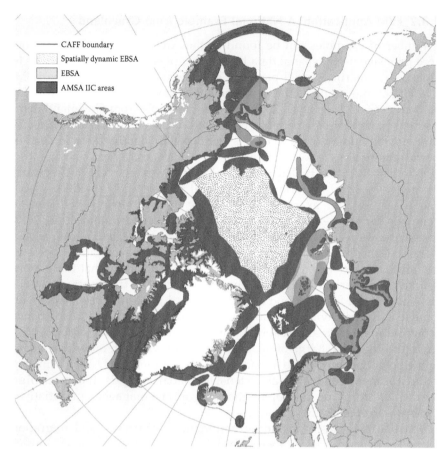

FIGURE 11.4
Arctic EBSA and Arctic marine areas of heightened ecological and cultural significance.

as EBSAs (CDB, 2015). These data were compiled and analyzed,* and in cooperation with experts at the Arctic EBSA workshop were used to identify and define Arctic EBSAs (Figure 11.4). As the application of the EBSA criteria is a scientific and technical exercise, management issues were not addressed but the outcomes will be relevant in any subsequent steps selecting conservation and management measures by states and for intergovernmental organizations, in accordance with international law, including the United Nations Convention on the Law of the Sea (Paragraph 26 of Decision X/29).

---

* The ABDS functions as a repository for the datasets compiled for the Arctic EBSA: www .abds.is/geo.

11.3.2 EBM Application: A National Example from Greenland

A number of examples can be found in the Arctic that demonstrate how more intensive use of spatial data has been applied in a national context to implement a marine spatial planning exercise in support of marine EBM. A well-known example is the development of an EBM plan by Norway for the Barents Sea (Olsen et al., 2007). A recent example from Greenland demonstrates how different spatial layers including species and ecosystem distribution and human-induced effects were used to identify areas in need of special attention.

Disko Bay and Store Hellefiskebanke along the western coast of Greenland are areas of critical importance for several species and provide key ecosystem services, for example, Greenlandic halibut, shrimp and marine mammals for the people who live in the region (Boertmann et al., 2013; Christensen et al., 2012) (Figure 11.5). However, rising temperatures and changes to the extent and nature of Arctic sea ice are facilitating access to new areas and extending the navigation season, meaning that these areas are likely to experience increased shipping, including transport of passengers and freight, fisheries and activities related to natural resource development. Adverse environmental impacts could include noise, disturbance to marine mammals and seabirds, introduction of invasive species, accidental or illegal discharge of oil, chemicals and waste. In order to prepare for the potential cumulative impacts of such activities, the Danish Ministry of Environment conducted an extensive spatial analysis and modelling exercise to inform the development of appropriate management initiatives (Christensen et al., 2015).

Forty-one map layers including abundance, occurrence and migration routes for over 65 species focused on the spatial distribution of important marine species and ecosystem components. These were combined to identify the most biologically important areas according to a set of criteria incorporating those used by the CBD to identify EBSAs and by the IMO to identify PSSAs. This method was inspired by impact mapping approaches used in marine regions outside the Arctic, as described by Halpern et al. (2008). Each of the biological layers were assessed and ranked according to their specific sensitivity to potential environmental effects caused by shipping. This analysis found that a number of smaller areas around Disko Bay and Store Hellefiskebanke are sensitive or very sensitive to the environmental impacts that shipping may cause and five sub-areas were identified (Figure 11.5) where there may be a need for heightened awareness in relation to impacts from shipping. Based on the analysis it was recommended that EBM should be applied in the study area but that at present there was no immediate need for further regulation in relation to shipping (Christensen et al., 2015).

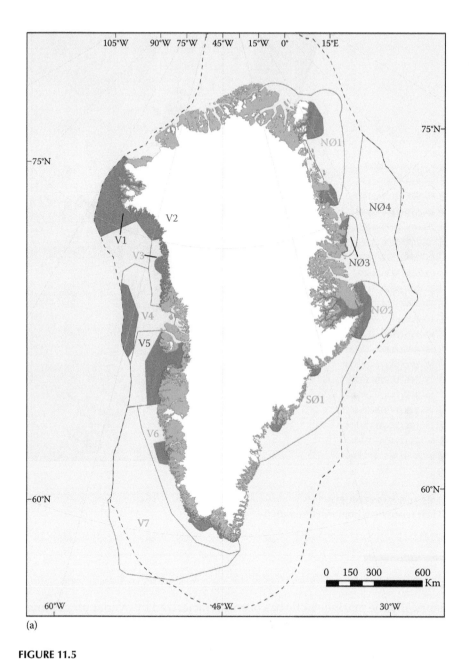

(a)

FIGURE 11.5

(a) Ecologically important marine areas in West Greenland. *(Continued)*

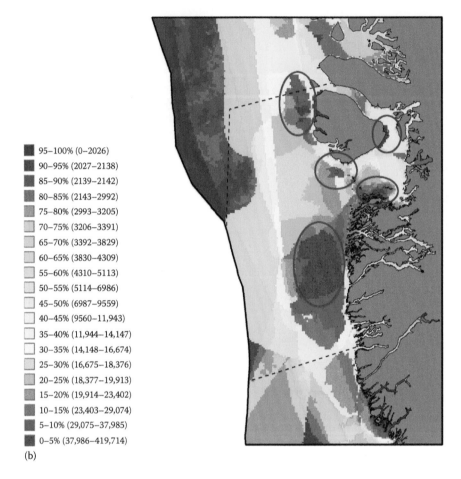

95–100% (0–2026)
90–95% (2027–2138)
85–90% (2139–2142)
80–85% (2143–2992)
75–80% (2993–3205)
70–75% (3206–3391)
65–70% (3392–3829)
60–65% (3830–4309)
55–60% (4310–5113)
50–55% (5114–6986)
45–50% (6987–9559)
40–45% (9560–11,943)
35–40% (11,944–14,147)
30–35% (14,148–16,674)
25–30% (16,675–18,376)
20–25% (18,377–19,913)
15–20% (19,914–23,402)
10–15% (23,403–29,074)
5–10% (29,075–37,985)
0–5% (37,986–419,714)
(b)

FIGURE 11.5 (CONTINUED)
(b) Their relative environmental sensitivity.

11.4 Conclusions

Historically the marine Arctic has been characterized by overlapping, sometimes conflicting and often uncoordinated efforts to establish a more reliable spatial framework to inform management. However, in recent years a range of initiatives has been embarked upon which is helping develop a more reliable framework to facilitate EBM implementation in the Arctic; examples include the CBMP, Arctic SDI and agreement on the LMEs.

In order for spatial data to reach its potential in effectively enabling EBM in the marine Arctic, access to data needs to be more flexible. Increased interoperability between data management frameworks is key to allowing spatial data to contribute more effectively towards implementing EBM, for example, through marine spatial planning. Key steps include an increased emphasis on improved interoperability between data holdings (e.g. the ABDS), avoiding duplication of data storage and efforts to establish common and/or compatible approaches to data collection, sharing and analysis, for example, via the Polar Data Forum. There is also a need for more focus on harmonization of datasets obtained with different methods to allow for scientifically valid data integration.

There has also been a tendency to develop online interactive mapping tools focused on providing the ability to interact with diverse data layers, for example, the Alaskan Ocean Observing System (www.aoos.org), the Circumpolar Seabird Information Network (axiom.seabirds.net/circumpolar_portal) and the University of the Arctic's Atlas (research.uarctic.org/resources/atlas). This has allowed freedom to analyse data but within defined environments which restricts the extent to which data can be applied to practical ends. While providing the means to interact online with spatial data is laudable within the context of EBM, resources might be better served in facilitating increased data interoperability whereby scientists, managers and planners can directly access standardized data in a manner more suited to their needs, for example, direct download of data or mapping services.

The above challenges have been widely acknowledged and a range of initiatives undertaken which focus on improving the manner in which data is collected, stored and accessed; and developing tools to enable EBM.

References

AMAP/CAFF/SDWG. 2013. Identification of Arctic Marine Areas of Heightened Ecological and Cultural Significance: Arctic Marine Shipping Assessment (AMSA) IIc. Conservation of Arctic Flora and Fauna, Iceland.

Arctic Biodiversity Data Service (ABDS). Conservation of Arctic Flora and Fauna (CAFF), Arctic Council Working Group. www.abds.is, accessed November 2015.

Arctic Council. 2013. Ecosystem Based Management in the Arctic, Arctic Council.

Arctic Ocean Observing Viewer (AOOV). http://www.arcticobservingviewer.org, accessed January 2015.

Arctic SDI. 2013. Memorandum of understanding between the mapping agencies of the eight Arctic states.

Arkema, K. K., Abramson, S. C., and Dewsbury, B. M. 2006. Marine ecosystem-based management: From characterization to implementation. *Frontiers in Ecology and the Environment*, 10(10), 525–532.

Barry, T., Christensen, T., Payne, J., and Gill, M. 2013. Circumpolar Biodiversity Monitoring Program Strategic Plan, 2013-2017: Phase II Implementation of the CBMP. CAFF Monitoring Series Report Nr. 8. CAFF International Secretariat. Akureyri, Iceland. ISBN 978-9935-431-27-1.

Boertmann, D., Mosbech, A., Schiedek, D., and Dünweber, M. (Eds.). 2013. *Disko West. A Strategic Environmental Impact Assessment of Hydrocarbon Activities.* Aarhus University, DCE – Danish Centre for Environment and Energy. Scientific Report from DCE – Danish Centre for Environment and Energy No. 71.

CAFF. 2010. Arctic Biodiversity Trends 2010: Selected Indicators of Change. CAFF Secretariat, Akureyri.

Conservation of Arctic Flora and Fauna (CAFF). 2011. *Circumpolar Marine Biodiversity Monitoring Plan*, CAFF Monitoring Series Report No. 3, April 2011, CAFF International Secretariat, Akureyri, Iceland.

CAFF. 2013. Arctic Biodiversity Assessment, Status and Trends in Arctic Biodiversity. Conservation of Arctic Flora and Fauna, Akureyri.

Conservation of Arctic Flora and Fauna (CAFF). 2015. *CAFF Work Plan 2015–2017.*

Christensen, T., Mosbech, A., Geertz-Hansen, O., Johansen, K. L., Wegeberg, S., Boertmann, D., Clausen, D. S. et al. 2015. *Analyse af Mulig Økosystembaseret Tilgang Til Forvaltning Af Skibstrafik i Disko Bugt og Store Hellefiskebanke.* Aarhus Universitet, DCE – Nationalt Center for Miljø og Energi, 103 s. – Teknisk rapport fra DCE – Nationalt Center for Miljø og Energi nr. Xxx. http://dce2.au.dk/pub/TRxxx.pdf.

Christensen, T., Falk, K., Boye, T., Ugarte, F., Boertmann, D., and Mosbech, A. 2012. *Identifikation af Sårbare Marine Områder i Den Grønlandske/Danske Del af Arktis.* Aarhus Universitet, DCE – Nationalt Center for Miljø og Energi.

Convention on Biological Diversity (CBD). 2012. *Decision IX/20* (paragraph 36 of decision X/29).

Convention on Biological Diversity (CBD). 2012. Ecologically or Biologically Significant Marine Areas (EBSAs) Scientific Collaboration among Dedicated Experts to Better Understand Marine Biodiversity and Support Country Efforts to Achieve the Aichi Biodiversity Targets.

Convention on Biological Diversity (CBD). 2014. Arctic Regional Workshop to Facilitate the Description of Ecologically or Biologically Significant Marine Areas, UNEP/CBD/EBSA/WS/2014/1/5.

Det Norske Veritas. 2014. *Specially Designated Marine Areas in the Arctic High Seas report.* Report for Norwegian Environment Agency.

Gill, M. J., Crane, K., Hindrum, R., Arneberg, P., Bysveen, I., Denisenko, N. V., Gofman, V. et al. 2011. *Arctic Marine Biodiversity Monitoring Plan (CBMP-MARINE PLAN).* Monitoring Series Report No. 3. Conservation of Arctic Flora and Fauna (CAFF).

Halpern, B. S., Walbridge, S., Selkoe, K. A., Kappel, C. V., Micheli, F., D'Agrosa, C., Bruno, J. F. et al. 2008. A global map of human impact on marine ecosystems. *Science.* 319.

International Council for the Exploration of the Sea (ICES). 2015. http://www.ices.dk/community/groups/Pages/WKICA.aspx, accessed November 2015.

IPCC. 2007. Fourth Assessment Report: Climate Change 2007 (AR4).

O'Boyle, R. and Jamieson, G. 2006. Observations on the implementation of ecosystem-based management: Experiences on Canada's east and west coasts. *Fisheries Research*, 79(1–2), 1–12, ISSN 0165-7836, http://dx.doi.org/10.1016/j.fishres.2005.11.027.

Olsen, E., Gjøsæter, H., Røttingen, I., Dommasnes, A., Fossum, P., and Sandberg, P. 2007. The Norwegian ecosystem-based management plan for the Barents Sea. *ICES Journal of Marine Science*, 64, 599–602.

OSPAR. 2015. ICG-MAQ. Intersessional Correspondence Group to manage preparation and publication of the Intermediate Assessment 2017 and the QSR21.

Ottersen, G., Kim, S., Huse, G., Polovina, J.J., and Stenseth, N.C. 2010. Major pathways by which climate may force marine fish populations. *Journal of Marine Systems*, 79(3–4), 343–360, ISSN 0924-7963, http://dx.doi.org/10.1016/j.jmarsys.2008.12.013.

Ottersen, G., Olsen, E., van der Meeren, G. I., Dommasnes, A., and Loeng, H. 2011. The Norwegian plan for integrated ecosystem-based management of the marine environment in the Norwegian Sea. *Marine Policy*, 35(3), 389–398, ISSN 0308-597X, http://dx.doi.org/10.1016/j.marpol.2010.10.017.

Polar Data Forum. www.polar-data-forum.org, accessed December 2015.

Protection of the Arctic Marine Environment (PAME). 2013. *Large Marine Ecosystems (LMEs) of the Arctic Area Revision of the Arctic LME Map*, 2nd ed. Protection of the Arctic Marine Environment (PAME), Akureyri, Iceland.

Protection of the Arctic Marine Environment (PAME). 2016. *Working Group Meeting Report*, PAME Secretariat, Portland, Maine, September 19–21.

Sustained Arctic Observation Network (SAON). March 2014. SAON board meeting report. http://www.arcticobserving.org/images/pdf/Reports/10_SAON_Board_Meeting_06MAR2014_Minutes.pdf, accessed January 2016.

12

Geospatial Technologies for Siting Coastal and Marine Renewable Infrastructures

David R. Green

CONTENTS

12.1 Introduction – A Contextual Setting

Renewable energy production has gained considerable attention worldwide in recent years, as a means to replace the rapid depletion of fossil fuels and to meet growing demands for energy, predicted to increase by up to three times by 2050. Furthermore, with growing concerns around the world being expressed about rising levels of carbon dioxide (CO_2) and other greenhouse gases in the atmosphere, and the consequent potential impacts of climate change, renewable energy and the associated infrastructures are rapidly becoming an essential part of many countries' future energy provision and supply plans (www.renewableuk.com/en/renewable-energy/wind-energy /offshore-wind) (Figure 12.1). Increasing the proportion of power generated from renewable energy sources is now seen as playing a vital part in helping to reduce current global emissions.

Many countries are already committed to reducing greenhouse gas emissions. In Europe, for example, the EU Renewable Energy Directive (EU RED) (European Union, 2009) requires member states to increase the percentage contribution of electricity generation from renewables such as wind to more than 30% by 2020. This will require significant investment in new low-carbon and renewable energy infrastructure by 2030 (Natural England, 2010; Scottish Borders Council, 2011).

Renewables will initially be used to supplement our current demand for, and continuing reliance on, fossil fuels (coal, oil and gas), and nuclear power for the foreseeable future. Ultimately though, they will gradually take an increasingly larger role in meeting our energy demands. In the United Kingdom, more electricity is already being generated from offshore wind than in any other country, and the renewable sector is now meeting around 5% of annual electricity requirements, a figure that is expected to grow to 10% by 2020 (www.thecrown estate.co.uk/energy-minerals-and-infrastructure/offshore-wind-energy).

At present most of the operational examples of marine renewables worldwide take the form of wind farms or wind parks and, as technology continues

FIGURE 12.1
Offshore renewables (windmills at the windmill farm Middelgrunden outside Copenhagen). (From www.awea.org [https://www.flickr.com/photos/andjohan/1022097482/].)

to advance, wind energy will likely be further expanded with improved efficiency to meet growing energy needs (Miller and Li, 2014). It is estimated that approximately 10 million megawatts (MW) of wind energy are available to use around the globe (Miller and Li, 2014). Compared with fossil fuels, wind energy also has an advantage in that it can be generated and used at the local level with virtually no transportation costs, aside from those associated with bringing it onshore. Aside from wind turbines, there are also many other possible sources of coastal and marine renewable energy that are being trialled around the world, notably wave and tidal energy.

Not surprisingly, renewable energy technologies have rapidly been transforming landscapes and seascapes all around the world, particularly over the last 10 years. Many coastlines already have both nearshore and offshore wind farms in place, or are currently in the process of planning for their future development, whilst marine test hubs for wind and other renewable energy sources are under construction in many locations, for example, EMEC (www.emec.org.uk) in Scotland, and FaB Test in Cornwall in the United Kingdom (www.fabtest.com).

Deciding where to place marine renewable energy installations relies upon an intimate knowledge and understanding of the coastal and marine environment, the so-called renewable climate and many other physical and social factors. Many of the criteria and analyses required for locating and evaluating potential sites are inherently spatial in nature, and GIS and other geospatial technologies have been used for a number of years to help map and analyse relevant data, as well as in providing objective and repeatable methods for assessing where renewable energy developments can be sited.

This chapter will examine some of the coastal and marine renewable energy options now available. It will also consider how renewable energy is currently being located and sited, with an emphasis on the role that geospatial technologies such as GIS, remote sensing, environmental modelling and the Internet are now playing in planning and decision making. Examples of the methods and approaches commonly used for suitability and constraint mapping and the siting of renewable energy sites will be used to illustrate these points. The need for data and information as the future basis for siting renewables in an ever more widely utilized marine and coastal environment will also be considered within the context of marine spatial planning (MSP).

12.2 Coastal and Marine Renewables

Although the term 'offshore' usually refers to the marine environment, strictly it can also include inshore or nearshore water areas such as seas, fjords, as well as sheltered coastal areas and even lakes. Offshore energy resources are also increasingly being located further from the coast, requiring new and innovative approaches, for example, from floating platforms, and in the engineering and delivery of the offshore network infrastructure (www .ppaenergy.co.uk/Insights/Offshore_Renewable_Energy/).

Whether operational or experimental, the location of marine renewable installations depends upon a very detailed multidisciplinary knowledge and understanding of the marine environment, the renewable technologies themselves, the marine renewable climates and offshore engineering, as well as economics, and proximity to market.

There are currently three main types of coastal and marine renewable technologies in use or under development around the world. Including both operational and experimental devices, these seek to extract energy from wind (wind turbines); from surface waves (wave energy converters) and from tidal currents (kinetic hydropower technologies). These, plus the entire infrastructure also associated with marine renewables including the transmission lines, are referred to collectively as offshore renewable energy installations (OREIs) (www.gov.uk/government/publications/offshore -renewable-energy-installations-orei).

12.2.1 Wind Energy: Offshore Wind Farms

Some of today's renewable energy technologies have long historic anteced-ents. In twelfth-century China (Zhang, 2009), as well as in many parts of pre-industrial Europe, the landscape was once dotted with mills that utilized wind power for the milling of grain (grist mills) and the pumping of water. Later developments of this technology in the United States, the USSR and Denmark were the precursor of modern wind turbines. With the industrial revolution came new sources of energy, such as steam power and the inter-nal combustion engine, and windmills soon fell into disuse. However, the landscape, especially in areas exposed to wind such as hilltops, and now the coast and offshore, is being taken up once again with the modern equivalent of windmills, namely the single wind turbine, and multiple wind turbine arrays known as wind parks and wind farms (Drews, 2012; www.wwindea .org/technology/ch01/en/1_3_1.html).

There are two types of wind turbine currently in use: horizontal axis and vertical axis turbines (www.wwindea.org/technology/ch01/en/1_3_1.html). Wind turbines comprise the rotor blades, the generator and the structural component, for example, a tower (see, e.g. Malhotra, 2011). In addition, there may be additional components required, such as boat landing platforms, maintenance platforms, decks and cable tubes (www.wind-energy-the-facts .org/offshore-support-structures.html). Some offshore platforms are fixed to the seabed, whilst others can be placed on floating platforms. As the technol-ogy has advanced so too have the possibilities for placement further offshore and, with the development of structures strong enough to withstand the greater extremes of the marine environment and the force of the wind, this now includes both shallow and deeper water areas. Current research also seeks to reduce the cost of the electricity generated, as well as to improve the reliability and performance of the turbines, particularly where they are to be sited in high-risk areas (energy.gov/eere/energybasics/articles/wind-energy -technology-basics and energy.gov/eere/next-generation-wind-technology).

12.2.2 Wave Energy

Like wind energy, the harnessing of water-based energy has precedents that predate the industrial revolution, and had its origins with water-driven wheels to power mills, weaving looms and other installations. Hydro-electric power (HEP) subsequently provided cheap electricity supplies for many areas of the world. At the present day, however, marine water-based renewable energy technologies are relatively less well-developed than wind energy, and are still at the experimental stage.

The origins of modern wave energy technology lie mainly in the work of Professor Stephen Salter at the University of Edinburgh in the mid-1970s based on his wave energy generation tank and the 'Salter Duck' experimental wave energy device that he developed (Salter, 1974; Salter et al., 2002). Building on these foundations, a variety of designs of marine wave energy converters (WEC) are currently under development, with several full-scale prototypes being tested in the open sea. These include the Oscillating Wave Surge Converter (Folley et al., 2004); Pelamis Wave Energy Convertor (Yemm et al., 2011); LIMPET 500 (Land Installed Marine Power Energy Transformer) energy converter (Czech and Bauer, 2012); Archimedes Wave Swing (de Sousa Prado et al., 2006) and several others. Each device captures and stores wave energy in a slightly different way. Many rely on the conversion of the wave motion into electricity, whilst others utilize overtopping or a pressure differential. However, most WECs remain largely experimental, whilst others are no longer in existence, having failed to achieve and demonstrate long-term reliability for continued operation, or the current financial climate precluding further research and development.

12.2.3 Tidal Energy

Compared to wind or wave power, tidal power seeks to harness a highly predictable and potentially reliable source of renewable energy. However, exploiting this potential is only possible in relatively limited areas of the coast, or at specific sites. Tidal systems capture the energy generated by tidal flows in a similar way to wind and wind-turbines, and are of three types: tidal stream, tidal range barrage schemes and tidal range lagoon schemes (Crown Estate, 2012). Current technology includes horizontal and vertical axis turbines, oscillating hydrofoil, enclosed tips, Archimedes screw and tidal kites (www.emec.org.uk). These devices are either attached to the seabed or via a piling, a floating mooring or with the aid of a hydrofoil.

To date, the best-known operational tidal power system is the 25-year-old barrage scheme in France on the Rance Estuary, Brittany (Andre, 1978; Frau, 1993), but despite the long-term demonstration of its potential at this site, tidal technology is still regarded as being 'immature', and in its infancy. There are also now a number of other smaller projects in operation around the world, with new ones under construction, and more proposed for the future. The Marine Current Turbines SeaGen device at Strangford Lough in Northern Ireland is the world's first commercial-scale, grid-connected tidal stream turbine (Douglas et al., 2008).

12.3 Siting Criteria

Planning and development for any offshore renewable energy installation requires the consideration of many factors. Site assessment can be very complex, as there is a need to meet many technical, economic, environmental and social acceptance-related criteria. This is further complicated by a changing policy framework, dynamic energy markets, a variety of stakeholders (particularly at the local level) and factors influencing investment decision making, all of which will inevitably affect potential profitability. Until quite recently, many decisions may have been based on the *ad hoc* recommendations and individual experiences of the decision maker or planner in charge (Sunak et al., 2015, p. 3). However, many of the criteria that have to be taken into account when siting renewable energy installations are inherently spatial in nature, lending themselves well to analysis and processing through the use of geoinformatics technologies.

The siting of marine renewables is primarily dependent on the availability of a reliable source of wind, wave or tidal energy. This naturally 'constrains' each type of renewable to a specific area of the coastal or marine environment. For example, stronger and steadier winds, and the large open areas at sea, allow the establishment of larger wind farms than can be developed in most inland areas. Marine locations also tend to offer more continuous and stronger sources of wind than is usually available on land, while offshore breezes may also be stronger in the afternoon than on land, and less turbulent, and rather fortuitously coincide with peak demand for electricity (en.wikipedia.org/wiki/Offshore_wind_power). Reduced turbulence also means a reduction in the fatigue load and an increase in the lifetime of the wind-turbines while, according to Henderson et al. (2002) and Mathew (2006), a reduced occurrence of wind shear also allows the use of shorter support towers. Wind farms also yield the highest net revenues if located in areas with high wind speeds (Archer, 2012). Understanding the wind climate characteristics over space and time, including wind shear, is therefore important (Beaucage et al., 2014).

To begin a renewable energy siting exercise, the general areas of the coast or marine environment that may potentially be suitable for the installation of marine renewables need to be identified. Prasad et al. (2009) note the need to carry out a wind, wave or tide resource assessment for any renewable energy project. In the case of wind, for example, a wind resource assessment exercise uses data and information on wind speed/wind power and wind density estimates, the prevailing wind direction, daily and seasonal variation, long-term consistency (climate cycles), turbulence and wake, temperature and the uncertainty of the wind at various heights above the Earth's surface to isolate areas where there is a potential source of wind energy to be harvested. Ozerdem et al. (2006) consider the wind resource to be the most important consideration for a wind energy plant (Lip-Wah et al., 2012). Corresponding assessments will similarly be undertaken where wave or tidal power resources are being investigated.

Where the potential energy supply criteria are met, then this will define those areas that may be deemed suitable for further assessment, and everything else will be unsuitable (Harrison, 2012). This is sometimes known as sieve mapping (the locations remaining after sieving are suitable) or, its inverse, masking (the areas that do not meet the criteria are removed from further consideration) (Green and Ray, 2003).

Once the general area of interest has been identified, a range of additional considerations then brings the focus of attention to the individual sites within this area that match or most closely match the set of criteria deemed essential for the type of renewable energy that is being targeted. Refinement of the siting process to the local level means identifying all of the factors that might constrain those sites or locations that are actually available for the proposed development. These may be physical, social, environmental, commercial or even military in origin. For example, a wind farm needs to be placed at a certain altitude, within a certain distance from the end-user market, and not in the proximity of airports and army testing ranges. Other criteria that may be relevant include some or all of the following: geology, topography, slope, ground suitability, land price, site access, land use/land cover/land ownership, wildlife habitat, protected areas and human factors (noise, settlement, aesthetics, visual, auditory disturbances), proximity to existing structures, glare, roads, transmission lines, public participation, safety (ice throw and broken blades) and cost-revenue analysis, as well as radar interference, energy storage and energy grid requirements (Fang, 2015; Scottish Borders Council, 2011; Crill et al., 2010; American Wind Energy Association, 2004; Dabiri, 2011; Ozerdem et al., 2006; Prasad et al., 2009).

Offshore, apart from the obvious need for the availability of wind, wave or tidal energy itself, some or all of the following factors may also need to be taken into account: water depth (bathymetry), coastal geomorphology, the topography and geological composition of the seabed, seismic areas, landscape (as seen from land to sea and in the opposite direction) and visual enjoyment of the natural environment. Climate and extreme meteorological conditions also need to be considered, as well as the potential impacts of future climate change. Aspects related to other infrastructures, such as ports, shipyards, production clusters, fleets of suitable ships and other craft for the installation and maintenance of offshore plants may also be relevant (Serri et al., 2012), as may rules, standards, legislation, permitting and financing, social aspects (e.g. public acceptance) (www.pgs.eonerc.rwth-aachen.de /go/id/fbzf/lidx/1), interaction with human activities such as tourism, fishing and navigation and the creation of new jobs (Serri et al., 2012). It has also been noted by Natural England (2010, p. 8) that in deciding where best to locate renewable energy it is important to 'separate (out) the identification of relevant site factors from the process of making judgements about them, either individually or in the context of the natural environment as a whole'.

One attraction to locating wind farms or parks offshore has been the need to avoid affecting the visual landscape through intrusion and noise. Being out of sight also means less likely opposition from the public: not everyone likes a land- or seascape covered by wind farms, and planning objections to a proposal have often arisen because of the potential visual impact of the structures, the number and spatial formation (the spatial combinations of multiple wind turbines), the flicker and movement effect of the blades, the colour of the turbines and their visibility from the shoreline. Concerns have also arisen about scale, distance, proportional visibility, noise, contrast, lighting, skylining and back-clothing, orientation, elevation and design (University of Newcastle, 2002; Bishop and Lange, 2005). To address these latter concerns, Lindquist et al. (2014) in Canada, and Roth and Gruehn (2014) in Germany, among others, have advocated the emerging new concept of geodesign (Dangermond, 2009; Steinitz, 2012) as a methodological framework to help inform the siting and design of wind turbine installations. Geodesign seeks to incorporate geographic concepts and principles (Dangermond, 2009), implemented through geoinformatics technologies, into the process of designing elements of the built environment.

Potentially adverse environmental impacts arising from placing renewable energy installations in the marine environment must also be addressed, and the impacts on biodiversity and ecosystem functions need to be considered in location choice (Punt et al., 2009), although many types of installation may have both positive or beneficial and negative or detrimental effects on ecology (Inger et al., 2009). For example, wind farms have been found to have negative effects on flora and fauna, including through avian collisions along bird migration routes (Exo et al., 2003; Bright et al., 2006; Drewitt and Langston, 2006), underwater noise (Koschinski et al., 2003; Wahlberg and Westerberg, 2005; Thomsen et al., 2006) and electromagnetic fields (Gill, 2005; Petersen and Malm, 2006). At the same time, some researchers (Petersen and Malm, 2006; Wilhelmsson et al., 2006; Fayram and De Risi, 2007; Punt et al., 2009; Jay, 2010) have concluded that offshore wind farms can also be beneficial and can act as artificial reefs and no-take zones.

The location, size and irregular shape of OREIs and wind farms in relation to the navigational hazards to shipping have also been identified as issues by leisure and sailing enthusiasts, port and harbour authorities and the emergency services. In the latter case, for example, the Maritime Coastguard Agency (MCA) in the United Kingdom cites difficulties that could arise from their location, potentially leading to injury, death or loss of property, either at sea or among the population ashore (www.gov.uk/guidance/offshore -renewable-energy-installations-impact-on-shipping).

Aside from the concerns raised by stakeholders regarding their placement, there are a number of economic advantages and disadvantages associated with the location of wind farms nearshore or offshore. According to Breton and Moe (2009) wind farms that are located in areas with low average water depths and at a short distance from the shore will yield the highest net revenues. For the investor this means lower costs. In direct contrast, however, Henderson et al. (2002) and Mathew (2006) observe that in the case of offshore renewables, investment costs for turbine foundations are higher, the distance to the main electrical grid is further and improved equipment is needed because of the harsher environment that leads to rapid corrosion and makes maintenance more difficult and more costly (Punt et al., 2009). Miller and Li (2014) also note that final site selection needs to consider the higher development costs associated with a proposed development offshore, in order to derive the greatest economic benefit for the investment.

In some respects trying to find places to install renewable energy is a Catch-22 situation: clean energy is deemed to be an essential part of a more environmentally sustainable future, and yet the optimum locations for siting of renewable energy installations may be significantly and irreversibly impacted by the development of these facilities. The trade-off between the economic or social benefits that might be gained, and the potential adverse ecological impacts to be avoided must therefore be thoroughly analyzed (Lip-Wah et al., 2012, p. 73). Arguably though, with more informed and objective approaches to planning and decision making, using GIS, sensitively located renewable energy can bring social and economic benefits to communities and to local businesses that outweigh the objections (Scottish Borders Council, 2011).

12.4 Spatial Datasets, Data and Information Sources

In order to carry out any GIS-based siting or mapping exercise, there is a need for spatial datasets for each of the criteria being considered. Whilst many of the datasets may already be available, there are often considerable gaps and new data may be required at an appropriate scale and resolution. In addition, not all datasets that are already available will be ideal for the task and many may have been collected for another purpose, and at the wrong

FIGURE 12.2

National marine plan interactive (NMPI) screen showing some of the datasets relevant to renewable energy siting held in this system. (Courtesy of the author – Marine Scotland – https://marinescotland.atkinsgeospatial.com/nmpi/.)

scale, resolution or even be out of date. The use of datasets that are not ideally fit for the purpose may lead to errors and poor decision making when using a GIS for site evaluation (Wright et al., 1998; Green and Ray, 2003).

Fortunately, thanks to recent spatial data infrastructure (SDI) initiatives (Burrough and Masser, 1998; Williamson et al., 2003) a growing number of geoportals (Bernard et al., 2005; Longley, 2014), web-based data catalogues, downloadable spatial datasets and other online resources are available that may be suitable for use in GIS suitability and siting exercises. In Scotland, for example, Marine Scotland has developed the Marine Atlas of Scotland, and with the evolution of marine spatial planning (MSP), the National Marine Planning Information (NMPi) system has been released. Figure 12.2 shows some of the datasets relevant to renewable energy siting held in this system (marinescotland.atkinsgeospatial.com/nmpi/and www .gov.scot/Topics/marine/science/atlas). Other sources of UK digital data-sets relevant to renewable energy siting can be found at the INSPIRE data website (aws2.caris.com/ukho/mapViewer/map.action), and numerous sim-ilar resources for other parts of the world either exist already or else are under development, many of them under the guidance and coordination of the International Coastal Atlas Network (http://www.iode.org/index.php ?option=com_content&view=article&id=335&Itemid=100065).

Despite the wide range of spatial datasets now available, there are still many gaps at the local scale. This is highlighted by Serri et al. (2012), who examined the available data sources for renewable energy siting as part of project ORECCA. They concluded that some data were not of a suit-able resolution to support commercial site selection decisions, and not well resolved close to the shore, especially for areas where detailed data would be required. The study also concluded that more collaboration was needed between regions, to produce regional resource atlases, marine spatial plan-ning maps and catalogues of available data (Serri et al., 2012).

Archer (2012, p. 2) mentions the potential role of datasets provided by stakeholders to provide additional local data and information. Data acquired during participatory planning and siting exercises (see, e.g. Alexander et al., 2011) can be very useful at the local level. However, Archer (2012) also cau-tions that participatory siting can be fraught with difficulties between stake-holders, and can lead to extended debates, conflicts and delays.

12.5 The Potential of Geospatial Technologies for Siting Marine Renewables

The use of geospatial technologies for siting exercises is not new, and it has long been recognized that GIS hardware and software provides a powerful spatial data toolbox for renewable energy management and planning. For example,

Sparkes and Kidner (1996) used GIS to site onshore windfarms, while Wright et al. (1998) and Green and Ray (2003) used GIS to seek the best sites for artificial reefs in the Moray Firth in Scotland, UK. In a similar way, GIS has also been used to generate wind, wave and tidal resources maps; (a) to determine wind, wave and tidal energy locations, (b) to determine experimental marine renewable technology test sites, (c) to visualize the appearance of turbines and windfarms in the landscape, (d) to connect offshore renewable energy to the land and infrastructure and (e) for participatory planning and decision making.

Especially when combined with additional geospatial tools such as remote sensing, digital image processing, mobile GIS and the Internet, GIS can support the full planning and decision-making workflow, from data input, storage and management, to analysis, modelling, mapping and visualization (Malczewski, 2004; Mariononi, 2004; Harrison, 2012). Before GIS, sites were selected mainly on the basis of economic and technical issues (Shouman, ND) as well as what is usually referred to as 'common sense'. Today, a higher degree of sophistication in the selection process is expected and, as outlined in Section 12.3, the number of potential factors that need to be taken into account when siting renewable energy installations has multiplied (Shouman, ND, p. 3). Geospatial technologies allow the development of decision-making systems that may be used to address planning and management issues and to support problem-solving in complex environments where many spatially referenced data have to be integrated (Carrion et al., 2008; Tegou et al., 2010).

In addition to applications built on general-purpose GIS software, a number of dedicated commercial software packages are now also available, based on GIS concepts and methods but specifically designed and developed by specialist companies working within and for the renewable energy sector. These include products such as the WindFarmer, WaveFarmer and TidalFarmer from the Norwegian company DNV-GL (www.dnvgl.com), WindPRO (www.emd.dk/windpro/) and WindFarm (ReSoft: www.resoft.co.uk/English/index.htm) (University of Newcastle, 2002). Besides excluding areas deemed non-viable for siting renewable energy installations due to constraints such as purchasable land and visibility and so on, some of these packages, such as the Farmer series from DNV-GL, also use computational fluid dynamics models to accurately predict the energy yield; the aim being to minimize the cost of energy produced from the site whilst meeting regulatory constraints (personal communication with Tom Levick, DNV-GL Energy).

12.6 GIS-Based Analytical Methods: MCA, OWA, WLC and AHP

Whilst some GIS siting exercises have approached the problem of where best to locate a feature in the environment by using simple masking and overlay techniques (e.g. Wright et al., 1998; Green and Ray, 2003), other more advanced analytical methods have been developed to help fine-tune the result. In particular, MultiCriteria Analysis (MCA), MultiCriteria Evaluation (MCE) and MultiCriteria Decision Analysis (MCDA) have been widely used. According to Harrison (2012) other terminology which is often used interchangeably includes MultiCriteria Decision Making (MCDM), MultiCriteria Decision Support Systems (MCDSS), Spatial MultiCriteria Decision Making (SMCDM) and Spatial MultiCriteria Analysis (SMCA).

MCA allows multiple information sources (e.g. layers of spatial information or factors) to be combined into an aggregated measure of suitability for decision-making purposes (Comber et al., 2010). Weighting procedures are used to determine the importance of each criterion to the process. Multi-objective decision making is common in environmental management and is designed to improve the chance of making a logical and well-structured decision where there are many possible criteria, or where there is a multidisciplinary team involved and a need to arrive at a jointly reached conclusion (Chen et al., 2010). Taha and Daim (2013) provide a review of MCA in renewable energy analysis.

There are a number of different types of MCA in use. Where there is a need to simply divide the study area into the binary categories of suitable and non-suitable areas, and create layers that are then combined using an overlay operation, then Boolean MCE is sufficient. However, in some cases there may be varying degrees of uncertainty or preference associated with identifying suitability and, for each criterion, the contribution (weight) to the decision-making process may be different (Comber et al., 2010). Fuzzy MCE approaches were developed to deal with the lack of sensitivity of the Boolean MCE approach in such situations. Deciding on how to assign weights to each criterion (in other words, establishing the importance of each criterion), and their relative contribution in the decision-making process can be difficult. Boolean overlay represents the extreme cases with no trade-off, Boolean 'AND' operator represents the MIN risk decision making and Boolean 'OR' operator represents the MAX risk decision-making strategy.

According to Jiang and Eastman (2000), weighted linear combination (WLC) MCE allows a trade-off between each criterion by weighting them according to their assigned importance when assessing suitability. In the WLC method, a factor with a high criterion weight can trade-off or compensate for poor weights given to other factors. According to Chen et al. (2010), WLC is an averaging risk decision making (situated at the mid-point between the MIN [Boolean 'AND' operator] and MAX [Boolean 'OR' operator]), that indicates full trade-off among criteria. However, limitations associated with both Boolean MCE and WLC, concerning the risk associated with making the wrong decision, have led to another MCE method known as ordered weighted analysis or averaging (OWA) (Jiang and Eastman, 2000). OWA provides the means to control the level of risk by using both criterion weights (the significance of a criterion) and order weights (Chen et al., 2010). Compared to WLC, OWA method can obtain any results along the continuum (between the MIN-MAX of risk), allowing any degree of trade-off among criteria between no trade-off and full trade-off according to the decision-making strategy (Chen et al., 2010).

The analytical hierarchy process (AHP) (Saaty, 2008) is another increasingly widely used type of MCA that has been used in renewables siting exercises. AHP involves five steps: (1) Defining the problem and alternatives to evaluate; (2) structuring the criteria for the decision-making process; (3) pairwise comparison of options; (4) weighting and calculation of relative priorities (how do options fare in their pairwise comparisons) and (5) aggregation of relative priorities to produce an overall ranking of options (www.weadapt.org/knowledge-base/climate-adaptation-training/module -ahp). AHP can be useful to help decide on the weights selected in the MCA process (Karapetrovic and Rosenbloom, 1999; Mariononi, 2004; Saaty, 2004; Boroushaki and Malczewski, 2008; Chen et al., 2010; Chandio et al., 2013) and can be used to make one choice from many alternatives; to rank and prioritize these alternatives and to involve the data, experience and judgement of stakeholders in the decision-making process. Examples include: Tegou et al. (2010) and Harrison (2012).

Chen et al. (2010) mention the need to include sensitivity analysis (SA) as part of the MCA approach, in order to gain some insight into the sensitivity of the outputs generated – namely the areas deemed suitable for development. The sensitivity analysis relates to errors, inaccurate assumptions, perturbations in the input values (criteria values or weights) and so on, that may be based upon the subjective or conditional perceptions of the stakeholders and decision makers involved in the decision-making process. Furthermore, SA helps to assess the precision and limitations of the model, and provides an assessment of uncertainties in the simulation results (Chen et al., 2009).

Other alternative approaches to MCA mentioned in the literature include the Dempster-Shafer Method, which combines multiple layers of spatial data from different sources, using weights of evidence (Comber et al., 2010). The objective here is to arrive at a degree of belief that the derived output is the appropriate one, rather than generating a binary yes/no choice.

12.7 Case Studies: GIS and Renewable Energy

12.7.1 GIS and Wind Energy

The recent upsurge of interest in renewable energy tended to focus initially on the potential of wind as a resource and this, along with the popularity and success of wind farms, means that a very large body of GIS literature has focused on their siting. Several of these studies have focused on the spatial analysis of wind data in order to assess potential sites for wind power parks and plants (e.g. Hillring and Kreig, 1998; Baban and Parry, 2001; Lejeune and Feltz, 2008; Palaiologou et al., 2011), or else looked at wind power planning issues to provide an overview of the overall renewable energy potential of an area (Belmonte et al., 2009; Tegou et al., 2010; Arnette, 2010; Arnette and Zobel, 2011). Moller (2011), for example, used GIS and a supply cost curve analysis to develop a spatially continuous resource economic assessment model (SCREAM), which was used as a tool for analysing cumulative off-shore wind energy production and its marginal costs.

Other studies have focused on the application of GIS as a (spatial) decision support system ((S)DSS) and environmental decision support system (EDSS) for site investigations (Malczewski, 1999; Lejeune and Feltz, 2008; Aydin et al., 2010), many of which have used GIS-based MCA in the decision-making process. This includes work by Karapetrovic and Rosenbloom (1999), Baban and Parry (2001), Mariononi (2004), Boroushaki and Malczewski (2008), Lee, Chen and Kang (2009), Chen, Yu and Khan (2010), Tegou, Polatidis and Haralambopoulos (2010), van Haaren and Fthenakis (2011), Harrison (2012), Hofer et al. (2014) and Sunak et al. (2015).

12.7.2 GIS and Tidal Energy

With growing use of the coastal and marine environment for renewable energy, there has been a need to manage the multiple interests and potential conflicts through marine spatial planning (MSP). Alexander et al. (2011) demonstrate the use of a novel GIS-based approach to examine a potential site for renewable tidal energy off the Mull of Kintyre in Scotland, where there are potential conflicts of interest between multiple stakeholders. They used an interactive touch-table mapping device in stakeholder workshops to gather data and to facilitate negotiation of trade-offs. The result was to reveal conflicts between users and the use preferences and concerns of stakeholders, spatial issues and gaps in the data. The workshops developed a degree of consensus between conflicting users on the best areas for potential development suggesting that this approach should be adopted during MSP.

12.7.3 GIS and Wave Energy

Nobre et al. (2009) developed a GIS-based multi-criteria model for selecting the coastal sites best suited to wave energy farms. An exclusion mask was developed, restricting the available area using the following criteria: areas outside 12 nautical miles of the coastline, military exercise areas, marine protected areas, 500 m buffer zones around underwater cables, wave shadow areas, harbour entrances and navigational channels and areas <30 m and >200 m depths. Following this a set of weighted features including ocean depth, bottom type, distance to ports, distance to shoreline and power grid and wave climate or resource availability were reclassified to a common scale within the available area. The final step was to use geospatial algebra to create a suitability layer revealing the areas where wave energy farms would be most successful. Although dealing with the physical aspects of siting, this model does not take into account the aesthetic impacts of the development.

The volume by Cascales et al. (2014) also contains a number of chapters on the use of MCDA and fuzzy AHP in relation to marine renewables such as WECs.

12.7.4 GIS Siting of Experimental Marine Renewables and Connection to the Coast

Offshore wind-farm developments also need to take into account connection to the grid as well. The high costs associated with taking the energy generated via a cable to the sub-station and linking to the grid have sometimes been excluded from the siting criteria and oversimplified (Prest et al., 2007). The paths for siting the cables can be very complicated, and considerations need to include physical, environmental and political features such as national parks and marine protected areas (MPA), the location of existing undersea cables and even steep cliffs.

Graham (2006) developed a GIS-based model using a wide range of spatial data to determine the best sites for the location of experimental WECs. Network access is a major constraint on the development of marine energy and selecting areas that make the best use of the resource and network capacity available to get the first generation of devices into the water are essential. Siting considerations included identifying suitable areas of marine energy resource as well as determining the route to network, and locating network capacity for the new generation (Graham, 2006). Using bathymetric, seabed, environmental and economic data, the model developed utilized a least-cost path analysis to position the optimal route of the submarine cable from the WEC to the network. In addition, power flow analysis on the Scottish electricity network was used to determine the location and amount of wave energy resource available.

In South Australia, Prest et al. (2007) developed a GIS method and tool to optimize the cable route between wave-farm parks and the existing electricity infrastructure. Using GIS and MCDA they modelled the effects of different exclusion zones on the cost of transmission paths for individual site development assessment. The goal was to provide policy makers with important information about the available economic feasibility of wave farm locations as well as reaching a compromise between conservation and protection designations and the need to reduce emissions.

12.7.5 Landscape, Visualization, Participatory GIS and Web-Based Resources

The visual impacts of renewable energy developments have also been increasingly considered as society has become more aware of the impact renewable energy installations can have on the landscape and seascape (see, e.g. Soerensen and Hansen, 2001; Miller et al., 2007; Natural England, 2010). Such studies in Germany and Denmark have considered visibility in relation to the siting distance from the shoreline and under different atmospheric conditions that may or may not be negligible.

Another role for GIS and the geospatial technologies in renewable energy developments involves the potential to visualize the impacts of installations on landscape aesthetics and the need to involve communities in the decision-making process through the use of 2D and 3D visualization tools, participatory GIS and access to information through the Internet or web-GIS. Such tools help to identify the visual impact, and the scale and cumulative impacts on different fields of view (FOV), and can offer useful and valuable insights for communities into proposals in relation to landscape and seascape change prior to planning permission for development being granted (Natural England, 2010). In the context of planning applications for both onshore and offshore renewables, the generation of a semi-realistic landscape visualization of a proposed site can be very useful for presenting a visual idea of the development to a wider audience, and may provide opportunities to involve stakeholders in the planning process, sometimes in an interactive way, through participatory GIS. Over time, lower-cost technologies have evolved to provide higher resolution and more realistic graphics, interactive and immersive tools (e.g. virtual reality [VR] and headsets such as Oculus Rift), as well as facilitating the use of familiar free software such as Google Earth (GE) and Sketchup (http://www.sketchup.com/), allowing individuals to create, use and share visualizations of proposed developments.

Computing power and data projection technology now also allow for very realistic computer-generated visuals using GIS and remote sensing data as a backdrop to a graphic representation of a wind-farm, wave or tidal energy development. Aside from being able to see what impact a development will have on the landscape, the ability to 'fly' through the landscape, which can either be controlled by an operator or the stakeholder, can provide unique opportunities for individuals to explore a development proposal, and to interact with the visualization, providing invaluable feedback from the public and communities. In Scotland, the James Hutton Institute (JHI) has been involved in the development of a number of interactive marine renewable landscape visualizations including wind and tidal energy (www.hutton.ac.uk/learning/exhibits/vlt). With the aid of interaction handsets, stakeholders are able to view the development from a number of different locations, to fly through the proposed development and to vote on questions raised about the potential impact of the development on the landscape.

Two types of landscape visualization theatres have been developed: a static screen and projection facility; and a mobile equivalent that has been used to take the visualization to the public. Advantages of the visualization theatres are the realistic imagery and the 3D visuals. Controls allow the operator to change the appearance of the turbines, the visibility, lighting and weather conditions, and even enable the provision of animated graphics with rotating turbine blades. One limitation of this approach, however, is that the graphic nature of the visuals, along with the projections systems and computer generated displays, can distract the participant from the purpose of the visualization.

Other approaches, albeit on a much smaller scale, have used specialized software, for example, Windfarmer from Garrad Hassan (Figure 12.3) to present the user with a wide range of different map and 3D visuals. Another alternative has been to use a combination of Google Earth (GE) and Sketchup to create a visualization of a proposed wind farm (Figure 12.4). Using the functions of GE it is also possible to generate a flythrough of the site, as well as to add additional GIS layers.

All of these provide opportunities for stakeholders to engage with the planning process and to acquire a more informed basis for decision making. These types of approaches can also provide invaluable insight into proposed wind-farm sites that can be informative as well as being influential in winning over opponents of wind-farm developments. In a participatory situation, the use of GIS technology can be valuable in the planning process.

Other approaches combine GIS and photographs, wireframes and photomontages along with procedures for calculating zones of visual influence (ZVI) (University of Newcastle, 2002; Riddington et al., 2008). Most of these studies have utilized digital terrain model (DTM) data from various sources, increasingly including LiDAR data collected from aerial platforms, in relation to the ZVI. Basic modules needed to calculate the ZVI are now a standard feature of much GIS software, and integrated links to other software programmes for producing wireframes and photomontages are widely available.

Animated computer simulations now provide the means to compare landscape scenes with and without a wind turbine in order to test the ability of respondents to first detect, then recognize and judge the impact of turbines in relation to distance, contrast and atmospheric conditions (Shang and Bishop, 2000).

A slightly different study, by a consortium comprising academics from the Glasgow Caledonian University and commercial consultancy company Cogent Strategies International (Riddington et al., 2008), used GIS, questionnaires and an Internet survey to explore the impact of existing wind farms in the landscape on the behaviour of tourists seeking accommodation or when travelling in Scotland. Simao et al. (2009) have used a conceptual web-based GIS for a participatory approach to wind farm planning incorporating a MCSDSS (MultiCriteria Spatial Decision Support System), followed by a proof-of-concept web-based participatory wind energy planning (WePWEP) system.

Early landscape and seascape visualization studies were somewhat limited by the available technology and the ability to create realistic landscapes and features. Whilst useful and informative, stakeholders could easily be

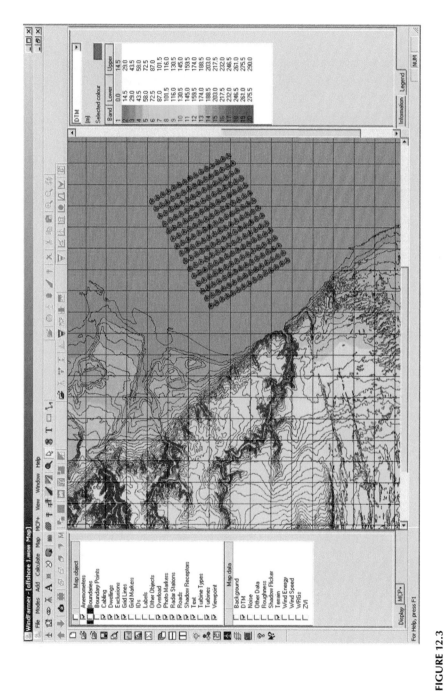

FIGURE 12.3
Garrad Hassan's WaveFarmer software. (From www.dnvgl.com.)

FIGURE 12.4

The use of Google Earth and Google Sketchup to create a coastal wind farm visualization. (Courtesy of the author.)

distracted by the shortcomings of the technology to produce realistic scenes, including the lack of a convincing landscape and real-looking wind turbines, as well as the resolution of the display, the capability to include convincing animation and the ease with which the viewer can navigate the scene. With faster computer processors, improved screen resolutions, higher resolution spatial data and new developments such as VR, more realistic results can now be achieved. In addition, the availability of GIS interfaces in the form of touch-tables (Alexander et al., 2011) and haptic or tangible interfaces that allow navigation through a virtual landscape using gestures and movements (Beleboni, 2014; Petrasova et al., 2014) have also aided the potential for stakeholders to become more actively involved with spatial data in siting exercises.

12.8 Discussion

As is evident from the foregoing sections, GIS and other geospatial technologies are now widely used as the basis for more detailed and objective siting of wind-farms and other renewable energy sources both on- and offshore. Over time these methods have become progressively more sophisticated, and have evolved to support a more objective and participatory planning and decision-making process. The once relatively limited availability of appropriate spatial datasets at the right scale, coverage and resolution to undertake the siting of renewable energy is also changing rapidly. While the siting of renewable energy is complex, and involves consideration of many physical, social, economic, legislative and landscape factors, geoinformatics methods and technologies facilitate approaches to the task that allow these to be taken into more effective account. This in turn ensures that the social and economic benefits of renewable energy are maximized, while hopefully avoiding long term and irreversible change to the environment and conflicts with other uses of marine and coastal resources.

Past experience suggests that there is also a need for up-to-date legislation and policy instruments at the local level to ensure proper siting of offshore energy that minimizes environmental impacts and prevents proposals being only led by market forces (Nadai, 2007; Jay, 2010; Sperling et al., 2010). The available space to develop renewable energy in the coastal and marine environment is ultimately clearly limited. To this end, maritime (or marine) spatial planning (MSP) will be needed to ensure best use of the space in light of multiple other users and the potential for conflicts of use particularly in the coastal zone. It has been argued (Jacques et al., 2011) that MSP and the definition of zones for offshore wind and ocean energy could reinforce existing legislation, and thereby encourage successful renewable energy development while also supporting investment decisions in the sector. MSP is already being implemented in many countries around the world (Archer, 2012) but it has been suggested (Jacques et al., 2011) that for many EU countries the effectiveness of available MSP instruments at international or regional level is often constrained at the national level by limitations of policy and related legal frameworks, coupled with complex permitting and incentives procedures.

In addition, as noted by Gubbay et al. (2006), although GIS can provide the means to identify the optimum locations for renewable energy sites, the final location will still require a formal environmental impact assessment (EIA) (CEFAS, 2004), and the incorporation of the results of that assessment into an environmental statement (ES). MSP does not negate the need for an EIA.

12.9 Summary and Conclusion

Geospatial technologies clearly provide a valuable and wide-ranging set of tools and techniques to aid in developing a more informed and objective decision-making approach to the location of onshore and offshore renewable energy developments. However, these tools should perhaps only be seen as one piece of the toolbox, and siting is in fact far more complex in practice. Beyond the physical constraints, there is also a need to consider their location from a much wider perspective that includes building into the process an intimate understanding of the marine and coastal environment at a variety of different scales and resolutions, the business case and economics of choosing an optimum location, marine legislation (both past and present), the developing area of MSP, as well as established approaches such as environmental impact assessments.

Finally, whilst current GIS and the geospatial technologies are a valuable component of siting coastal and marine renewable infrastructures, there is clearly a need in the future to build into any analyses the social aspects of renewable energy siting, including the potential visual impacts and scenic intrusions upon the landscape and seascape, noise and acceptance by the public, as well as developing increasingly interactive GIS-based participatory approaches in the future.

References

Alexander, K.A., Janssen, R., Arciniegas, G., O'Higgins, T.G., Eikelboom, T., and Wilding, T.A., 2011. Interactive marine spatial planning: Siting tidal energy arrays around the Mull of Kintyre. *PLos ONE*. 7(1):e30031. doi: 10.1371/journal .pone.0030031.

American Wind Energy Association, 2004. AWEA electrical guide to utility scale wind turbines. *AWEA Grid Code White Paper. Interim Report.* 31 p.

Andre, H., 1978. Ten years of experience at the 'La Rance' tidal power plant. *Ocean Management.* 4(2–4):165–178.

Archer, T., 2012. GIS and wind turbine siting. Annotated Bibliography. 6 p.

Arnette, A.N., and Zobel, C.W., 2011. The role of public policy in optimizing renewable energy development in the Greater Southern Appalachian mountains. *Renewable and Sustainable Energy Reviews.* 15(8):3690–3702.

Aydin, N.Y., Kentel, E., and Duzgun, S., 2010. GIS-based environmental assessment of wind energy systems for spatial planning: A case study from western Turkey. *Renewable Sustainable Energy Review.* 14:364–373.

Baban, S.M.J., and Parry, T., 2001. Developing and applying a GIS-assisted approach to locating wind farms in the UK. *Renewable Energy.* 24:59–71.

Beaucage, P., Brower, M.C., and Tensen, J., 2014. Evaluation of four numerical wind flow models for wind resource mapping. *Wind Energy.* 17:197–208.

Beleboni, M.G.S., 2014. A brief overview of Microsoft Kinect and its applications. Interactive Mulitmedia Conference, University of Southampton, UK. Retrieved from http://mms.ecs.soton.ac.uk/2014/papers/2.pdf, accessed 20 April 2016.

Belmonte, S., Núñez, V., Viramonte, J.G., and Franco, J., 2009. Potential renewable energy resources of the Lerma Valley, Salta, Argentina for its strategic territorial planning. *Renewable and Sustainable Energy Reviews.* 13(6–7):1475–1484.

Bernard, L., Kanellopoulos, I., Annoni, A., and Smits, P., 2005. The European geoportal – One step towards the establishment of a European spatial data infrastructure. *Computers, Environment and Urban Systems.* 29(1):15–31, http://dx.doi.org/10.1016/j.compenvurbsys.2004.05.009.

Bishop, I., and Lange, E. (Eds.), 2005. *Visualization in Landscape and Environmental Planning: Technology and Applications.* London: Taylor & Francis.

Boroushaki, S., and Malczewski, J., 2008. Implementing an extension of the analytical hierarchy process using ordered weighted averaging operators with fuzzy quantifiers in ArcGIS. *Computers and Geosciences.* 34(4):399–410.

Breton, S.-P., and Moe, G., 2009. Status, plans and technologies for offshore wind turbines in Europe and North America. *Renewable Energy.* 34(3):646–654

Bright, J.A., Langston, R.H.W., Bullman, R., Evans, R.J., Gardner, S., Pearce-Higgins, J., and Wilson, E., 2006. Bird Sensitivity map to provide locational guidance for onshore wind farms in Scotland. *RSPB Research Report No 20.* RSPB. June 2006. 140 p.

Burrough, P., and Masser, I. (Eds.), 1998. *European Geographic Information Infrastructures: Opportunities and Pitfalls.* London: Taylor & Francis.

Carrion, A.J., Estrella, E.A., Dols, A.F., Toro, Z.M., Rodriquez, M., and Ridao, R.A., 2008. Environmental decision-support systems for evaluating the carrying capacity of land areas: Optimal site selection for grid-connected photovoltaic power plants. *Renewable and Sustainable Energy Reviews.* 12:2358–2380.

Cascales, M. del S.G., Arredondo, A.D.M., Corona, C.C., and Lozano, J.M.S. (Eds.), 2014. *Soft Computing Applications for Renewable Energy and Energy Efficiency.* IGI Global.

CEFAS, 2004. Offshore Wind Farms. Guidance note for Environmental Impact Assessment. In respect of FEPA and CPA requirements. Version 2 – June 2004. Prepared by the Centre for Environment, Fisheries and Aquaculture Science (CEFAS) on behalf of the Marine Consents and Environment Unit (MCEU). 48 p.

Chandio, I.A., Matori, A.N.B, WanYusof, K.B., Mir Aftab Hussain Talpur, M.A.H., Balogun, A.-L., and Lawal, D.U., 2013. GIS-based analytic hierarchy process as a multicriteria decision analysis instrument: A review. *Arabian Journal of Geosciences.* 6(8):3059–3066.

Chen, Y., Yu, J., and Khan, S., 2010. Spatial sensitivity analysis of multi-criteria weights in GIS-based land suitability evaluation. *Environmental Modelling and Software.* 25(12):1582–1591.

Chen, Y., Yu, J., Shahbaz, K., and Xevi, E., 2009. A GIS-based sensitivity analysis of multi-criteria weights. *18th World IMACS/MODSIM Congress*, Cairns, Australia. 13–17 July 2009. pp. 3137–3143.

Comber, A., Carver, S., Fritz, S., McMorran, R., Washtell, J., and Fisher, P., 2010. Different methods, different wilds: Evaluating alternative mappings of wildness using fuzzy MCE and Dempster-Shafer MCE. *Computers, Environment and Urban Systems.* 34:142–152.

Crill, C., Gillman, W., Malaney, J., and Stenz, T., 2010. A GIS-driven approach to siting a prospective wind farm in South Central Wisconsin. Unpublished Report. Geography 565. 26 p.

Crown Estate, 2012. UK Wave and Tidal Key Resource Areas Project: Summary Report. Crown Estate. October 2012. 10 p.

Czech, B., and Bauer, P., 2012. Wave energy converter concepts: Design challenges and classification. *IEEE Industrial Electronics Magazine*, 6(2):4–16. doi: 10.1109/MIE .2012.2193290.

Dabiri, J.O., 2011. Potential order-of-magnitude enhancement of wind farm power density via counter-rotating vertical-axis wind turbine arrays. *Journal of Renewable Sustainable Energy.* 3:043104 (2011). 12 p.

Dangermond, J., 2009. GIS: Designing our future, *ArcNews* (2009, summer).

de Sousa Prado, M.G., Gardner, F., Damen, M., and Polinder, H., 2006. Modelling and test results of the Archimedes wave swing. *Proceedings of the Institution of Mechanical Engineers, Part A: Journal of Power and Energy.* 220:855–868, doi:10.1243/09576509JPE284.

Douglas, C.A., Harrison, G.P., and Chick, J.P., 2008. Life cycle assessment of the Seagen marine current turbine. *Proceedings of the Institution of Mechanical Engineers, Part M: Journal of Engineering for the Maritime Environment.* 222:1–12, doi:10.1243/14750902JEME94.

Drewitt, A., and Langston, R., 2006. Assessing the impacts of wind farms on birds. *Ibis.* 148:29–42.

Drews, L.V., 2012. Multi-criteria GIS analysis for siting of small wind power plants – A case study from Berlin. LUMA-GIS Thesis nr 17. Unpublished Masters Thesis. Lund University, Sweden. 134 p.

European Union, 2009. Directive 2009/28/EC of the European Parliament and of the Council of 23 April 2009 on the promotion of the use of energy from renewable sources and amending and subsequently repealing Directives 2001/77/EC and 2003/30/EC (Text with EEA relevance). OJ L 140, 5.6.2009, pp. 16–62 (EN).

Exo, K., Hüppop, O., and Garthe, S., 2003. Birds and offshore wind farms: A hot topic in marine ecology. *Wader Study Group Bulletin.* 100:50–53.

Fang, B., 2015. A GIS-multicriteria approach to analyzing noise and visual impacts of wind farms. Unpublished MSc Thesis. University of Waterloo, Canada. 154 p.

Fayram, A.H., and De Risi, A., 2007. The potential compatibility of offshore wind power and fisheries: An example using bluefin tuna in the Adriatic Sea. *Ocean & Coastal Management.* 50(8):597–605.

Folley, M., Whittaker, T., and Osterried, M., 2004. The oscillating wave surge converter. *The Fourteenth International Offshore and Polar Engineering Conference,* 23–28 May, Toulon, France. (Available at https://www.onepetro.org/conference-paper/ISOPE-I-04-073.)

Frau, J.P., 1993. Tidal energy: Promising projects: La Rance, a successful industrial-scale experiment. *IEEE Transactions on Energy Conversion,* 8(3):552–558. doi: 10.1109/60.257073.

Gill, A.B., 2005. Offshore renewable energy – Ecological implications of generating electricity in the coastal zone. *Journal of Applied Ecology.* 42:605–615.

Graham, S., 2006. Optimisation of the network delivery of marine energy using a geographical information system. Unpublished Doctor of Philosophy thesis. University Of Edinburgh. 274 p.

Green, D.R., and Ray, S.T., 2003. Using GIS for siting artificial reefs – Data issues, problems and solutions: 'real world' to 'real world'. In Green, D.R., and King, S.D. (Eds.) *Coastal and Marine Geo-Information Systems.* Volume 4 of the Series – Coastal Systems and Continental Margins. Springer. pp. 113–131.

Gubbay, S., Earll, R., Gilliland, P., and Ashworth, J., 2005. Case study: Marine spatial planning pilot. Scenario – Tidal stream energy. (Taken from Gubbay, S., Earll, R., Gilliland, P.M. & Ashworth J. 2006. *Mapping European Seabed Habitats (MESH).* Workshop Report and Additional Case Studies. Report to Natural England).

Harrison, J.D., 2012. Onshore wind power systems (ONSWPS): A GIS-based tool for preliminary site-suitability analysis. Unpublished MSc Thesis. University of Southern California. 134 p.

Henderson, A.R., Morgan, C., Smith, B., Sørensen, H.C., Barthelmie, R.J., and Boesmans, B., 2003. Offshore wind energy in Europe – A review of the state-of-the-art. *Wind Energy.* 6:35–52.

Hillring, B., and Krieg, R., 1998. Wind energy potential in southern Sweden – Example of planning methodology. *Renewable Energy.* 13(4):471–479. http://dx.doi.org/10.1016/S0960-1481(98)00027-5.

Hofer, T.M., Sunak, Y., Siddique, H., and Madlener, R., 2014. Wind farm siting using a spatial analytic hierarchy process approach: A case study of the Städteregion Aachen. *FCN Working Paper. No. 16/2014.* 52 p.

Inger, R., Attrill, M.J., Bearhop, S., Broderick, A.C., Grecian, W.J., Hodgson, D.J., Mills, C. et al., 2009. Marine renewable energy: Potential benefits to biodiversity? An urgent call for research. *Journal of Applied Ecology.* 46(6):1145–1153.

Jay, S., 2010. Planners to the rescue: Spatial planning facilitating the development of offshore wind energy. *Marine Pollution Bulletin.* 60:493–499.

Jiang, H., and Eastman J.R., 2000. Application of fuzzy measures in multi-criteria evaluation in GIS. *International Journal of Geographical Information Sciences.* 14:173–184.

Karapetrovic, S., and Rosenbloom, E.S., 1999. Quality control approach to consistency paradoxes in AHP. *European Journal of Operational Research.* 119:704–718.

Koschinski, S., Culik, B.M., Henriksen, O.D, Tregenza, N., Ellis, G., Jansen, C., and Kathe, C., 2003. Behavioural reactions of free-ranging porpoises and seals to the noise of a simulated 2MW windpower generator. *Marine Ecology Progress Series.* 265:263–273.

Lee, A., Chen, H., and Kang, H., 2009. Multi-criteria decision making on strategic selection of wind farms. *Renewable Energy*. 34(1):120–126.

Lejeune, P., and Feltz, C., 2008. Development of a decision support system for setting up a wind energy policy across the Walloon Region (Southern Belgium). *Renewable Energy*. 33:2416–2422.

Lindquist, M., Lange, E., and Kang, J., 2014. An assessment of the potential of using visual abstraction and sound for inclusive geodesign. In Wissen Hayek, U., Fricker, P. and Buhmann, E. (Eds.), *Peer Reviewed Proceedings of Digital Landscape Architecture 2014 at ETH Zurich*. Herbert Wichmann Verlag, VDE VERLAG GMBH, Berlin/Offenbach.

Lip-Wah, H., Ibrahim, S., Sutarji, K., Omar, C.M.C., and Abdullah, A.M., 2012. Review of offshore energy assessment and siting methodologies for offshore wind energy planning in Malaysia. *American International Journal of Contemporary Research*. 2(12):72–85.

Longley, P., 2014. Geoportals (Editorial). *Computers, Environment and Urban Systems*, 29(1):1, http://dx.doi.org/10.1016/j.compenvurbsys.

Malczewski, J., 2004. GIS-based land-use suitability analysis: A critical overview. *Progress in Planning*. 62:3–65.

Malhotra, S., 2011. Selection, design and construction of offshore wind turbine foundations. In *Wind Turbines*, Al-Bahadly, I. (Ed.). InTech. pp. 231–264.

Mariononi, O., 2004. Implementation of the analytical hierarchy process with VBA in ArcGIS. *Computers and Geosciences*. 30:637–646.

Mathew, S., 2006. *Wind Energy Fundamentals Resources Analysis and Economics*. Berlin: Springer.

Miller, A., and Li, R., 2014. A geospatial approach for prioritizing wind farm development in Northeast Nebtraska, USA. *ISPRS International Journal of Geo-Information*. 3:968–979.

Moller, B., 2011. Continuous spatial modelling to analyse planning and economic consequences of offshore wind energy. *Energy Policy*. 39:511–517.

Nadai, A., 2007. Planning, siting and the local acceptance of wind power: Some lessons from the French case. *Energy Policy*. 35(5):2715–2726.

Natural England, 2010. Making Space for Renewable Energy: Assessing On-Shore Wind Energy Development. Natural England. 28 p.

Nobre, A., Pacheco, M., Jorge, R., Lopes, M.F.P., and Gato, L.M.C., 2009. Geo-spatial multi-criteria analysis for wave energy conversion system deployment. *Renewable Energy*. 34:97–111.

Ozerdem, B., Ozer, S., and Tosun, M., 2006. Feasibility study of wind farms: A case study for Izmir, Turkey. *Journal of Wind Engineering and Industrial Aerodynamics*. 94:725–743.

Palaiologou, P., Kalabokidis, K., Haralambopoulos, D., Feldas, H., and Polatidis, H., 2011. Wind characteristics and mapping for power production in the island of Lesvos, Greece. *Computers and Geosciences*. 37(7):62–972.

Petersen, J.K., and Malm, T., 2006. Offshore windmill farms: Threats to or possibilities for the marine environment. *AMBIO: A Journal of the Human Environment*. 35(2):75–80.

Petrasova, A., Harmon, B., Petrasa, V., and Mitasova, H., 2014. GIS-based environmental modeling with tangible interaction and dynamic visualization. *International Environmental Modelling and Software Society (iEMSs) 7th International Congress on Environmental Modelling and Software,* San Diego, CA, D.P. Ames, and N. Quinn (Eds.). (Available at http://www.iemss.org/society/index.php/iemss -2014-proceedings.)

Prasad, R., Bansal, R., and Sauturaga, M., 2009. Some of the design and methodology considerations in wind resources assessment. *IET Renewable Power Generation.* 3(1):53–64.

Prest, R., Daniell, T., and Stendorf, B., 2007. Using GIS to evaluate the impact of exclusion zones on the connection cost of wave energy to the electricity grid. *Energy Policy.* 35(9):4516–4528.

Punt, M.J., Groeneveld, R.A., van Ierland, E.C., and Stel, J.H., 2009. Spatial planning of offshore wind farms: A windfall to marine environmental protection? *Ecological Economics.* 69:93–103.

Riddington, G., Harrison, T., McArthur, D., Gibson, H., and Millar, K., 2008. *The Economic Impacts of Wind Farms on Scottish Tourism.* A Report for the Scottish Government. March 2008. 305 p. (Available at http://www.gov.scot/resource /doc/214910/0057316.pdf.)

Roth, M., and Gruehn, D., 2014. Digital participatory landscape planning for renewable energy – Interactive visual landscape assessment as basis for the geodesign of wind parks in Germany. In Wissen Hayek, U., Fricker, P., and Buhmann, E. (Eds.), *Peer Reviewed Proceedings of Digital Landscape Architecture 2014 at ETH Zurich.* Herbert Wichmann Verlag, VDE VERLAG GMBH, Berlin/Offenbach.

Saaty, T., 2008. Decision making with the analytic hierarchy process. *International Journal of Services Sciences.* 1(1):83–98. DOI: 10.1504/IJSSci.2008.01759.

Salter, S.H. 1974. Wave power. *Nature,* 249:720–724.

Salter, S.H., Taylor, J.R.M., and Caldwell, N.J., 2002, Power conversion mechanisms for wave energy. *Proceedings of the Institution of Mechanical Engineers, Part M: Journal of Engineering for the Maritime Environment* June 1, 2002, 216(1):1–27. doi: 10.1243/147509002320382103.

Scottish Borders Council, 2011. Wind energy. Scottish Borders Council: Supplementary Planning Guidance. May 2011. 52 p.

Serri, L., Sempreviva, A.M., Pontes, T., Murphy, J., Lynch, K., Airoldi, D., Hussey, J. et al., 2012. ORECCA – Resource data and GIS tool for offshore renewable energy projects in Europe. Report. Results of the FP7 ORECCA Project Work Package 2. 111 p.

Shang, H., and Bishop, I., 2000. Visual thresholds for detection, recognition and visual impact in landscape settings. *Journal of Environmental Psychology.* 20:125–140.

Shouman, M., ND. Industrial site selection: An intelligent GIS-based decision analysis approach. Unpublished Paper. 16 p. (Available at https:// www.academia.edu/1480307/INDUSTRIAL_SITE_SELECTION_AN _INTELLIGENT_GIS-_BASED_DECISION_ANALYSIS_APPROACH.)

Simao, A., Densham, P.J., and Haklay, M., 2009. Web-based GIS for collaborative planning and public participation: An application to the strategic planning of wind farm sites. *Journal of Environmental Management.* 90:2027–2040.

Soerensen, H.C., and Hansen, L.K., 2001. Social Acceptance, environmental impact and politics. Concerted action on offshore wind energy in Europe. Draft Report. NNE5-1999-00562.

Sparkes, A., and Kidner, D., 1996. A GIS for the environmental assessment of windfarms. *Proceedings 11th Esri European User Conference.* October 1996.

Sperling, K., Hvelplund, F., and Mathiesen, B.V., 2010. Evaluation of wind power planning in Denmark – Towards an integrated perspective. *Energy*. 35:5443–5454.

Steinitz, C., 2012. A Framework for Geodesign: Changing Geography by Design. Redlands, CA: Esri Press.

Sunak, Y., Hofer, T., Siddique, H., Madlener, R., and De Doncker, R.W., 2015. A GIS-based decision support system for the optimal siting of wind farm projects. *E.ON Energy Research Center Series*, 7(2). Rwth Aachen University. 71 p.

Taha, R.A., and Daim, T., 2013. Multi-criteria applications in renewable energy analysis, a literature review. In Daim, T.U., Oliver, T., and Kim, J. (Eds.). *Research and Technology Management in the Electricity Industry, Green Energy and Technology*. London: Springer-Verlag. pp. 17–30.

Tegou, L.I., Polatidis, H., and Haralambopoulos, D.A., 2010. Environmental management framework for wind farm siting: Methodology and case study. *Journal of Environmental Management*. 91:2134–2147.

Thomsen, F., Lüdemann, K., Kafemann, R., and Piper, W., 2006. Effects of offshore wind farm noise on marine mammals and fish. Report. Cowrie Ltd. Hamburg, Germany. 62 p.

University of Newcastle, 2002. Visual assessment of windfarms best practice. Scottish Natural Heritage. Commissioned Report F01AA303A. 79 p.

van Haaren, R.V., and Fthenakis, V., 2011. GIS-based wind farm site selection using spatial multi-criteria analysis (SMCA): Evaluating the case for New York state. *Renewable Sustainable Energy Review*. 15:3332–3340.

Wahlberg, M., and Westerberg, H., 2005. Hearing in fish and their reaction to sounds from offshore wind farms. *Marine Ecology Progress Series*. 288:295–309.

Wilhelmsson, D., Malm, T., and Ohman, M.C., 2006. The influence of offshore windpower on demersal fish. *ICES Journal of Marine Science*. 63:775–784.

Williamson, I., Rajabifard, A., and Feeney, M.-E., 2003. *Developing Spatial Data Infrastructures: From Concept to Reality*. London: Taylor & Francis.

Wright, R., Ray, S., Green, D., and Wood, M., 1998. Development of a GIS of the Moray Firth (Scotland, UK) and its application in environmental management (site selection for an 'artificial reef'). *The Science of the Total Environment*. 223:65–76.

Yemm, R., Pizer, D., Retzler, C., and Henderson, R., 2011. Pelamis: Experience from concept to connection. *Phil. Trans. R. Soc. A*. 370:365–380; doi: 10.1098/rsta.2011 .0312.

Zhang, B., 2009. Ancient Chinese windmills. In Yan, H-S. and Ceccarelli, M. (Eds.), *International Symposium on History of Machines and Mechanisms*, pp. 203–214. Netherlands: Springer. 10.1007/978-1-4020-9485-9_15.

13

Geoinformatics for Fisheries Management

Tom Nishida, Kiyoshi Itoh, Albert Caton and Darius Bartlett

CONTENTS

13.1 Introduction

Ever since the dawn of humanity, fish have played an important part in the diet of most coastal communities. Today, fish and fish products are the most extensively traded commodities in the global food sector (United Nations Environment Programme, [UNEP], 2013). Data from the Food and Agriculture Organization of the United Nations (FAO) suggest that per capita apparent fish consumption worldwide nearly doubled, from an average of 9.9 kg in the 1960s to 19.2 kg in 2012 (FAO, 2014) (see Table 13.1). Drivers for this increase include world population growth, rising incomes and urbanization, greater efficiencies of targeting and capture within the fisheries sector, major expansion of aquaculture worldwide and transformations in the processing and distribution of fisheries produce.

As well as being of direct importance to human society as a food source and a basis for economic activity, fish populations also deliver important ecosystem services (Holmund and Hammer, 1999) including several that are fundamental to the health of the planet and ultimately to human society. Overexploitation of fish resources, and the large number of nontarget

TABLE 13.1

World Fisheries and Aquaculture, 2007–2012

	2007	2008	2009	2010	2011	2012
Production						
Capture						
Inland	10.1	10.3	10.5	11.3	11.1	11.6
Marine	80.7	79.9	79.6	77.8	82.6	79.7
Total capture	90.8	90.1	90.1	89.1	93.7	91.3
Aquaculture						
Inland	29.9	32.4	34.3	36.8	38.7	41.9
Marine	20.0	20.5	21.4	22.3	23.3	24.7
Total aquaculture	49.9	52.9	55.7	59.0	62.0	66.6
Total World Fisheries	140.7	143.1	145.8	148.1	155.7	158.0
Utilization						
Human consumption	117.3	120.9	123.7	128.2	131.2	136.2
Non-food uses	23.4	22.2	22.1	19.9	24.5	21.7
Population (billions)	6.7	6.8	6.8	6.9	7.0	7.1
Per capita food fish supply (kg)	17.6	17.9	18.1	18.5	18.7	19.2

Source: Food and Agriculture Organization of the United Nations (FAO) 2014. *The State of World Fisheries and Aquaculture*, Vol. 2014. doi:92-5-105177-1.

Note: Utilization data for 2012 based on FAO provisional estimates.

species being caught worldwide along with the intended catch, puts these ecosystem services at risk and also reduces fish production, which may ultimately force the closure of otherwise profitable fisheries (Gaines and Costello, 2013: 15859) with consequent negative social and economic repercussions (UNEP, 2013).

The potential of geographical information systems (GIS) and remote-sensing technologies for improved fisheries management has been recognized since the early 1990s (FAO, 1985; Meaden and Kapetsky, 1991; Simpson, 1992; Meaden, 1996). FAO was an important early adopter and champion of fisheries-related GIS (FAO, 1985; Meaden and Aguilar-Manjarrez, 2013), leading Meaden (2000) to observe that 'progress in applying GIS to fisheries management has been rapid, diverse and imaginative' and that 'it will be most surprising if this does not continue in the immediate future' (Meaden, 2000: 222).

This chapter draws on papers presented at two recent symposia on GIS/ Spatial Analyses in Fishery and Aquatic Sciences (the 5th Symposium held in Wellington, New Zealand, in 2011; and the 6th, held in Tampa, Florida, in 2014), supplemented by other relevant studies, to outline some of the current ways in which geoinformatics are contributing to improved fisheries management and the quest for sustainability in the sector. Readers seeking more technical guidance on the construction and use of GIS and remote-sensing applications for fisheries management are referred to recent publications by Morales et al. (2011) and by Meaden and Aguilar-Manjarrez (2013).

For the purposes of the present chapter, we define *fisheries management* as including both fisheries *resource* management and also *fishing-activity-related* management, and examine these from the interrelated perspectives of 'tools', 'habitats' and 'fisheries management' (Figure 13.1).

Responsibility for the management of fisheries usually rests with governments either independently or, where resources extend beyond national boundaries, consequential to international agreements. Governments may manage unilaterally or may establish formal internal entities that, for example, may include industry or scientific representatives. Normally a fisheries

FIGURE 13.1
Relations among spatial data, tools, habitats, fishing activities and fisheries management.

manager will be a government official with legislated responsibility to manage fisheries resources and the activities of fishers and industry. Hence, managers will be the primary users of geoinformatics. However, fisheries scientists are users, as they need to conduct scientific spatial analyzes to assist managers; and fishers are also users, applying geoinformatics to search for fishing grounds. Thus, in a fisheries management context, geoinformatics provide integrated tools for managers, scientists and fishers alike.

13.2 Tools

The field of geoinformatics provides a support for fisheries management in a number of ways. Of particular relevance are hands-on tools for spatial data collection, and the systems to support the processing and dissemination of these data. This section examines some of these.

13.2.1 Spatial Fisheries Data Collection Systems

Spatial fisheries data (such as catch and effort) are essential for identifying habitats and for fisheries management. These data are normally collected by means of, *inter alia*, logbooks, observers, dockside samplings and national statistics-collection systems. One recent development has been the increasing availability and use of electronic logbooks, which have demonstrated effectiveness for collecting real-time data.

In addition, especially in developing countries where there are difficulties collecting basic spatial fisheries (catch and effort) data, global positioning system (GPS) data loggers have been developed and applied (Alabsi, 2014). These devices provide easy-to-use and efficient means of collecting spatial fisheries data along with sea temperature and depth information (Figure 13.2) (Itoh and Nishida, 2013a).

13.2.2 Remote Sensing

Information from remote sensing has been actively used in the fisheries sector for the last three decades to collect information about fish habitats, behaviour and distribution. Each of these is described in more detail below.

13.2.2.1 Acoustic and Satellite Remote Sensing

Sea water is opaque to most forms of electromagnetic energy, including in the optical and microwave parts of the spectrum, which raises significant challenges for the use of remote sensing techniques based on these (see Morales et al., 2011 and Chapter 4 by Lück-Vogel of the present volume). Two

Consideration on coastal and small-scale fisheries resource management
using the GPS data logger and Marine Explorer

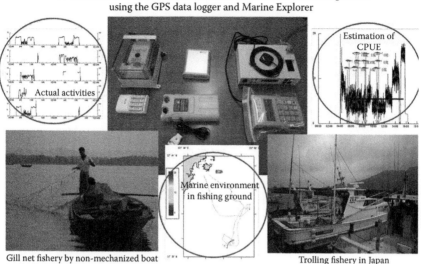

Gill net fishery by non-mechanized boat
in Chilika Lagoon

Trolling fishery in Japan

Environmental Simulation Laboratory, Inc.

FIGURE 13.2
The use of a GPS data logger in India. (From Itoh, K., and Nishida, T., Coastal and small-scale fisheries resource management using RealMC, a real-time GPS data-logger and marine-monitoring system. In *GIS/Spatial Analyses in Fishery and Aquatic Sciences* [Vol. 5], Nishida, T., and A.E. Caton, Eds., 2013a, pp. 93–96. Saitama, Japan: International Fishery GIS Society.)

approaches in particular have been developed to overcome these difficulties. The first is to use acoustic remote sensing, based on the transmission of soundwaves through the water column, instead of using light. The use of sidescan and multibeam sonar has already been discussed in the context of seabed mapping in Chapter 2 by Scott et al. and Chapter 3 by Murphy and Wheeler of the present volume. The same principles and family of techniques may also be applied to detecting and assessing phenomena that occur within the water column itself. Sasakura et al. (2016) and Itoh et al. (2016), for example, have demonstrated the use of real-time acoustic remote-sensing devices for measuring environment data such as sea temperature and depth, and for transmitting data to boats ultrasonically (Figure 13.3).

The second approach is to use remote sensing of the sea surface in order to infer information about habitats through one or more proxy indicators. Yan (2014), Yan et al. (2014) and Alabsi (2014), for example, developed data-collection systems for sea-surface temperature (SST) and productivity (chlorophyll-a) using satellite remote sensing with GPS. Alabsi (2014) found Indian mackerel (*Rastrelliger kanagurta*) concentrated in high-chlorophyll waters in the Red Sea off Yemen. Some of the important habitat parameters examined included SST, chlorophyll-a, sea surface salinity, depth and sea surface currents. Escudero (2016) integrated data on chlorophyll-a, SST, sea level anomaly and salinity,

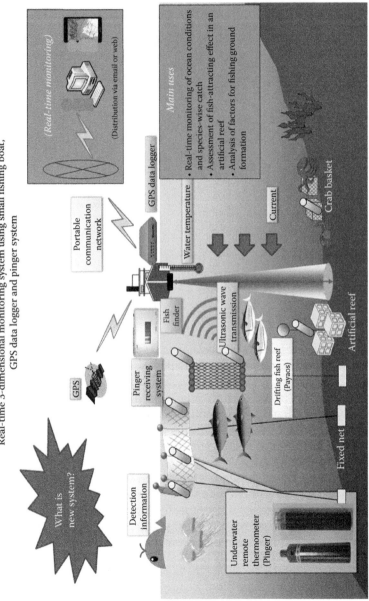

FIGURE 13.3

Real-time monitoring system for underwater environments using ultrasonic wave technology. (From Itoh, K., Nishida, T., and Sasakura, T., Real-time monitoring system for underwater environments using ultrasonic wave technology. In *GIS/Spatial Analyses in Fishery and Aquatic Sciences* [Vol. 6], Nishida, T., and A.E. Caton, Eds., 2016, pp. 209–214. Saitama, Japan: International Fishery GIS Society.)

obtained through both acoustic and satellite remote sensing, within a geographical information system to estimate the spatial distributions and biomass of anchovy off the coast of Peru (see also Escudero and Rivera, 2011).

13.2.2.2 Acoustic Telemetry and Video Systems

Acoustic telemetry and video systems have been used in a number of studies, to monitor and collect fish behaviour information. Using these devices, the position of fish can be determined in three dimensions, and through the use of GIS technologies, compared to habitat distributions. For example Dutka-Gianelli et al. (2014) used kernel density estimates and other techniques within the ArcGIS software to analyze data captured by means of acoustic telemetry tags to examine the movement of largemouth bass (*Micropterus salmoides*) and common snook (*Centropomus undecimalis*), two species of gamefish commonly found together within coastal river waters in Florida. Similarly, Hollensead (2014) monitored southern flounder (*Paralichthys lethostigma*) in the New River estuary in North Carolina; Brinton (2014) studied patterns of Atlantic stingray (*Dasyatis sabina*) in tidal creeks near Savannah, Georgia and Taylor (2014) used video systems to monitor and collect information on fisheries resources in north-eastern waters in the United States (Figure 13.4).

13.2.3 Spatial-Data Processing and Dissemination Systems

Spatial data on fisheries and habitats may be processed by various GIS software packages and/or through computer languages (C, C++, R, Visual Basic, etc.) before visualization and dissemination. Web-GIS and Web Atlases are especially powerful tools for disseminating visualization products, in a simple way, for sharing common information globally (see Wright et al., 2011 for a detailed examination of coastal web atlases, their construction and uses). A growing number of coastal atlases are under development around the world, many of which include fisheries information as a key element. Examples include the European Atlas of the Seas (http://ec.europa.eu/maritimeaffairs /atlas/maritime_atlas), the Gulf Atlas (http://gulfatlas.noaa.gov/) (Figure 13.5) and the African Marine Atlas (http://www.africanmarineatlas.org/).

13.2.4 Visualization

Visualization (geovisual analyzes) for fisheries management, as a part of geoinformatics, refers to the generation of maps and other graphics output from GIS, both in hard (printed) copy, and in the form of digital and sometimes interactive maps presented on the screen of a computer or other display device. While the topic of effective map design and visualization output from GIS has received much attention in the general geoinformatics literature (e.g. Kraak and Ormeling, 1996, 2010), in many cases the maps produced in fisheries GIS studies would merit significant improvement. It seems certain that future moves

Sixth International Symposium on GIS/Spatial Analyses in Fishery and Aquatic Sciences 2014
HABCAM Instrument System

Operational Surveys   HABCAM V2 vehicle   2007–2014

(a)

Sixth International Symposium on GIS/Spatial Analyses in Fishery and Aquatic Sciences 2014
HABCAM Instrument System

Operational Surveys–Results   HABCAM V2 Vehicle   2007–2014

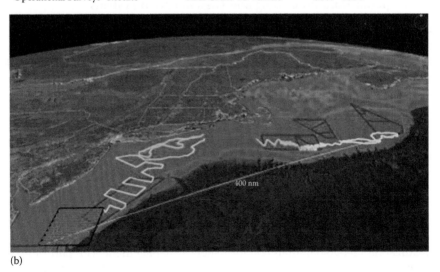

(b)

FIGURE 13.4
Habitat mapping camera (HABCAM) video monitoring system (a) and surveyed area (b)
in northeast United States. (From Taylor, R., Ten years of development and operation of the
HABCAM instrument system [HABCAM: HABitat mapping CAMera system]. Paper pre-
sented at *6th International Symposium on GIS/Spatial Analyses in Fishery and Aquatic Sciences*,
Tampa, FL, 2014.)

gulfatlas.noaa.gov

2 ways to browse and search for data

FIGURE 13.5

Gulf of Mexico Data Atlas (Web-GIS). (From Rose, K., Gulf of Mexico Data Atlas: Data discovery and access. Paper presented at *6th International Symposium on GIS/Spatial Analyses in Fishery and Aquatic Sciences*, Tampa, FL, 2014.)

will be made within the industry towards making maps that are more readily interpreted, particularly where the users of the maps may have had no formal training in map-reading, and where the conditions of use (e.g. on the bridge of a fishing vessel at sea) may raise particular challenges not encountered elsewhere.

13.2.5 Instrumentation

A number of recent developments have been directed towards the use of a wider range of digital recording devices, as a means of securing data to be used as basic information for fisheries management. These vary from types of acoustic underwater image detection through to data buoys and electronic tags inserted into fishes and even whales (see Zhang et al., Chapter 5, this volume). Because of the urgent need to find ways of increasing data inputs to fisheries-oriented geoinformatics applications, it is likely that instrumentation will receive a lot of attention in the coming decades. This is even more important for dealing with four-dimensional data in aquatic environments that are constantly mobile – as are the occupants of these environments.

13.2.6 Big Data

Until fairly recently, most GIS applications were hampered by a lack of available data. In recent years the situation has become reversed, and many sectors of geoinformatics application are now grappling with the problem of 'Big Data', defined here as datasets whose volume, complexity and dynamic nature preclude storage and analysis through conventional means.

While most datasets relating to fisheries do not normally fit these descriptions as yet, the emergence and growing use of sensors, improvements in satellite and other remote-sensing technologies, the availability of autonomous vessels and improved infrastructures for data transmission and integration, all suggest that Big Data may become a challenge for the fisheries geoinformatics developer or user in the foreseeable future. Furthermore, the fisheries sector is increasingly seeing a diversity of separate and often independently developed geoinformatics technologies being used in combination. This raises numerous challenges, especially those relating to technical and semantic interoperability, and the effective use of data that are increasingly detailed in scale and regularity of update.

13.3 Habitats

Habitat information is essential in the application of geoinformatics for fisheries management. With knowledge about habitats we can understand fish distributions and area-specific biology, ecology, abundances and life history. As well as being a key parameter for marine spatial planning and the designation of marine protected areas, habitat information also enables decisions on how to manage fisheries resources by area. For example, through knowing the availability and status of habitats, managers can more easily establish evidence-based and area-specific conservation measures when the status of a fish stock is compromised: one such measure might be time-and-area-specific no-fishing zones to protect juveniles.

As well as being interested in the nature and distribution of fish and habitats at the present day, fisheries managers must also consider the projected impacts of future climate change on aquatic species distributions. Such information is essential for forward planning and longer-term management decisions. The Intergovernmental Panel on Climate Change provides access to different scenarios where projected changes of sea temperature are computed (Silva et al., 2015). In a fisheries context, the development of more accurate long-term predictions of different oceanographic and climate parameters will have major implications and potential benefits for analyzing changes in habitats, fish population distributions and their dynamics and other related aspects.

Four particular aspects of habitats are important in fisheries management, namely essential fish habitat, habitat-suitability modelling, habitat abundance modelling and spatial prediction. The role of geoinformatics in each of these is discussed in the following sections.

13.3.1 Essential Fish Habitat

The concept of an essential fish habitat is a fundamental component of ecosystem-based fisheries management. In the United States, the 1996 Sustainable Fisheries Act defines essential fish habitat as 'those waters and substrate necessary to fish for spawning, breeding, feeding and/or growth to maturity' (Levin and Stunz, 2005; see also Rosenberg et al., 2000 and Peterson et al., 2000).

Geoinformatics technologies can greatly facilitate the collection, synthesis, analysis and visualization of data that identify and support the management of essential fish habitat (Valavanis et al., 2008). They can also enable ecosystem data to be related to catch, catch rate (i.e. 'catch per unit of fishing effort'; CPUE), biological and ecological information and so on, to better understand the fundamental characteristics of fish habitat that have relevance for fisheries resource management. The ultimate goal of these studies is to improve our understanding of the nature and dynamics of fish populations, which is pivotal to informed fisheries management decisions.

While many studies of essential fish habitat focus on distribution and abundance of fisheries resources, they can also provide useful information about the species of fish involved, and in the interactions between fish and fishers. Nóbrega and Lessa (2016), for example, have described unique work on biological and mortality aspects of the spatial variation in age structure and survival of greater amberjack (*Seriola dumerili*); and Montes et al. (2012) analyzed spatio-temporal co-occurrences of users (recreational fishers) and North Atlantic right whales (*Eubalaena glacialis*) in southeast United States in order to assess and manage potential conflicts of interest.

13.3.2 Habitat-Suitability Modelling

Unlike essential fish habitat studies, which deal with one or two simple parameters to represent habitats, and where an association between a fish population and its essential habitat is known, it is possible to use GIS, remote sensing and spatial analysis to create models of fish habitat suitability (Morris and Ball, 2006; Valavanis et al., 2008). There are several models available. Rubec et al. (2014a,b, 2016b), for example, demonstrated how upwelling and current velocity enhanced the abundance of pink shrimp (*Farfantepenaeus duorarum*) on the southeast coast of the United States. One conclusion drawn was that applying habitat-suitability modelling to the West Florida Shelf and Florida estuaries was more cost-effective than undertaking a mapping of benthic conditions in the study area. Elsewhere, Song (2014) presented an integrated habitat index, modelling CPUEs using quantile regression to determine potential habitat for albacore (*Thunnus alalunga*) near the Cook Islands in the South Pacific; while Yen (2014) used habitat-suitability index models to map skipjack tuna (*Katsuwonus pelamis*) in the Central and Eastern Pacific, and related their distributions to the influence of El Niño–Southern Oscillation events.

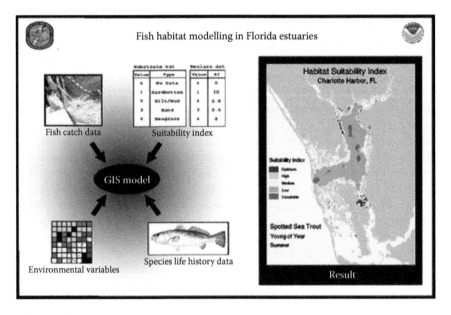

FIGURE 13.6
Habitat suitability modelling (HSM) in Florida. (From Rubec, P.J., Lewis, J., Reed, D. and Westergren, C., An evaluation of transferability of habitat suitability models between Tampa Bay and Charlotte Harbor, Florida. Paper presented at *6th International Symposium on GIS/ Spatial Analyses in Fishery and Aquatic Sciences*, Tampa, FL, 2014a.)

The transferability study in Tampa Bay and Charlotte Harbor, Florida, by Rubec et al. (2016a) used habitat-suitability modelling in which CPUEs in suitability functions were weighted according to their relative magnitude across environmental gradients (Figure 13.6). The study demonstrated that models with fewer effective factors could estimate predictive parameters more accurately than the one using all parameters. This implies that habitat-suitability modelling can be done at less cost (fewer parameters to collect) once a few effective parameters are identified. Elsewhere, Good and Peel (2007) have used maximum-entropy methods to estimate area-specific abundances for prawn stocks in Australia, while Worthington (2014) applied MaxEnt to freshwater habitats, and effectively modelled and mapped distributions of Arkansas River shiner, concluding that changes in flow and river fragmentation explained its decline.

13.3.3 Habitat Abundance

Habitat abundance estimates are derived from catch rates that are analyzed using approaches such as CPUE standardization, with the focus being on analysis of availability of a specific habitat or range of habitats within a given area, with this information then being linked to fish populations.

Both habitat-suitability modelling and CPUE provide population abundance. However, the modelling will provide abundance estimates that are more dependent upon marine ecosystem information, whereas CPUE analysis takes less account of spatial ecological data. Normally, habitat-suitability modelling uses CPUE and ecological information to identify suitable locations of fish concentration, CPUE being fundamental to habitat-suitability modelling. Hence, in some cases the two approaches are highly interrelated and the distinction is blurred.

13.3.4 Spatial Prediction

Integration of the knowledge obtained through essential-fish-habitat determination, habitat-suitability modelling and habitat-abundance estimation enables alternative management scenarios to be evaluated, and their outcomes predicted. GIS is an ideal tool to facilitate this. The three following case studies are examples of the many that demonstrate this approach.

In the first, Itoh and Nishida (2010) successfully developed pin-point fishing-ground prediction techniques (Figure 13.7) using a specialist GIS package called Marine Explorer, developed specifically for fisheries and oceanic applications (Itoh and Nishida, 2001). In the second, Farmer and Karnauskas (2013) developed habitat-suitability modelling that predicted the presence of speckled hind (*Epinephelus drummondhayi*) and Warsaw grouper (*Hyporthodus nigritus*) along the shelf edge off the south-eastern United States for possible creation of marine protected areas. And thirdly, Silva et al. (2015) used the Idrisi GIS and remote-sensing software developed by Clark Labs (Worcester, MA) to predict changes in the spatial distributions and abundances of swordfish (*Xiphias gladius*) and common sardine (*Strangomera bentincki*) off the coast of Chile, under the Intergovernmental Panel on Climate Change A2 scenarios of climate change. Predictions of this nature can provide important inputs to planning and managing fisheries activities, particularly over the longer term.

13.4 Spatially Based Fisheries Management

13.4.1 Fisheries Resource Management Based on Habitats

In the past, fisheries resource management has generally concentrated on using stock assessment results to decide total allowable catch (TAC), catch and effort limits, quota allocation and so on, without regard to spatial aspects. This has been routine practice in, for example, tuna and demersal regional fisheries management organizations such as, for tunas, the Inter-American Tropical Tuna Commission, the Indian Ocean Tuna Commission,

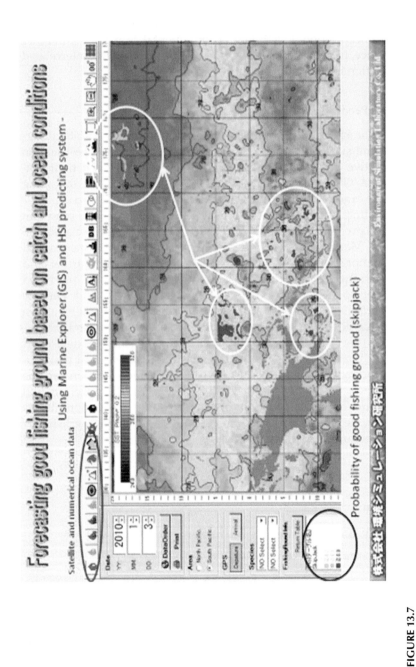

FIGURE 13.7
Pinpoint fishing ground prediction system using Marine Explorer GIS software and habitat-suitability-index models. (From Itoh, K., and Nishida, T., Pinpoint prediction of good fishing grounds using Jack Pot Skipper. In *GIS/Spatial Analyses in Fishery and Aquatic Sciences* [Vol. 5], Nishida, T., and A.E. Caton, Eds., 2013b, pp. 81–84, Saitama, Japan: International Fishery GIS Society.)

the Western and Central Pacific Fisheries Commission and the Commission for the Conservation of Southern Bluefin Tuna and, for demersal species, the Northwest Atlantic Fisheries Organization, the Commission for the Conservation of Antarctic Marine Living Resources and the North East Atlantic Fisheries Commission.

However, by taking spatial considerations into account, there is scope to focus on particular zones such as spawning grounds, 'hot spots', juvenile areas and so on, facilitating conservation according to habitats. Geoinformatics can be used to generate a structured TAC scheme that can improve resource management. For example, in British Columbia, Canada, Yamanaka and Flemming (2013) used fisheries and environmental data to identify habitat areas for small quillback rockfish (*Sebastes maliger*). This enabled the operation of a marine protected area for a limited period, and succeeded in protecting the stock from overfishing.

13.4.2 Spatial Monitoring and Management of Fishing Activities for Compliance

In Chapter 9 of the present volume, Saunders has discussed the role of geoinformatics technologies in policing the seas in general, and in ensuring compliance with the many national, regional and international laws and regulations that apply to maritime spaces. Fishing is one activity that is increasingly being subject to regulations governing, *inter alia*, the equipment the vessels may carry, the numbers and species of fish they may harvest and the locations in which they may operate. However the small size of nearshore fishing vessels in particular, along with the characteristics of the sea surface on which they operate, present numerous challenges for effective monitoring of their activities (Kourti et al., 2005).

A wide and growing range of spatially enabled technologies is contributing to this effort. One of these is the vessel monitoring system. It consists of three interoperating elements (FAO, 2016), namely

- **The ship-board component** comprising an aerial (antenna), transceiver and power source and often an integrated GPS unit, that are installed on the vessel and assigned a unique identifier;
- **A communications component**, which typically uses satellites such as the polar orbiting US Argos or the geostationary Inmarsat C and D+ services to relay locational information and
- **A fisheries monitoring centre** operated by one or more national or regional fisheries management organizations.

Under European Union legislation, all fishing vessels longer than 15 m that are operating in European waters are obliged to be fitted with a monitoring unit, and to report their positions at regular (usually hourly) intervals

(European Council, 1992, 2002). Similar requirements have been introduced in other jurisdictions, and have given enormous insight (Kourti et al., 2005) into the dynamics and *modus operandi* of fisheries in many parts of the world. Devillers (2014), for example, demonstrated the increasing utilization and availability of VMS data in areas managed by the North Atlantic Fisheries Organization and by Canada, respectively. However, the use of VMS is largely dependent on the cooperation of the owners or operators of a ship and, therefore, particularly when illegal, unreported and unregulated operations are concerned, the system is most effective when combined with remote sensing (Kourti et al., 2005).

Soon after the launch of Landsat in the 1970s, McDonnell and Lewis (1978) investigated the ability of that platform's multi-spectral scanner to assist in ship detection, while Burgess (1993) explored the suitability of both Landsat Thematic Mapper (TM) and SPOT data for the purpose. However, optical remote sensing, especially from space-borne platforms, is only effective during daylight hours, and is also impacted by cloud cover. Furthermore, the narrow swath-width of high-resolution optical remote sensing makes it difficult to apply to broad-scale surveillance of wide expanses of ocean (Corbane et al., 2010). Microwave radar imagery is not affected by these impediments, and has been widely used in recent years to provide additional information alongside the VMS for fishing-vessel detection and monitoring. Gower and Skey (2000), for example, demonstrated the potential of Radarsat ScanSAR synthetic-aperture-radar imagery for detecting small boats (lengths between 18 and 25 m) such as are typically used for nearshore fishing, and O'Shea (2007) has outlined how these data may be used in fisheries control to detect illegal, unreported and unregulated vessels.

More recently, Corbane et al. (2010) have revisited the potential of optical imagery for ship detection, noting that '[c]ompared to the large amount of investigations on the feasibility of satellite-based SAR for ship detection purposes, far less research and development activity has taken place in automatic detection and classification of vessels using optical imagery than using SAR imagery' (Corbane et al., 2010: 5838). They conclude that a combination of high-resolution optical imagery, as is now available from modern platforms, along with synthetic-aperture radar, can offer synergies and advantages for ship detection and monitoring that are lacking when one or another method is used alone.

13.4.3 Fisheries Management Incorporating Socioeconomics

Effective fisheries management needs to be based on a sound knowledge and appreciation of human factors as well as environmental ones. This is another area where the integrative and decision-support abilities of modern geoinformatics can make significant contributions.

A number of examples and case studies have addressed spatially based fisheries management incorporating socioeconomics, and illustrate this point well.

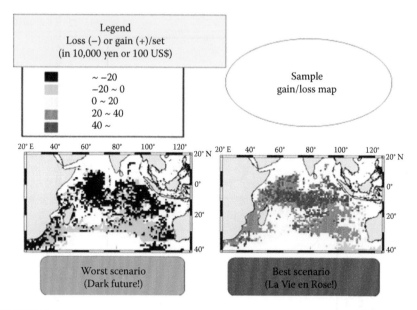

FIGURE 13.8
Integrated ecosystem approach for sustainable tuna longline fisheries, considering fishing conditions, socioeconomics (running costs, fish and fuel prices) and ecosystem (depredation and bycatch). (From Nishida, T., Murakami, T., Shiba, Y., Ooyama, M., Miura, N. and Itoh, K., Integrated ecosystem approach for sustainable tuna longline fisheries [Case study: Tropical tuna in the Indian Ocean]. *Fourth International Symposium on GIS/Spatial Analyses in Fishery and Aquatic Sciences, Abstract Proceedings*. Universidade Santa Úrsula, Rio de Janeiro, Brazil, 25–29 August 2008.)

Nishida et al. (2008), for example, describes an integrated ecosystem approach for sustainable tuna longline fisheries in the Indian Ocean, in which interrelationships between fishing conditions, socioeconomics (running costs, fish and fuel prices) and ecosystem (depredation and bycatch) factors were examined (Figure 13.8). Meanwhile, Cruz et al. (2016) demonstrated the value of participatory GIS for resource management in Belize, and suggested that engagement with the community through an ecosystem framework might widen the economic opportunities available in the study area. Interesting examples of the role of participatory GIS in fisheries management have also been provided from the Seychelles (Koike, 2014), Yemen (Alabsi, 2014) and Morocco (Wahbi et al., 2014), all three of these being countries where difficulties might be faced when trying to involve stakeholders in the management process (see also Chapter 7 by Goldberg, D'Iorio and McClintock in this volume).

13.4.4 Ecosystem-Based Fisheries Management

Whereas traditional fisheries management normally focused on maximizing the catch of a single target species, ecosystem-based management presents a

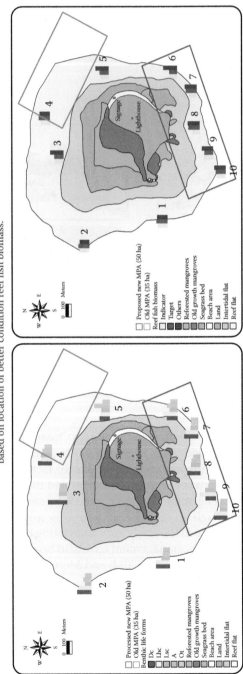

FIGURE 13.9

Coastal habitat mapping of Nogas Island, Philippines, for conservation and management (MPA: marine protected areas). (Adapted from Aguilar, G.D., and Villamor S., Habitat mapping for conservation and management of Nogas Island, Philippines. In *GIS/Spatial Analyses in Fishery and Aquatic Sciences* [Vol. 4], Nishida, T., and A.E. Caton, Eds, 2010, pp. 307–318. Saitama, Japan: International Fishery GIS Society.)

more holistic and longer-term approach that starts with consideration of the entire ecological assemblage (Hall and Mainprize, 2004; Pikitch et al., 2004). It is now the dominant paradigm for sustainable fisheries worldwide, and is the subject of numerous international agreements (Curtin and Prellezo, 2010), even though there is still considerable debate in the literature regarding the appropriate methods for achieving it, or even the indicators, reference points and parameters needed to gauge progress (Hall and Mainprize, 2004; Pitcher et al., 2009).

The potential contribution of geoinformatics to ecosystem-based fisheries management, and especially the value of GIS as an integrating technology for data management, analysis and output of data products, was recognized early on (see, e.g. Nishida and Booth, 2001). More recently, Aswani (2011) gives examples and case studies from Oceania, and particularly the Solomon Islands, to show how GIS may be used effectively as a means of eliciting indigenous ecological knowledge that might then be incorporated into hybrid customary and ecosystem-based management strategies for artisanal and other fisheries. Similar work has been documented by Aguilar and Villamor (2010) in the Philippines (Figure 13.9) and numerous recent studies in other parts of the world.

13.4.5 Cooperation with Fishers

There is evidence that better cooperation is being achieved between fishers and management, with geoinformatics technologies and methods playing a key role in this development. Increasingly, this also involves the use of the World Wide Web (e.g. Bay of Bengal Program at http://www.bobpigo.org), as access to the Internet expands around the world. It seems reasonable to expect that, as the situation with respect to the availability of fish for commercial extraction significantly declines, fishers will see the need for increasing management as a way to rationalize the situation. This better cooperation will make it easier to get more reliable information and data on social and economic aspects of fisheries, including catch and effort data.

13.5 Competition for Marine Space and Marine Spatial Planning

The fisheries industry represents just one group of stakeholders, among many, with interests in the management of marine space. Together, these activities spread across more than 70% of the planet's surface and can, and frequently do, have potential for conflicts and competition. Marine spatial planning seeks to provide frameworks for reconciling these differences (see Chapter 8 by McWhinnie and Gormley, this volume). The need for such planning is crucial to the survival not just of fisheries but of many world marine ecosystems. Geoinformatics approaches will be central to this process. Tools, data and methods developed specifically for fisheries management will also increasingly provide key inputs to wider-ranging decision making, and must in the future be designed with this extended use in mind. This raises important issues of data compatibility and integration, system interoperability, the implementation of spatial data standards and infrastructures, as well as the political and administrative instruments and frameworks needed to support these goals.

13.6 Summary

As has been outlined in this chapter, geoinformatics play an increasingly important role in many aspects of fisheries management. Nevertheless, with growing human populations to feed, changes to the marine environment due to climate change and other issues, and multiple potential or actual conflicts between users of marine resources a paradigm shift is needed to support the spatial assessment of fish populations, ecosystem-based fisheries management and marine spatial planning. There is a continued need for interdisciplinary research using geoinformatics and GIS wherein fishers, oceanographers, marine geologists, GIS analysts and fisheries scientists

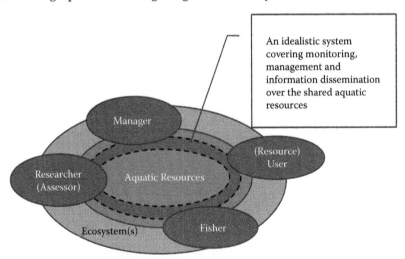

FIGURE 13.10
Summary of the integration required for implementing fisheries management that incorporates geoinformatics and GIS, taking into account stakeholders, aquatic resources, fisheries monitoring, management and information dissemination.

collaborate to support management of marine ecosystems. Furthermore, new approaches are urgently needed to determine how fisheries species and practices are responding to climate change. Figure 13.10 summarizes the integration required for implementing fisheries management that incorporates geoinformatics, taking into account stakeholders, aquatic resources, fisheries monitoring, management and information dissemination.

References

Aguilar, G.D., and Villamor, S. 2010. Habitat mapping for conservation and management of Nogas Island, Philippines. In *GIS/Spatial Analyses in Fishery and Aquatic Sciences* (Vol. 4), Nishida, T., and A.E. Caton, Eds. pp. 307–318. Saitama, Japan: International Fishery GIS Society.

Alabsi, N. 2014. Habitat preferences of Indian mackerel from the Red Sea coast of Yemen (satellite remote sensing and GPS data logger are applied). Paper presented at *6th International Symposium on GIS/Spatial Analyses in Fishery and Aquatic Sciences*, Tampa, FL.

Aswani, S. 2011. Socioecological approaches for combining ecosystem-based and customary management in Oceania. *Journal of Marine Biology* 2011: 1–13. doi:10.1155/2011/845385.

Brinton, C.P. 2014. Movement patterns of the Atlantic stingray in tidal creeks near Savannah, Georgia (using acoustic telemetry). Paper presented at *6th International Symposium on GIS/Spatial Analyses in Fishery and Aquatic Sciences*, Tampa, FL.

Burgess, D.W. 1993. Automatic ship detection in satellite multispectral imagery. *Photogrammetric Engineering and Remote Sensing* 59: 229–237.

Corbane, C., Najman, L., Pecoul, E., Demagistri, L., and Petit, M. 2010. A complete processing chain for ship detection using optical satellite imagery. *International Journal of Remote Sensing* 31(22): 5837–5854. doi:10.1080/01431161.2010.512310.

Cruz, S., Robinson, J., and Tingey, R. 2016. Integrating participatory planning in the design of Belize's marine replenishment zones. In *GIS/Spatial Analyses in Fishery and Aquatic Sciences* (Vol. 6), Nishida, T., and A.E. Caton, Eds. pp. 135–152. Saitama, Japan: International Fishery GIS Society.

Curtin, R., and Prellezo, R. 2010. Understanding marine ecosystem based management: A literature review. *Marine Policy* 34(5): 821–830. doi:10.1016/j.marpol.2010.01.003.

Devillers, R. 2014. Interactive mapping and analysis of VMS data in support of fisheries enforcement activities in Eastern Canada. Paper presented at *6th International Symposium on GIS/Spatial Analyses in Fishery and Aquatic Sciences*, Tampa, FL.

Dutka-Gianelli, J., Struve, J., Lorenzen, K., and Crandall, C.C. 2014. Integrating acoustic telemetry and GIS techniques in fisheries research. Paper presented at *6th International Symposium on GIS/Spatial Analyses in Fishery and Aquatic Sciences*, Tampa, FL.

Escudero, L. 2016. Evaluation of pelagic resources using GIS in Peruvian waters (using acoustics and satellite remote sensing). Paper presented at *6th International Symposium on GIS/Spatial Analyses in Fishery and Aquatic Sciences*, Tampa, Florida.

Escudero, L., and Rivera, V. 2011. Distribution of Peruvian anchovy fleets in relation to oceanic parameters. In *Handbook of Satellite Remote Sensing Image Interpretation: Applications for Marine Living Resources Conservation and Management*, Morales J. et al. Eds. Chapter 14. Dartmouth, Canada: EU PRESPO Project and IOCCG (International Ocean-Colour Coordinating Group).

European Council. 1992. Council Regulation (EEC) No 3760/92 of 20 December 1992 establishing a community system for fisheries and aquaculture. *Official Journal* L389, 1–14, 31 December 1992.

European Council. 2002. Council Regulation (EC) No 2371/2002 of 20 December 2002 on the conservation and sustainable exploitation of fisheries resources under the Common Fisheries Policy. *Official Journal* L358: 59–80, 31 December 2002.

Farmer, N.A., and Karnauskas, M. 2013. Spatial distribution and conservation of speckled hind and Warsaw grouper in the Atlantic Ocean off the Southeastern U.S. *PLoS ONE* 8(11): e78682. doi:10.1371/journal.pone.0078682.

Food and Agriculture Organization of the United Nations (FAO). 1985. Report on the Ninth International Training Course on Applications of Remote Sensing to Aquaculture and Inland Fisheries. RSC Series 27. Rome: FAO.

Food and Agriculture Organization of the United Nations (FAO). 2014. *The State of World Fisheries and Aquaculture*, Vol. 2014. doi:92-5-105177-1.

Food and Agriculture Organization of the United Nations (FAO). 2016. Fishing Vessel Monitoring Systems (VMS). Online at http://www.fao.org/fishery/vms/en [accessed 1 March 2016].

Gaines, S.D., and Costello, C. 2013 Forecasting fisheries collapse. *Proceedings of the National Academy of Sciences of the United States of America* 110(40): 15859–15860. doi:10.1073/pnas.1315109110.

Good, N.M., and Peel, D. 2007. Using maximum entropy methods to calculate an index of abundance for prawn stocks, based on vessel monitoring system derived catch and effort data. In *GIS/Spatial Analyses in Fishery and Aquatic Sciences* (Vol. 3), Nishida, T., P.J. Kailola, and A.E. Caton, Eds. pp. 161–180. Saitama, Japan: International Fishery GIS Society.

Gower, J., and Skey, S. 2000. Wind, slick, and fishing boat observations with Radarsat ScanSAR. *Johns Hopkins APL Technical Digest (Applied Physics Laboratory)* 21(1): 68–74.

Hall, S.J., and Mainprize, B. 2004. Towards ecosystem-based fisheries management. *Fish and Fisheries* 5(1): 1–20. doi:10.1111/j.1467-2960.2004.00133.x.

Hollensead, L. 2014. Monitoring movement of southern flounder to elucidate residency time and migration patterns in a North Carolina estuary (using acoustic telemetry). Paper presented at *6th International Symposium on GIS/Spatial Analyses in Fishery and Aquatic Sciences*, Tampa, FL.

Holmund, C.M., and Hammer, M. 1999. Ecosystem services generated by fish population, *Ecological Economics* 29(2): 253–268.

Itoh, K., and Nishida, T. 2001. Marine explorer: Marine GIS software for fisheries and oceanographic information. In *Proceedings of the First International Symposium on Geographic Information Systems (GIS) in Fishery Science*, Nishida, T., P.J. Kailola, and C.E. Hollingworth, Eds. pp. 427–437. Saitama, Japan: Fishery GIS Research Group.

Itoh, K., and Nishida, T. 2010. Marine Explorer (marine GIS): A platform for spatial data processing and analysis in fisheries oceanography. In *GIS/Spatial Analyses in Fishery and Aquatic Sciences* (Vol. 4), Nishida, T., and A.E. Caton, Eds. pp. 451–468. Saitama, Japan: International Fishery GIS Society.

Itoh, K., and Nishida, T. 2013a. Coastal and small-scale fisheries resource management using RealMC, a real-time GPS data-logger and marine-monitoring system. In *GIS/Spatial Analyses in Fishery and Aquatic Sciences* (Vol. 5), Nishida, T., and A.E. Caton, Eds. pp. 93–96. Saitama, Japan: International Fishery GIS Society.

Itoh, K., and Nishida, T. 2013b. Pinpoint prediction of good fishing grounds using Jack Pot Skipper. In *GIS/Spatial Analyses in Fishery and Aquatic Sciences* (Vol. 5), Nishida, T., and A.E. Caton, Eds. pp. 81–84. Saitama, Japan: International Fishery GIS Society.

Itoh, K., Nishida, T., and Sasakura, T. 2016. Real-time monitoring system for underwater environments using ultrasonic wave technology. In *GIS/Spatial Analyses in Fishery and Aquatic Sciences* (Vol. 6), Nishida, T., and A.E. Caton, Eds. pp. 209–214. Saitama, Japan: International Fishery GIS Society.

Koike, H. 2014. Can market price prevent overfishing? Spatial bio economic simulation in the Seychelles sea cucumber fishery. Paper presented at *6th International Symposium on GIS/Spatial Analyses in Fishery and Aquatic Sciences*, Tampa, FL.

Kourti, N., Shepherd, I., Greidanus, H. et al. 2005. Integrating remote sensing in fisheries control. *Fisheries Management and Ecology* 12: 295–307.

Kraak, M.-J., and Ormeling, F. 1996. *Cartography: Visualization of Spatial Data*. Harlow, Essex, England: Longman.

Kraak, M.-J., and Ormeling, F. 2010. *Cartography: Visualization of Geospatial Data*. Harlow, New York: Prentice Hall.

Levin, P.S., and Stunz, G.W. 2005. Habitat triage for exploited fishes: Can we identify essential 'essential fish habitat'? *Estuarine, Coastal and Shelf Science* 64(1 SPEC. ISS.): 70–78. doi:10.1016/j.ecss.2005.02.007.

McDonnell, M.J., and Lewis, A.J. 1978. Ship detection from LANDSAT imagery. *Photogrammetric Engineering and Remote Sensing* 44: 297–301.

Meaden, G.J. 1996. Potential for geographical information systems (GIS) in fisheries management. In *Computers in Fisheries Research*, Megrey, B.A. and E. Moksnes, Eds. pp. 41–79. London: Chapman and Hall.

Meaden, G.J. 2000. Applications of GIS to fisheries management. In *Marine and Coastal Geographical Information Systems*. Wright, D.J., and D.J. Bartlett, Eds. pp. 205–226. London: Taylor & Francis.

Meaden, G.J., and Aguilar-Manjarrez, J. Eds. 2013. Advances in geographic information systems and remote sensing for fisheries and aquaculture. *Fisheries and Aquaculture Technical Paper No. 552*. 425 pp. Rome, Italy: FAO. CD–ROM version.

Meaden, G.J., and Kapetsky, J.M. 1991. Geographical information systems and remote sensing in inland fisheries and agriculture. *FAO Fisheries Technical Paper No 318*. Rome, Italy: FAO.

Montes, N., Swett, B., Sidman, C., and Zoodsma, B. 2012. Spatial and temporal analyses of encounters between the North Atlantic right whale (*Eubalaena glacialis*) and recreational vessels in the Southeastern United States. *Florida Sea Grant's 2012 Coastal Science Symposium*, October 17, 2012. Smathers Library, University of Florida, Gainesville, FL. http://nsgl.gso.uri.edu/flsgp/flsgpw12003.pdf, accessed 29 February 2016.

Morales, J., Stuart, V., Platt, T., and Sathyendranath, S., Eds. 2011. *Handbook of Satellite Remote Sensing Image Interpretation: Applications for Marine Living Resources Conservation and Management*. Dartmouth, Canada: EU PRESPO Project and IOCCG (International Ocean-Colour Coordinating Group).

Morris, L., and Ball, D. 2006. Habitat suitability modelling of economically important fish species with commercial fisheries data. *ICES Journal of Marine Science* 63(9): 1590–1603. doi:10.1016/j.icesjms.2006.06.008.

Nishida, T., and Booth, A. 2001. Recent approaches using GIS in the spatial analysis of fish populations. In *Spatial Processes and Management of Marine Populations*, G.H. Kruse et al. Eds. Proceedings of the *Symposium on Spatial Processes and Management of Marine Populations*, October 27–30, 1999, Anchorage, Alaska. 1–18. University of Alaska Sea Grant, AK-SG-01-02, Fairbanks.

Nishida, T., Murakami, T., Shiba, Y., Ooyama, M., Miura, N., and Itoh, K. 2008. Integrated ecosystem approach for sustainable tuna longline fisheries (Case study: Tropical tuna in the Indian Ocean). *Fourth International Symposium on GIS/Spatial Analyses in Fishery and Aquatic Sciences, Abstract Proceedings*. Universidade Santa Úrsula, Rio de Janeiro, Brazil, 25–29 August 2008.

Nóbrega, M., and Lessa, R. 2016. Spatial variation in age structure and survival of the greater amberjack, *Seriola dumerili* (Risso, 1810), caught off northeastern Brazil. In *GIS/Spatial Analyses in Fishery and Aquatic Sciences* (Vol. 6), Nishida, T., and A.E. Caton, Eds. pp. 23–40. Saitama, Japan: International Fishery GIS Society.

O'Shea, B. 2007. Fisheries enforcement – Unlock the potential benefits of satellite surveillance. In *GIS/Spatial Analyses in Fishery and Aquatic Sciences* (Vol. 3), Nishida, T., P.J. Kailola, and A.E. Caton, Eds. pp. 275–284. Saitama, Japan: International Fishery GIS Society.

Peterson, C.H., Summerson, H.C., Thomson, E. et al. 2000. Synthesis of linkages between benthic and fish communities as a key to protecting essential fish habitat. *Bulletin of Marine Science* 66(3): 759–774.

Pikitch, E.K., Santora, C., Babcock, E.A. et al. 2004. Ecosystem based fishery management. *Science*, 16 July 2004. 305(5682): 346–347. doi:10.2236/science.10982222.

Pitcher, T., Kalikoski, D., Short, K., Varkey, D., and Pramod, G. 2009. An evaluation of progess in implementing ecosystem-based management of fisheries in 33 countries. *Marine Policy* 33: 223–232.

Rose, K. 2014. Gulf of Mexico Data Atlas: Data discovery and access. Paper presented at *6th International Symposium on GIS/Spatial Analyses in Fishery and Aquatic Sciences*, Tampa, FL.

Rosenberg, A., Bigford, T.E., Leathery, S., Hill, R.L., and Bickers, K. 2000. Ecosystem approaches to fishery management through essential fish habitat. *Bulletin of Marine Science* 66(3): 535–542.

Rubec, P.J., Lewis, J., Reed, D., and Westergren, C. 2014a. An evaluation of transferability of habitat suitability models between Tampa Bay and Charlotte Harbor, Florida. Paper presented at *6th International Symposium on GIS/Spatial Analyses in Fishery and Aquatic Sciences*, Tampa, FL.

Rubec, P.J., Kiltie, R., McEachron, L., Flamm, R., and Leone, E. 2014b. Development of zero-inflated models to support habitat suitability modeling and mapping of species distributions in Tampa Bay, Florida. Paper presented at *6th International Symposium on GIS/Spatial Analyses in Fishery and Aquatic Sciences*, Tampa, FL.

Rubec, P.J., White, M., Ashbaugh, C.F., Lashley, C., and Versaggi, S. 2016a. Development of electronic logbooks linked to GPS, VMS and data loggers to support collection of geo-referenced catch, effort and environmental data on shrimp-fishing vessels. In *GIS/Spatial Analyses in Fishery and Aquatic Sciences* (Vol. 6), Nishida, T., and A.E. Caton, Eds. pp.183–208. Saitama, Japan: International Fishery GIS Society.

Rubec, P.J., Lewis, J., Reed, D., Westergren, C., and Baumstark, R. 2016b. An evaluation of transferability of habitat suitability models between Tampa Bay and Charlotte Harbor, Florida. In *GIS/Spatial Analyses in Fishery and Aquatic Sciences* (Vol. 6), Nishida, T., and A.E. Caton, Eds. pp. 51–70. Saitama, Japan: International Fishery GIS Society.

Sasakura, T., Itoh, K., and Yamaguchi, A. 2016. Underwater remote thermometer FRTD-600. In *GIS/Spatial Analyses in Fishery and Aquatic Sciences* (Vol. 6), Nishida, T., and A.E. Caton, Eds. pp. 215–244. Saitama, Japan: International Fishery GIS Society.

Silva, C., Yáñez, Barbieri, M.A., Bernal, C., and Aranis, A. 2015. Forecasts of swordfish (*Xiphias gladius*) and common sardine (*Strangomera bentincki*) off Chile under the A2 IPCC climate change scenario. *Progress in Oceanography*, 134: 343–355.

Simpson, J.J. 1992. Remote sensing and geographical information systems: Their past, present and future use in global marine fisheries. *Fisheries Oceanography* 1: 238–280.

Song, L. 2014. An integrated habitat index for albacore tuna in waters near the Cook Islands based on the quantile regression method. Paper presented at *6th International Symposium on GIS/Spatial Analyses in Fishery and Aquatic Sciences*, Tampa, FL.

Taylor, R. 2014. Ten years of development and operation of the HABCAM instrument system (HABCAM: HABitat mapping CAMera system). Paper presented at *6th International Symposium on GIS/Spatial Analyses in Fishery and Aquatic Sciences*, Tampa, FL.

United Nations Environment Programme (UNEP). 2013. Fisheries & Aquaculture. In *Green Economy and Trade–Trends, Challenges and Opportunities*. pp. 93–123 (Available at: http://www.unep.org/greeneconomy/GreenEconomyandTrade).

Valavanis, V.D., Pierce, G.J., Zuur, A.F. et al. 2008. Modelling of essential fish habitat based on remote sensing, spatial analysis and GIS. *Hydrobiologia* 612(1): 5–20. doi:10.1007/s10750-008-9493-y.

Wahbi, F., Tojo, N., Ramzi, A., Somoue, L., Manchih, K., and Errhif, A. 2014. Real-life spatial platform for the sustainability of coastal fisheries and communities with GIS. Paper presented at *6th International Symposium on GIS/Spatial Analyses in Fishery and Aquatic Sciences*, Tampa, FL.

Worthington, T. 2014. Combining GIS data and species distribution models to highlight the role of landscape-scale factors in the decline of an endemic Great Plains Cyprinid (using maximum entropy methods). Paper presented at *6th International Symposium on GIS/Spatial Analyses in Fishery and Aquatic Sciences*, Tampa, FL.

Wright, D., Dwyer, N., and Cummins, V. 2011. *Coastal Informatics: Web Atlas Design and Implementation*, pp. 1–344. Hershey, PA: IGI Global. doi:10.4018/978-1-61520-815-9.

Yamanaka, K.L., and Flemming, R. 2013. Development of spatial management tools to address fisheries conservation concerns for quillback rockfish (*Sebastes malinger*) in British Colombia Canada. In *GIS/Spatial Analyses in Fishery and Aquatic Sciences* (Vol. 5), Nishida, T., and A.E. Caton, Eds. pp. 199–214. Saitama, Japan: International Fishery GIS Society.

Yan, Y. 2014. Research and development of South China Sea fisheries information dynamic collection and real-time automatic analysis system based on GPS, GIS and (satellite) remote sensing integration. Paper presented at *6th International Symposium on GIS/Spatial Analyses in Fishery and Aquatic Sciences*, Tampa, FL.

Yan, Y., Wang, F., Guo, X., Feng, B., and Lu, H. 2014. Research and development of South China Sea fisheries information dynamic collection and real-time automatic analysis system based on GPS, GIS and remote sensing integration. *Journal of Fisheries of China* 2014(05).

Yen, 2014. Contrasting the habitat (suitability index) and fishing conditions of Skipjack in the western and central Pacific between two types of El Niño. Paper presented at *6th International Symposium on GIS/Spatial Analyses in Fishery and Aquatic Sciences*, Tampa, FL.

14

Geoinformatics in Hydrography and Marine Navigation

Adam Weintrit

CONTENTS

14.1 Introduction

Electronic charting for the safe and efficient navigation of shipping began life in the late 1970s and early 1980s (Astle and van Opstal, 1990; Ward et al., 2000; Ternes et al., 2008). From these beginnings, the international hydrographic community has responded to the introduction of modern mapping tools such as geographical information systems (GIS), digital terrain models (DTM), video plotters, raster charts, digital maps and global positioning system (GPS) by departing from the traditional paper chart to the development and use of a range of digital spatial data products that meet their specialized needs. Electronic charting technologies are primarily designed for safe navigation. Through their ability to display information selectively and relate it spatially, they may be considered a real-time GIS application in the marine and inland waterways. This chapter describes the potential of electronic chart display and information system (ECDIS) for maritime, coastal, offshore and inland applications and especially the databases upon which these systems are built.

The latest state of-the-art information technology, comprising computers and efficient communication networks and route-finding algorithms has significantly changed the concept of navigation. Ship owners, vessel traffic service centres, river information systems and individual vessels can all be connected through spatially enabled computer networks, leading to safer and more efficient operation in ocean, coastal and inland waters (Weintrit, 2010).

Many GIS applications are significant at the regional or local level. Electronic charting, on the other hand, demonstrates the implementation of GIS-based capabilities at the global level, specifically in terms of:

- International projects and tests, leading to system development, enhancement and implementation
- The worldwide database that provides the geospatial underpinnings for the technology
- International legal and regulatory frameworks
- International standards for systems and data
- The user community, both in terms of their ports or bases of origin and the locations that the users might frequent during the course of their working activities

Early research into electronic charting focused on the 'soft copy' display of conventional charts, maps and mapping techniques on cathode ray tube (CRT) devices. Although the resolution of the screen is the immediately noticeable difference between the paper and the video display medium, the true value of the electronic chart is not in simply imitating the paper nautical chart, but in providing a dynamic display which successfully combines the chart with real-time location of the ship and other information.

To maintain the visual simplicity of this more complex display, the data format, organization and type of chart features shown and the way they appear on the screen, must reflect the relative importance of the information to safe navigation. Unlike the static paper chart, the electronic chart can change the display and emphasis of symbols, based on actual real-time events and the viewing scale chosen. The computer-based algorithms of the electronic chart allow it to always include the least number of symbols that are most relevant to a given situation. However an electronic chart is more than simply a computer display: it is a real-time navigation system that integrates a variety of information for use by the navigator; and an automated decision aid capable of continuously determining a vessel's position in relation to land, charted objects, aids-to-navigation, other vessels and unseen hazards. The electronic chart represents an entirely new approach to maritime navigation that provides significant benefits in terms of navigation safety and improved operational efficiency.

14.2 Electronic Charting Systems and ECDIS

There are two basic types of electronic chart systems:

- Those that comply with the International Maritime Organization (IMO) requirements for Safety of Life at Sea (SOLAS) class vessels, known as the electronic chart display and information system (ECDIS)
- All other types of electronic chart systems, regarded generically as electronic chart systems (ECS)

Some authors consider ECDIS to be a special class of ECS, and this is the approach adopted in the present chapter, while other authors believe ECS and ECDIS are entirely different kinds of systems (see Figures 14.1 and 14.2, and Weintrit, 2009a).

All ECSs use computers and other electronic systems, along with digital chart data, to plot and track a vessel's position. They will consist of, at least, a central computer, a library of electronic charts, a position input such as global positioning system (GPS) and a display screen. The electronic charts stored in the library may be in either raster or vector formats and they may be official or unofficial charts. Standards are being developed for ECS by the International Standards Organisation (ISO) and the International Hydrographic Organization (IHO), building on earlier work by the Radio Technical Commission for Maritime Services. These included the IHO Transfer Standard for Digital Hydrographic Data S-57; and IHO Specifications for Chart Content and Display aspects of ECDIS, S-52.

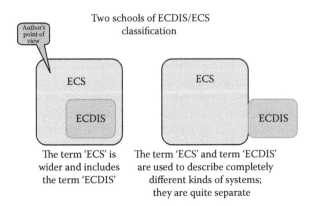

FIGURE 14.1
Two approaches to ECDIS/ECS classification. (From Weintrit, A. 2009a. (Ed.) *Marine Navigation and Safety of Sea Transportation*. Boca Raton, FL: CRC Press, Taylor & Francis.)

ECDIS equipment must conform to a more exacting suite of international standards, specifically: the IMO ECDIS Performance Standards (IMO Resolution A.817(19), as amended by Resolution MSC.64(67) – Adoption of New and Amended Performance Standards (adopted in 1996), Annex 5 – Amendment to Resolution A.817(19) Performance Standards for ECDIS; as amended by Resolution MSC.86(70) – Adoption of New and Amended Performance Standards for Navigational Equipment (adopted in 1998), Annex 4 – Amendments to the Recommendation on Performance Standards for ECDIS and as amended by Resolution MSC.232(82) – Adoption of the Revised Performance Standards for ECDIS (adopted in 2006). These Revised Performance Standards for ECDIS state that '(an) electronic chart display and information system (ECDIS) means a navigation information system which, with adequate back up arrangements, can be accepted as complying with the up-to-date chart required by regulation V/2, V/19 and V/27 of the SOLAS Convention, as amended, by displaying selected information from a system electronic navigational chart (SENC) with positional information from navigation sensors to assist the mariner in route planning and route monitoring, and by displaying additional navigation-related information' (Hecht et al., 2011).

The true ECDIS system displays information from electronic navigational charts (ENC) and integrates position information from the GPS/GNSS and other navigational sensors, such as radar/ARPA, Navtex, echosounder and automatic identification systems (AIS). It may also display additional navigation-related information, such as sailing directions, tide tables and so on (see Figure 14.3).

The main functions of ECDIS are:

- To contribute to safe navigation
- To come to assistance in decision making

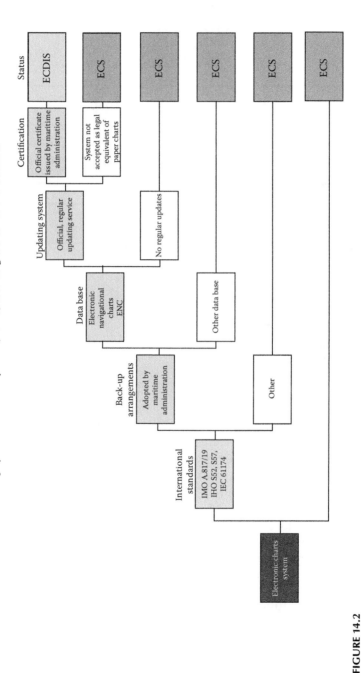

FIGURE 14.2

Classification scheme for electronic chart systems (ECS) and electronic chart display and information systems (ECDIS). (From Weintrit, A. 2001. *The Electronic Chart Systems and Their Classification.* Annual of Navigation No. 3/2001. Polish Academy of Sciences, Polish Navigation Forum, Gdynia.)

ECDIS components

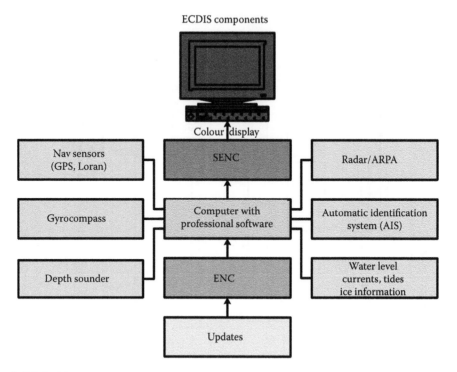

FIGURE 14.3
A typical ECDIS configuration. (From Weintrit, A. 2001. *The Electronic Chart Systems and Their Classification*. Annual of Navigation No. 3/2001. Polish Academy of Sciences, Polish Navigation Forum, Gdynia.)

- To reduce the navigational workload
- To execute in convenient and timely manner at least all navigational routines currently maintained on the paper chart
- To be the legal equivalent of the paper nautical chart required by regulation V/27 of the SOLAS Convention

Within the ECDIS, the ENC database stores the chart information in the form of structured, vector geographic objects represented by point, line and area shapes, carrying individual attributes, which make any of these objects uniquely identifiable. Appropriate mechanisms are built into the system to query the data, and then to use the obtained information to perform certain navigational functions (e.g. the anti-grounding surveillance). The presentation of ENCs on the screen is specified in another IHO standard, the 'Colours and Symbols Specifications for ECDIS IHO S-52', that is, in its

Appendix 2, called the 'ECDIS Presentation Library'. This style of presentation is mandatory.

As noted above, ECDIS meets the requirements of SOLAS, whereas non-ECDIS ECSs do not meet all the IMO, IHO and IEC standards for ECDIS. SOLAS, Chapter V regulation 2.2 states that '(a) Nautical chart or nautical publication is a special-purpose map or book, or a specially compiled database from which such a map or book is derived, that is issued officially by or on the authority of a Government, authorized Hydrographic Office or other relevant government institution and is designed to meet the requirements of marine navigation'. Therefore only when official ENCs are run in a compliant ECDIS system can it be called an ECDIS. Use of any other chart data immediately downgrades the system to an ECS, and is non-compliant, under the terms of the SOLAS regulations for use of electronic charts, as a primary means of navigation for merchant shipping. Furthermore, only the ENCs produced by HOs may be used by commercial vessels of more than 500 GRT. However, non-compliant systems (ECSs) are nonetheless in widespread use around the world and are characterized by being physically smaller, less sophisticated and less expensive than fully compliant ECDIS. ECS may also display a wider range of chart data types (vector or raster) than is allowed by ECDIS, and these data may variously be provided by hydrographic offices, commercial manufacturers or end-users. Electronic charting systems are intended for *use in conjunction with* a current, updated paper chart, but cannot function as *a legal substitute for* paper charts whereas a fully compliant ECDIS can. This distinction is often overlooked by would-be purchasers, but lawyers may not be quite so ready to ignore the regulations.

ECS is specified in ISO 19379 as follows: ECS is a navigation information system that electronically displays vessel position and relevant nautical chart data and information from an ECS database on a display screen, but does not meet *all* the IMO requirements for ECDIS and is not intended to satisfy the SOLAS Chapter V requirements to carry a navigational chart (ISO, 2014).

14.2.1 IMO MSC Approved Mandatory ECDIS

In July 2008, the IMO Safety of Navigation Sub-Committee agreed to implement the mandatory carriage of ECDIS and exclusive use of ENCs produced by official hydrographic offices; this was approved by the IMO Maritime Safety Committee at its meeting in December 2008.

A comprehensive phase-in schedule began on 1 July 2012 with ECDIS mandatory for newly built passenger ships of 500 gross tonnage (gt) and above, and for tankers of 3000 gt and above. Installation and use of ECDIS in existing ships is to be phased in from 1 July 2014 to 1 July 2018, as shown in Figure 14.4.

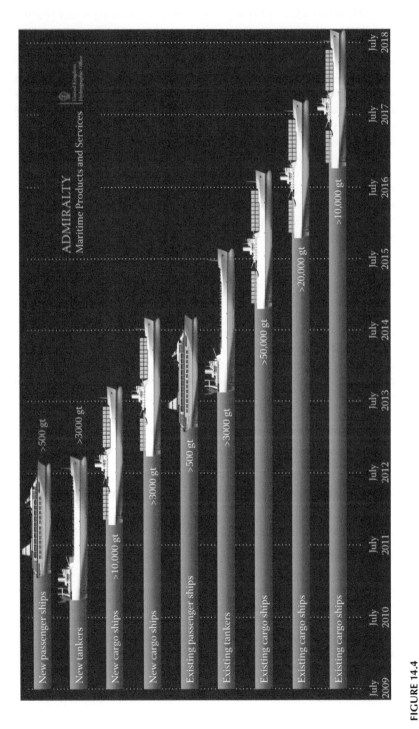

FIGURE 14.4
ECDIS implementation schedule: July 2012–July 2018. (From UKHO. British Crown Copyright 2016. All rights reserved.)

14.3 Databases for Electronic Charting

Many different formats exist for electronic charts, which may be classified in a variety of ways according to originating organization, degree of conformity with official standards and intended purpose or use (see Figure 14.5). Some of the main options and variants are described in this section, starting with official electronic navigational charts that conform to the IHO S-57 Standard.

14.3.1 Official Electronic Navigational Charts (ENCs)

14.3.1.1 Vector Charts

Electronic navigational charts (ENCs) are official specialist vector GIS datasets, intended for navigation purposes. They are created by national hydrographic offices for use with an ECDIS. An ENC must conform to standards stated in the IHO Special Publication S-57, and only ENCs can be used within ECDIS to meet the International Maritime Organisation (IMO) performance standard.

ENCs are made available through Regional Electronic Navigational Chart Coordinating Centers (RENCs) and National Electronic Chart Centers: for example, the Primar service, operated as a collaboration between the Hydrographic Service and the Electronic Chart Centre in Norway (www .primar.org) for seamless (non-overlapping) ENC cells, or the International Centre for Electronic Navigational Charts (www.ic-enc.org) for British-style ENCs where the coverage of individual sheets is variable and may overlap with others. The Hydrographic Office of Portugal (Instituto Hidrographico) maintains a web-GIS based graphical catalogue of ENC coverage, based on regularly-updated data, that can be accessed at http://websig.hidrografico .pt/website/icenc). Distributors such as the United Kingdom Hydrographic Office then distribute these to chart agents. Chart data are captured based on standards stated in IHO Special Publication S-57, and displayed according to a display format stated in IHO Special Publication S-52 to ensure consistency of data rendering between different systems. To guard against copyright infringement and data piracy, and thereby ensure the integrity of data and information published as ENC, an ENC Security Scheme was developed by the IHO Data Protection Security Working Group, based on an earlier Primar Security Scheme that had already become a *de facto* standard for ENC protection (IHO S-63, 2015). This is used to commercially encrypt and digitally sign ENC data.

An ENC contains in digital format all the chart information necessary for safe navigation, as well as supplementary information required to plan voyages and avoid groundings (route planning and route monitoring). The data are structured as vector objects, with point, line and polygon geometry and links between these objects and their attributes made possible through

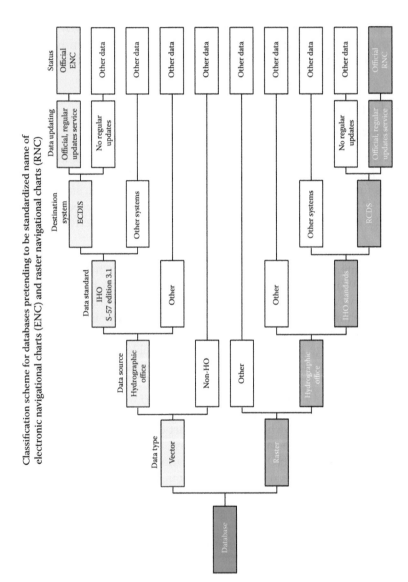

FIGURE 14.5

Classification scheme of databases of electronic navigational charts (ENC) and raster navigational charts (RNC). (From Weintrit, A. 2001. *The Electronic Chart Systems and Their Classification.* Annual of Navigation No. 3/2001. Polish Academy of Sciences, Polish Navigation Forum, Gdynia; Weintrit, A. 2009a. (Ed.) *Marine Navigation and Safety of Sea Transportation.* Boca Raton, FL: CRC Press, Taylor & Francis.)

TABLE 14.1

Comparison of Paper Charts versus Electronic Navigational Charts

Paper Charts	Electronic Charts
Fixed-scale sheet	Variable display scale
Fixed north-up orientation (usually)	Variable orientation with respect to north
Fixed symbol definition and arrangement	Variable symbol definition and arrangement
Limited paper size	Fixed display size
Limited types and amount of information	Variable types and amounts of information
Limited number of colours and colour combinations	Variable number and use of colour
Updated information requires manual annotation or replacement of the entire sheet	May allow dynamic or selective update of information

unique identifiers. This enables an ENC to boast electronic features and capabilities that paper charts lack. For instance, a navigator can integrate GPS data – which tells a navigator his or her precise latitude and longitude – with ENC data. The navigator can also integrate data from geographic information systems (GIS), as well as real-time tide, current and wind data, to enhance the capabilities of the ENC (see Table 14.1).

Incorporating these features can create a fuller, more accurate picture of the marine environment. A vessel using ENCs can detect an obstruction in advance and check planned travel routes to avoid crossing hazardous areas. The electronic charting systems used to view ENCs can also display warnings of impending danger in relation to the vessel's position and movement, and can sound an alarm if the vessel's projected course veers too close to a dangerous feature, as well as displaying any specific regulations that pertain to areas in which a vessel transits.

Official ENCs have the following attributes:

- ENC content is based on source data or official charts of the responsible Hydrographic Office
- ENCs are compiled and coded according to international standards
- ENCs are referred to World Geodetic System 1984 Datum (WGS84)
- ENC content is the responsibility of the issuing Hydrographic Office
- ENCs are issued only by the responsible Hydrographic Office
- ENCs are regularly updated with official update information distributed digitally

14.3.1.2 Raster Charts

Raster navigational charts (RNCs) also conform to IHO specifications, and are produced by digitally scanning a paper chart image (Figure 14.6). They contain all of the information found in the source document, but being raster images

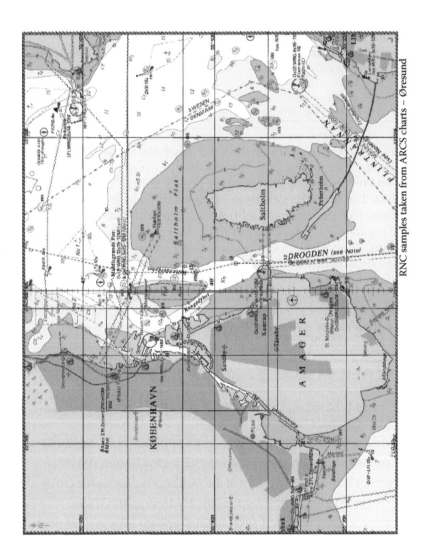

FIGURE 14.6
Raster navigational chart (RNC). (From Weintrit, A. (Ed.) *Marine Navigation and Safety of Sea Transportation*. Boca Raton, FL: CRC Press, Taylor & Francis, 2009a.)

RNC samples taken from ARCS charts – Øresund

of published analogue maps, they lack the structure or flexibility of the vector equivalent. The RNC digital file may be displayed in an electronic navigation system to give geographic context, while the vessel's position, generally derived from electronic position fixing systems, can be shown in superposition. Since the displayed data are merely a digital photocopy of the original paper chart, the image has no intelligence and, other than visually, cannot be interrogated.

IHO Special Publication S-61 provides guidelines for the production of raster data: IMO Resolution MSC.86(70) permits ECDIS equipment to operate in a raster chart display system (RCDS) mode in the absence of ENC.

The main features and determining criteria for official ENCs and RNCs are shown in Figure 14.5.

14.3.1.3 Electronic Navigational Chart (ENC) with Additional Military Layers (AML) for WECDIS Use

The concept of Additional Military Layers (AML) was introduced in 1995 with the intent to define a standardized format for non-navigational data (Figure 14.7). Since 1995, various North Atlantic Treaty Organization (NATO) standardization agreement documents concerning AML data and Warship Electronic Chart Display and Information Systems (WECDIS) have been created. As stated in the WECDIS standard, a WECDIS means 'an ECDIS as defined by the International Maritime Organization (IMO), with additional functionality for navigation and conduct of warfare onboard warships'.

Special features of WECDIS

Additional Military Layers (AML)

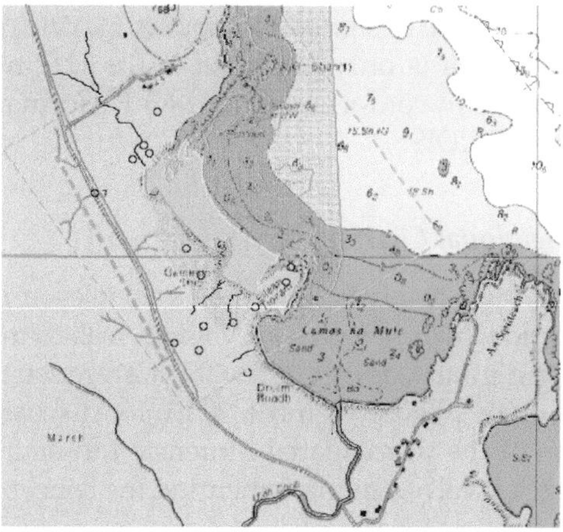

CLB:
Contour line bathymetry

ESB:
Environment, seabed and beach

LBO:
Large bottom objects

SBO:
Small bottom objects

MFF:
Marine foundations and facilities

RAL:
Routes, areas and limits

FIGURE 14.7
Electronic navigational chart (ENC) with Additional Military Layers (AML) for WECDIS use (on the left) compared to a standard ENC (on the right). (From Weintrit, A. 2009a. (Ed.) *Marine Navigation and Safety of Sea Transportation*. Boca Raton, FL: CRC Press, Taylor & Francis.)

NATO has since endorsed six AML product specifications, and completed sea trials using AML datasets. However, as more nations move toward AML data production, little is known about how the data will perform as overlays within a WECDIS adhering to NATO WECDIS standards.

14.3.2 Vector Charts Produced by the Commercial Sector

The existence of privately manufactured data is a fact of life. It is there, its volume is still increasing and it has proved to be meeting a demand of the maritime market (Malie, 2003). It is cost-effective, economically viable and it will not disappear. The major data manufacturers (e.g. Transas, Jeppesen, Navionics) offer a high quality and affordable means of world-wide navigation, including an easy to access update service, sold through reliable global networks offering a round-the-clock service. There is no doubt that in the years to come the volume of ENC will increase. However, the production rate is still too slow to provide the (minimum) necessary coverage, particularly of the major shipping routes, in an acceptable time. Moreover, it is very unlikely that ENC will ever have a world-wide coverage. In fact, it is questionable whether this is necessary.

14.3.2.1 Digital Nautical Chart (DNC)

The largest of the non-S-57 format databases is the digital nautical chart (DNC). The National Imagery and Mapping Agency (NIMA, now National Geospatial-Intelligence Agency – NGA) in the United States produced the content and format for the DNC according to a military specification.

The DNC is a vector-based digital product that portrays significant maritime features in a format suitable for computerized marine navigation. The DNC is a general purpose global database designed to support marine navigation and GIS applications. DNC data is only available to the U.S. military and selected allies. It is designed to conform to the IMO Performance Standard and IHO specifications for ECDIS.

14.3.2.2 Three-Dimensional Navigational Charts

In a number of offshore industries, such as offshore oil and gas, telecommunications, fishing, aggregate extraction, hydrographic surveying, seabed mapping and diving, three-dimensional digital nautical charts (3DNCs) are starting to be used (see Ford, 2000, Ternes et al., 2008 or Weintrit, 2009b for discussion on the arguments for the addition of the third spatial dimension to nautical charting, and some of the challenges involved in implementing the concept).

14.3.3 Filling the Gaps?

Not all sea areas have official vector ENC coverage. When operating in these areas, navigators have the choice either of using official raster data or employing privately manufactured vector data instead. The first option offers the advantage that the number of paper charts carried may be considerably reduced, whereas the second requires a full set of paper charts to be carried and used (since the paper charts have the legal authority whereas the non-official vector data do not) while also retaining full ECDIS functionality (including the alarm functions) that can only be obtained using vector charts.

This latter option is at present preferred by several ship-owners, particularly those operating their ships globally. Although most ECDIS do support privately sourced data produced by the major manufacturers, many users prefer (for cost reasons) ECS, particularly as many of these systems nowadays (also) meet the software requirements laid down in IEC 61174 (ECDIS Operational and Performance Requirements) and are less expensive. In this case, of course, paper charts are used for primary navigation because of their legal status. The time-consuming (IMO) mandatory passage planning however can be done using the ECS and, where applicable, copied to the paper chart.

A number of non-official charting data products are available, and some providers, for example, Transas, Navionics and Jeppesen, have almost worldwide coverage of vector charts, the data for which are based on existing paper charts. Unfortunately, these charts have not obtained official status because of the frequency of the updates and the lack of a controlling authority to approve the contents. In practice, the acceptability of these non-official chart products more or less depends on the flag of the ship and thus the flag state administration, and some national authorities have already considered accepting privately manufactured data meeting ISO 19379 as paper chart equivalent for certain (non-SOLAS) vessels. The U.S. and Italian governments, for example, have already amended the law to allow fishing vessels and leisure craft fitted with ECS and electronic navigational data that meets the ISO standard, to sail without paper charts in their waters.

14.4 Admiralty Vector Chart Service (AVCS)

The Admiralty Vector Chart Service (AVCS) is the most comprehensive global, official, digital maritime chart service in the world; with over 14,000 ENCs from hydrographic offices around the world, packaged and quality assured by the UKHO into a single value-added service. AVCS provides the widest official coverage, allowing ships to navigate on ECDIS for the entirety of most major routes on a single chart service. This coverage includes 4000 of the biggest and busiest ports worldwide, of which we offer unique coverage for over 150. All ENCs in AVCS satisfy the mandatory chart carriage requirements of SOLAS Chapter V, with certificates provided for local inspection. All AVCS customers will also receive weekly ENC updates by Internet download, email or disc. Installing the latest updates on a regular basis supports safe navigation and compliance with flag and port state requirements. AVCS is delivered in industry standard S-63/S-57 formats, guaranteeing compatibility with all ECDIS sold today – even in a mixed fleet. AVCS also meets the latest mandated data protection standard from the International Hydrographic Organization (IHO), designed to reduce the risk of both data piracy and the distribution of unofficial, dangerous ENCs to the mariner.

To help bridge crews identify areas of possible navigational uncertainty and risk at the crucial passage planning stage, all AVCS customers receive the unique Admiralty Information Overlay (AIO) at no extra cost. Combining all UKHO temporary and preliminary notices to mariners (T&P NMs) for all Admiralty standard nautical charts, and ENC preliminary notice to mariners (EP NMs), AIO is displayed as a single layer on top of an ENC. The AVCS, the world's leading ENC service, helps crews to navigate safely and efficiently.

14.5 ECDIS as Maritime Application of GIS Technology

ECDIS as a navigational information system is one of the examples of the use of GIS technology in marine applications. Like any working GIS, ECDIS integrates five key components: hardware, software, data, people and methods.

14.5.1 Hardware

Hardware is the computer on which an ECDIS operates, including high resolution computer screen of appropriate size, with adequate back up arrangements (usually second ECDIS). Today, ECDIS software runs on a wide range of hardware types, from centralized computer servers to personal computers used in stand-alone or networked configurations.

14.5.2 Software

ECDIS software provides the functions and tools needed to store, analyze and display ENC and additional navigation-related information, including user data. Currently on the market are dozens of manufacturers of ECDIS and each of them produces at least a few models of equipment. Before selling, each model should be checked and approved by the classification society according to IEC standards (IEC 61174, 2015).

14.5.3 Database

Possibly the most important and the most expensive component of an ECDIS is the data. ENCs are vector charts that conform to the requirements for the chart databases for ECDIS, with standardized content, structure and format, issued for use with ECDIS on the authority of government authorized hydrographic offices. ENCs are vector charts that also conform to IHO specifications stated in IHO Publication S-57. ENCs contain all the chart information necessary for safe navigation, and may contain supplementary information in addition to that contained in the paper chart. This supplementary information may be considered necessary for safe navigation and can be displayed together as a seamless chart. Systems using ENC charts can be programmed to give warning of impending danger in relation to the vessel's position and movement. The IHO developed the WEND (Worldwide Electronic Navigational Chart Database) concept to provide a timely, reliable worldwide uniform ENC data distribution service. An IHO concept, based on the set of WEND principles, designed specifically to ensure a worldwide consistent level of high-quality, updated official ENCs through integrated services that support chart carriage requirements of SOLAS and the requirements of IMO PS for ECDIS.

14.5.4 Navigator

ECDIS technology is of limited value without the educated, trained and well-experienced officers who manage the system and develop plans for applying it to real-world problems. ECDIS users range from technical specialists who design and maintain the system to those who use it to help them perform their everyday routine work onboard the vessels. We must remember that ECDIS is a decision support system adopted to assist navigators in route planning and route monitoring.

14.5.5 Methods and Procedures

A successful ECDIS operates according to international standards and well-designed implementation plan, policy and procedures (UKHO NP 232, 2014). According to STCW Convention all bridge officers on vessels that have ECDIS as the primary means of navigation are required to complete two types of ECDIS training:

- **Generic ECDIS training**, based on IMO Model Course 1.27 and approved by the relevant maritime administration. The required course length is 40 hours and is delivered by colleges, training centers or mobile training companies.
- **Type-specific ECDIS training**, given by the ECDIS manufacturer or their authorized representative. Also known as ECDIS familiarization, the regulatory requirements are covered by the ISM Code and the STCW Convention Regulation I/14, which require a company to establish procedures to ensure that new personnel and personnel transferred to new assignments related to safety and protection of the environment are given proper familiarization with their duties.

14.6 Inland ECDIS

The foregoing sections of this chapter have explained the origins and chief principles of ECDIS as the basic system for generation and presentation of digital charts for the global sea-going community. The development and use of ECDIS is regulated by a number of important supervisory bodies such as the IMO, the IHO and the ISO that, together, represent the interests of the worldwide maritime sector. Charts based on the appropriate maritime standards are officially equal to analogue charts in equipment duties.

In Europe, North America and elsewhere, extensive inland waterways comprising both natural river systems and artificial canals transect the continents, often crossing multiple jurisdictions and river basins in the process. These waterways carry large quantities of freight annually, and also have significant roles in passenger transport and in the leisure and tourism sectors. Recent figures (Eurostat, 2015) indicate that inland waterways transport of goods within Europe reached a peak in 2014 in terms of tonnage, taking into account both national and international traffic. Navigating these inland waterways presents many of the same challenges that arise from navigating coastal waters and the deep ocean, as well as posing additional ones that derive from their inland locations, the latter including generally narrower and more restricted space for manoeuver; potentially variable quantities and flow rates of the water and the frequent need to operate across multiple jurisdictions.

In the late 1990s, two European research and development projects, INDRIS (Inland Navigation Demonstrator for River Information Services), and a German pilot project on the river Rhine called ARGO, investigated the feasibility of extending ECDIS principles to these inland waters. Arising from these initiatives, an Inland ECDIS standard for ENC data and system requirements was passed by the Central Commission for the Navigation on the Rhine (CCNR) in summer 2001 (Inland ENC Harmonisation Group, 2010), and was subsequently adopted as the recommended standard for digital navigational charts by the Danube Commission, the United Nations European Commission for Economy (UN-ECE), the European RIS platform and the World Association for Waterborne Transport Infrastructure (INA/PIANC), making it currently the only standard accepted by all relevant inland navigation platforms. A similar initiative was undertaken by the U.S. Army Corps of Engineers in the United States, to facilitate the production and adoption of inland ENCs on the major navigable waterways of North America.

A North American–European Inland ENC Harmonisation Group (IEHG) was formed in 2003, comprising representatives of government, industry and academia, with the objective of developing standards that will meet the needs of Inland ENC applications worldwide (Inland ENC Harmonisation Group, 2010). In the course of discussions, it turned out that only an internationally agreed approach would be successful, since a boat master cannot be expected to employ different equipment in each country.

Since then, the Ministry of Transport of the Russian Federation, the Directorate of Hydrography and Navigation (DHN) of Brazil and the Water-borne Transportation Institute of the Ministry of Transport, Peoples Republic of China have all joined the IEHG. Close cooperation with the IHO also ensures that Inland ENC standards are compatible with those for maritime ECDIS applications. The idea was to adopt ECDIS for inland navigation and to supplement some distinct inland features, but not to change the original ECDIS standard. In this way, it will be possible to have compatibility between the original – maritime – ECDIS and Inland ECDIS. This is important for the estuaries of the rivers, where sea vessels as well as inland vessels navigate (Sandler et al., 1992). Figure 14.8 shows an example of an Inland ENC display, with radar information overlaid.

Cooperation between the North American, Russian and the (European) Inland ECDIS Expert Group led to improved and harmonized encoding rules for uniformly encoded Inland ENCs, which are written down in the Inland ENC Encoding Guide (Inland ENC Harmonisation Group, 2010). Copies of this and all other IENC-related standards are available online at http://ienc.openecdis.org, while further discussion of these initiatives, and their relationship to the ocean-going ECDIS/ENC may be found in Weintrit (2009b) (see also Scheid and Kuwalek, 2005; Gevers, 2006).

14.7 Towards a Universal Hydrographic Data Model

The IHO Transfer Standard for Digital Hydrographic Data S-7 standard has been in force for more than a decade, and has successfully been used for official ENCs adopted by hydrographic offices around the world and by navigation equipment manufacturers. Additionally S-57 has been used for many additional purposes. However S-57, and especially the administration of the standard, has also experienced limitations. In 2010, IHO released the next generation hydrographic standard called the S-100 Universal Hydrographic Data Model. In 2015, IHO released the second edition of the document (IHO S-100, 2015), in a move that will open the door to new possibilities to existing S-57 users and potentially broaden the use of IHO standards in the hydrographic community (Astle and Schwarzberg, 2013).

During the years that IHO S-57 has been in use, many people have come to regard the S-57 standard and the ENC product specification as the same thing. In reality, the ENC product specification is, in effect, a specific implementation of S-57 for the purpose of producing an ENC for use in ECDIS. This misconception resulted in a conclusion by many within the ECDIS and ENC community that the work on a new or revised edition of S-57 would radically

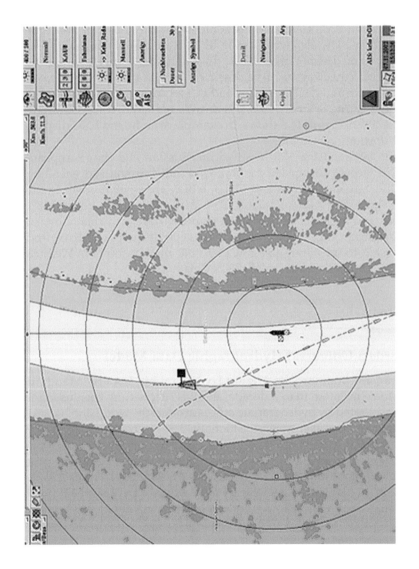

FIGURE 14.8
Inland ENC with radar picture overlay; north-up orientation of ECDIS. (From Weintrit, A. 2010. Six in one or one in six variants. Electronic navigational charts for open sea, coastal, off-shore, harbour, sea-river and inland navigation. *TransNav, the International Journal on Marine Navigation and Safety of Sea Transportation,* Vol. 4, No. 2, June.)

change the current ENC, thus affecting existing ENC production and ECDIS implementation. This was never the intention. In order to avoid a direct connection between S-57 edition 3.1 ENCs and any subsequent new IHO data transfer standard, the IHO decided in 2005 that the development of an Edition 4.0 of S-57 would henceforth be designated as S-100: Universal Hydrographic Data Model. Any product specifications developed using S-100 would then follow in an S-10n series. S-100 is the document that explains how the IHO will use and extend the ISO 19000 series of geospatial standards for hydrographic, maritime and related issues. S-100 extends the scope of the existing S-57 Hydrographic Transfer standard. Unlike S-57, S-100 is inherently more flexible and makes provision for such things as the use of imagery and gridded data types, enhanced metadata and multiple encoding formats. It also provides a more flexible and dynamic maintenance regime via a dedicated online registry. S-100 provides the data framework for the development of the next generation of ENC products, as well as other related digital products required by the hydrographic, maritime and GIS communities.

14.8 ECDIS – Guidance for Good Practice

The IMO's Maritime Safety Committee in 2015 approved the ECDIS – Guidance for Good Practice, drawing together relevant guidance from seven previous ECDIS circulars into a single, consolidated document. The undeniable safety benefits of navigating with ECDIS were recognized through formal safety assessments submitted to the IMO and experience gained by the voluntary use of ECDIS for many years. ECDIS was mandated for carriage by high-speed craft (HSC) as early as 1 July 2008. Subsequently, the mandatory carriage of ECDIS for ships other than HSC (depending on the ship type, size and construction date, as required by SOLAS regulation V/19.2.10) commenced in a phased manner from 1 July 2012 onwards.

ECDIS is a complex, safety-relevant, software-based system with multiple options for display and integration. The ongoing safe and effective use of ECDIS involves many stakeholders including seafarers, equipment manufacturers, chart producers, hardware and software maintenance providers, ship owners and operators and training providers. It is important that all these stakeholders have a clear and common understanding of their roles and responsibilities in relation to ECDIS.

ECDIS was accepted as meeting the chart carriage requirements of SOLAS regulation V/19 in 2002. Over the years, IMO Member States, hydrographic offices, equipment manufacturers and other organizations have contributed to the development of guidance on a variety of ECDIS-related matters. Over the years, IMO has issued a series of complementary circulars on ECDIS. While most useful IMO guidance on ECDIS was developed in this incremental manner, the information needed to be consolidated, where possible, to have ECDIS-related guidance within a single circular, which could be easily kept up to date without duplication or need for continual cross-referencing. Such consolidation of information offers clear and unambiguous understanding of the carriage requirements and use of ECDIS. The consolidated guidance termed 'ECDIS – Guidance for Good Practice' was published, and ship operators, masters and deck officers on ECDIS-fitted ships are encouraged to use this guidance to improve their understanding and facilitate safe and effective use of ECDIS.

14.9 Maintenance of IHO Standards for ECDIS

A major achievement in 2015 was the completion of a revised and mutually consistent set of the key standards underpinning ECDIS in accordance with the schedule presented at IMO's Sub-Committee on NCSR (Navigation, Communication and Search and Rescue). This activity was closely coordinated between the IHO and the International Electrotechnical Commission (IEC). The IHO contribution includes Edition 6.1 of Publication S-52 – Specifications for Chart Content and Display Aspects of ECDIS, Edition 4.0 of S-52 Annex A – IHO Presentation Library for ECDIS, Edition 3.0 of Publication S-64 – IHO Test Data Sets for ECDIS and Edition 1.2.0 of Publication S-63 – IHO ENC Data Protection Scheme. Following the publication of the 4th Edition of IEC 61174 – Maritime Navigation and Radiocommunication Equipment and Systems – Electronic Chart Display and Information System (ECDIS) – Operational and Performance Requirements, Methods of Testing and Required Test Results in August 2015, the revised editions are now the normative IHO references for the type approval of all new ECDIS equipment.

The plan was to have withdrawn the previous editions of S-52 and S-64 on 31 August 2016 but concerns were raised in November 2015 that this 12-month transition period would be too short to enable ship owners and operators to update existing systems to conform to the provisions of IMO circular MSC.1/Circ.1503 concerning the maintenance of ECDIS software.

The ENC/ECDIS Data Presentation and Performance Check referred to in the previous section will no longer be applicable to ECDIS equipment conforming to the revised set of ECDIS standards. The IHO ENCWG has been tasked to investigate the need to develop a new or revised check dataset, in the perspective of the development by CIRM of industry standards for software maintenance of shipborne equipment and the possible introduction of an annual performance test for ECDIS.

The revision of IHO Publication S-66 – Facts about Electronic Charts and Carriage Requirements initiated in 2014 has been delayed due to conflicting priorities. The draft new edition, reflecting the changes that have occurred since the first edition (January 2010), is now expected to be available for review in 2016.

TABLE 14.2

Comparison of ENC Coverage with Corresponding Paper Chart Coverage

	May 2009	May 2011	May 2013	April 2014	Dec 2014	Dec 2015
Small-scale ENCs (planning charts)	~100%	~100%	~100%	~100%	~100%	~100%
Medium-scale ENCs (coastal charts)	77%	88%	90%	90%	91%	92%
Large-scale ENCs (top 800 ports)	84%	94%	96%	97%	97%	97%

Source: Data from IMO NCSR 3/28, 2015. Report on monitoring of ECDIS issues by the IHO, submitted by the International Hydrographic Organization (IHO). Sub-Committee on Navigation, Communications and Search and Rescue, International Maritime Organization, London, 18 December 2015.

14.10 ENC Coverage

In parallel, the IHO continues to monitor the status of ENC coverage. Table 14.2 summarizes the global availability of ENCs by comparing the availability of paper charts intended for international voyages with the availability of corresponding ENCs. The figures are based on data available as of 1 December 2015.

IHO Working Group on the Worldwide ENC Database (WEND) continues to monitor gap and overlap issues in ENC coverage in liaison with the relevant Regional Hydrographic Commissions. Actions have been agreed to develop best practices for risk assessment. The Regional ENC Coordinating Centers (RENC) have developed and implemented tools and policies to detect and address overlapping ENCs as part of their quality control procedures. The pilot project led by Singapore on technical solutions to resolve the unpredictable performance of ECDIS caused by overlapping ENC coverage is expected to be completed in 2016.

14.11 Conclusions

ENCs have improved the safety of navigation and the efficiency of operations for mariners who have welcomed digital technology wholeheartedly.

The major advantages of Marine and Inland ECDIS electronic navigational charts are:

- Provision of information for all objects in text, graphical or video format
- Detailed and concise charts presentation in all resolutions and cut-out scales

- Simple and quick update of data (digital notices to skippers)
- Presentation in various levels of detail (e.g. depth) adapted to the needs of the skippers
- Provision of further information beyond shore and border zones
- Adoption to the requirements of skippers, for example, customizing the chart display brightness to the lighting conditions in the wheelhouse, dynamic objects like locking status
- Possibility of linking with the radar/ARPA/AIS display, route planning and route monitoring applications and so on

The expectation is that IHO S-100 will provide solutions needed by the growing market for hydrographic products, and will allow for the flexibility to grow as new needs are identified. This will not happen without significant effort and involvement by all branches of the hydrographic community. It has been evident with the developments of S-100 thus far that coordinated and combined input from and involvement by all stakeholders, including producing agencies, system manufacturers, governing bodies and end users are necessary for success.

References

Astle, H., and van Opstal, L.H. 1990. ECMAN. A standard database management system for electronic chart display and information systems (ECDIS). *International Hydrographic Review*, Monaco, Vol. 67, No. 2.

Astle, H., and Schwarzberg, P. 2013. Towards a universal hydrographic data model. *TransNav, the International Journal on Marine Navigation and Safety of Sea Transportation*, Vol. 7, No. 4.

Eurostat, 2015. Eurostat statistics explained: Inland waterways freight transport – Quarterly and annual data. Online at http://ec.europa.eu/eurostat/statistics -explained/index.php/Inland_waterways_ freight_transport_-_quarterly_and _annual_data (accessed 25 January 2016).

Ford, S.F. 2000. The first three-dimensional nautical chart. *Proceedings of the Esri User Conference*.

Gevers, K. 2006. Inland ECDIS – The unsung ECDIS success story. Admiralty World Series. *ECDIS Today*, Issue Three.

Inland ENC Harmonisation Group, 2010. *Encoding Guide for Inland ENCs*, Edition 2.2.0, February 2010. Available online at http://ienc.openecdis.org/files /Inland_ENC_Encoding_Guide_2_2_0.pdf.

Hecht, H., Berking, B., Jonas, M., and Alexander, L. 2011. *The Electronic Chart. Fundamentals, Functions and other Essentials: A Textbook for ECDIS Use and Training*, 3rd ed. Lemmer, The Netherlands: Geomares Publishing.

IEC 61174, 2015. Maritime Navigation and Radiocommunication Equipment and Systems – Electronic Chart Display and Information System (ECDIS) – Operational and Performance Requirements, Methods of Testing and Required Test Results, Edition 4. International Electrotechnical Commission, Genève, August.

IHO S-52, 2014. *Specifications for Chart Content and Display Aspects of ECDIS.* International Hydrographic Organisation, Monaco, Edition 6.1, October.

IHO S-52, Annex A, 2014. *IHO Presentation Library for ECDIS.* International Hydrographic Organisation, Monaco, Edition 4.0.1, October.

IHO S-63, 2015. *IHO Data Protection Scheme.* International Hydrographic Organisation, Monaco, Edition 1.2.1, January.

IHO S-64, 2015. *IHO Test Data Sets for ECDIS.* International Hydrographic Organisation, Monaco, Edition 3.0.1, June.

IHO S-66, 2010. *Facts about Electronic Charts and Carriage Requirements.* International Hydrographic Organisation, Monaco, Edition 1, January.

IHO S-100, 2015. *Universal Hydrographic Data Model.* International Hydrographic Organization, Monaco, Edition 2.0.0, June.

IMO. MSC.1/Circ.1503, 2015. *ECDIS – Guidance for Good Practice.* International Maritime Organization, London, 24 July.

IMO. MSC.232(82), 2006. Adoption of the Revised Performance Standards for Electronic Chart Display and Information Systems (ECDIS). International Maritime Organization, London, 8 December.

IMO NCSR 3/28, 2015. Report on monitoring of ECDIS issues by the IHO, submitted by the International Hydrographic Organization (IHO). Sub-Committee on Navigation, Communications and Search and Rescue, International Maritime Organization, London, 18 December 2015.

IMO Resolution A.817/19, 1995. *Performance Standards for Electronic Chart Display Systems (ECDIS).* International Maritime Organization, London, November.

IMO SOLAS, 2014. International Convention on Safety of Life at Sea, Consolidated Edition. International Maritime Organization, London.

ISO, 2014. ISO 19379:2003 Ships and marine technology – ECS databases – Content, quality, updating and testing (en). (Revised 2014). International Standards Organisation Technical Committee ISO/TC 8, Ships and marine technology, Subcommittee SC 6, *Navigation.* 14 pp.

Malie, C. 2003. ENC or privately manufactured data. One or the other or both? *Hydro International*, Vol. 7, No. 2, March.

Sandler, M., Kabatek, U., and Gilles, E.D. 1992. Application of an Electronic Chart in an Integrated Navigation System for Inland Ships. EURNAV'92. Digital Mapping and Navigation. The International Conference of the Royal Institute of Navigation (RIN) and the German Institute of Navigation (DGON), London.

Scheid, R.A., and Kuwalek, E. 2005. Trends for USACE Inland electronic navigational charts and the use of CARIS hydrographic production database. 6th International Symposium on Navigation, organised by Gdynia Maritime University, Faculty of Navigation and the Nautical Institute, Gdynia, June/July.

Ternes, A., Knight, P., Moore, A., and Regenbrecht, H. 2008. A user-defined virtual reality chart for track control navigation and hydrographic data acquisition. In Moore, A., and Drecki, I. (Eds.), *Geospatial Vision: New Dimensions in Cartography.* Springer Lecture Notes in Geoinformation and Cartography (selected papers from the 4th National Cartographic Conference GeoCart' 2008, New Zealand).

UKHO NP 231, 2012. *Admiralty Guide to the Practical Use of ENCs.* 1st ed. United Kingdom Hydrographic Office, Taunton.

UKHO NP 232, 2014. *Admiralty Guide to ECDIS Implementation, Policy and Procedures.* 1st ed. United Kingdom Hydrographic Office, Taunton.

Ward, R., Roberts, C., and Furness, R. 2000, Electronic chart display and information systems (ECDIS): State-of-the-art in nautical charting. In *Marine and Coastal Geographical Information Systems*, Wright, D.J., and D.J. Bartlett, Eds., pp. 149–161. Boca Raton, FL: Taylor & Francis.

Weintrit, A. 2001. *The Electronic Chart Systems and Their Classification.* Annual of Navigation No. 3/2001. Polish Academy of Sciences, Polish Navigation Forum, Gdynia.

Weintrit, A. 2009a. (Ed.) *Marine Navigation and Safety of Sea Transportation.* Boca Raton, FL: CRC Press, Taylor & Francis.

Weintrit, A. 2009b. *The Electronic Chart Display and Information System (ECDIS): An Operational Handbook.* Boca Raton, FL: CRC Press, Taylor & Francis.

Weintrit, A. 2010. Six in one or one in six variants. Electronic navigational charts for open sea, coastal, off-shore, harbour, sea-river and inland navigation. *TransNav, the International Journal on Marine Navigation and Safety of Sea Transportation,* Vol. 4, No. 2, June.

15

The Use of Geoinformatics by the Irish Naval Service in Maritime Emergency and Security Response

Brian Mathews and Cathal Power

CONTENTS

15.1 Introduction

Analogue geographical information technologies have been utilized to assist mariners at sea since the thirteenth century, through the use of charts and simple navigation aids such as the magnetic compass. Nowadays, geoinformatics is applied throughout a spectrum of services to assist maritime operations at sea, and exists in many forms such as cartography, bathymetry, radio, to satellite-based services, for instance. Commonly, ships' navigation systems rely on satellite-based systems such as GPS where the position of the vessel is mapped on an electronic chart display information system (ECDIS; Figure 15.1, see also Chapter 14, this volume) and updated automatically. Such systems have contributed to a significant recent growth in shipping activity, permitting 90% of all global trade to travel on the high seas (International Maritime Organisation, 2016), representing a 400% increase in shipping activity in the past 20 years (Oskin, 2014).

In addition to supporting the safe passage of shipping through the provision of advanced navigation systems since 1999, geoinformatics also enables

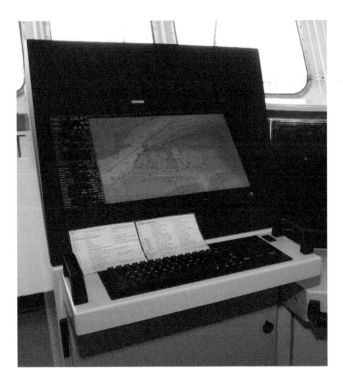

FIGURE 15.1
Electronic chart display information system (ECDIS) onboard an Irish Naval Ship. (From the Irish Naval Service.)

rescue services to determine the location of distressed persons at sea, through the implementation of the Global Maritime Distress and Safety System (GMDSS).

This chapter will briefly describe geoinformatics as it exists in daily maritime operations as introduced and will then focus on the application of geoinformatics to support maritime security and emergency rescue efforts at sea, culminating in a case study of the Irish Naval Service's involvement in the international humanitarian rescue effort in response to the Mediterranean Sea migrant crisis in 2015.

15.2 International Search and Rescue (SAR) and GMDSS

Under the provisions of the International Civil Aviation Organization (ICAO) and the International Maritime Organization (IMO), countries are obliged to provide a Search and Rescue (SAR) capability for the internationally designated area of responsibility. The goal of ICAO and IMO is to provide an effective worldwide (SAR) system, so that wherever people are in danger, in the air or sea, SAR services will be available if needed (Irish Coastguard, 2010; International Maritime Organization, 2013).

As a signatory or party to the International Convention for the Safety of Life at Sea 1982 (SOLAS), each country undertakes to provide certain SAR coordination and services. This requires the signatory states to have a certain capability to fulfil their international obligation.

To ensure a worldwide ability to respond to maritime emergencies, the IMO began considering the concept of a global distress system in the early 1970s. After a lengthy period of consultation at an international level, the Global Maritime Distress and Safety System (GMDSS) entered into force in February 1999 (UK Hydrographic Office, 2015). The legal compliance requirements are laid down in SOLAS Chapter IV, for all vessels (passenger and cargo) over 300 gross tonnes (GT), on international voyages.

The GMDSS regulations specify the technical carriage requirements for these vessels, to ensure that they can avail the services provided by GMDSS. Although warships do not have to comply with such regulations, the Irish Naval Service considers it best practice to adhere to the standards laid down for the equivalent sized merchant vessel for its ships. The principle of GMDSS aims to ensure that a signatory nation to SOLAS has a responsibility for a geographical area of the globe, thus ensuring that there will be a SAR response to an incident; these areas are referred to as Search and Rescue Regions (SRRs) (Figure 15.2). This global SAR plan has been considered one of the major achievements of the Global Maritime Distress and Safety System (UK Hydrographic Office, 2015).

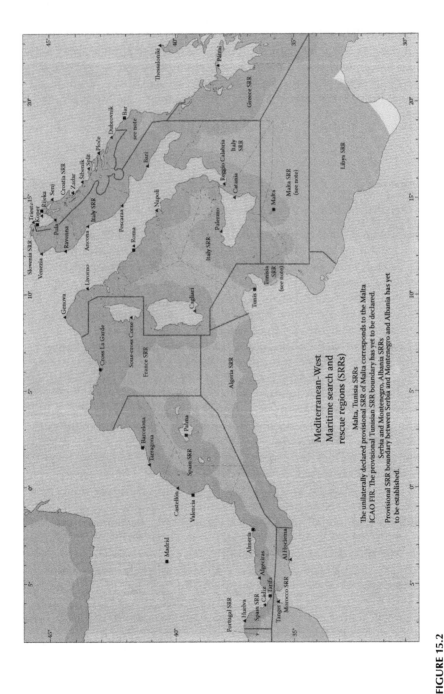

FIGURE 15.2

Search and rescue regions. (From the UK Hydrographic Office, *Admiralty List of Radio Signals*. Taunton, UK Hydrographic Office, 2015.)

15.3 Maritime Situational Awareness

In order to effectively monitor maritime activity for safety and security in a given jurisdiction or maritime domain, it is necessary to have a real time awareness of the maritime activity present and each vessel activity be it fishing, on passage, leisure and so on. Developing this maritime situational awareness requires access to multiple vessel monitoring systems and a capability to fuse the source information with intelligence, to provide a real-time picture of activity on a single display or common operating picture (COP) and broadcast to multiple ships as required. The COP is often one component of a Command and Control Information System (C2IS) and assists analysts in distinguishing between normal behaviour and anomalous behaviour which may warrant further investigation by authorities. Such analysis can be conducted manually or automatically within the C2IS depending on the complexity of the system. Complex algorithms exist to fuse vessel track information, and establish rules such as determining routine shipping routes between two locations. Essentially the C2IS consumes the data, converts it to geospatial information and, displays it on a COP allowing knowledge to be gleaned from the information, and where an anomaly is detected, this knowledge becomes wisdom which can be passed to the appropriate authority.

15.3.1 Generating Maritime Situational Awareness

The right command and control information system will depend upon organizational needs once the user requirements are captured. The Irish White Paper on Defence (2015) states, 'The Defence Forces must be able to operate jointly – that is to bring elements of the Army, Air Corps and Naval Service together to deliver effects in operations in a coordinated and cohesive manner. This is increasingly necessary for a broad range of operations at home and to be able to operate seamlessly with partners overseas. Jointness requires the capacity to develop and feed into a joint Common Operating Picture (JCOP)' (Government of Ireland, 2015).

In Defence Forces Ireland (DFI), of which the Naval Service is the maritime component, this single force C2IS approach means there exists full interoperability amongst the three components – land, sea and air (Figure 15.3). At the centre of the DFI C2IS is the joint COP (JCOP). Each service component of DFI promotes relevant information to the JCOP as necessary. The Naval Service component of the JCOP is the recognized maritime picture (RMP). Naval ships at sea on patrol are confronted by an unprecedented volume and velocity of data coming from ship sensors and external sources (Scott, 2016). Managing such volumes of information requires a suitable C2IS and the RMP delivers this capability. Information is managed in layers and can be shared to the JCOP as appropriate. As this is a military system, information security is also managed by layer access.

FIGURE 15.3
DFI C2IS JCOP approach. (From the Irish Naval Service.)

The single force platform approach requires a flexible solution that can share information both within and outside of the organization and so must conform to international agreements for information standardization amongst military forces in support of multinational or coalition military missions. The Multilateral Interoperability Programme referred to as MIP, is an interoperability organization established by national command and control information systems developers with a requirement to share relevant command and control information in a multinational or coalition environment (MIP, n.d.). The DFI C2IS is MIP-compliant.

15.3.2 Operational Capability Enhancement through RMP

Considering the vastness of the maritime domain and the fact that the majority of the world's oceans are international waters, these oceans have often been considered quite anarchic (Raymond and Morrien, 2009). The seminal naval strategist, Alfred Thayer Mahan, refers to the sea as the 'great highway' (Mahan, 1987, p. 25), considering the volume of commerce and activity in this domain. To this end, there has always been a requirement to conduct surveillance and patrol a state's claimed waters to ensure that such waters are not used for illicit or illegal activity. From an Irish perspective, this point is even more important, considering the scale of Ireland's maritime jurisdiction, 'taking our seabed area into account, Ireland is one of the largest EU states; with sovereign or exclusive rights over one of the largest sea to land ratios (over 10:1) of any EU State' (Government of Ireland, 2012, p. 3 – see also Chapter 2, this volume).

Where states have claims to maritime territory, it is essential that they are equipped with the capability to conduct surveillance of such a vast area of jurisdiction, and there is an imperative that they have a capacity to respond to an incident or interdict a vessel conducting activities contrary to International Maritime Law. The Chief of Staff of the Irish Defence Forces has reiterated this position on a number of occasions, stressing that 'sovereignty not upheld, is more imaginary than real' (Vice Admiral Mellett, 2015).

In Ireland, the White Paper on Defence 2015 stated that the Irish Naval Service will have a minimum of eight ships to conduct maritime, defence and security operations (Government of Ireland, 2015, pp. 66–67). However, the employment of ships is not the only tool to be used in developing an awareness of maritime activity. The utilization of surface ships to gather surveillance data is an age-old maritime tasking. This tasking has now been further enhanced by technological developments such as AIS* and LRIT.[†] Such technology is now referred to as the 'primary component in vessel monitoring, a ship's geospatial track or position' (Murphy, 2009, p. 22).

The collection, collation and dissemination of all this maritime data has always been a prerequisite of naval command decision making over the years. Naval Systems such as NATO Link 22[‡] have allowed for the sharing of tactical data in real time to allow for an understanding of what is occurring and realize effective decision making. Whether this process is conducted in an operations centre at sea or ashore is irrelevant; the crucial factor is that such information is time-dependent. The United States Coast Guard is credited in the 1990s as being the first agency to utilize the term 'Maritime Domain Awareness' (Murphy, 2009; see also Chapter 9 of the present volume).

MDA has since been defined as 'the effective understanding of anything associated with the global maritime domain that could impact the security, safety, economy or environment... It is critical to develop an enhanced capability to identify threats to the Maritime Domain as early and as distant from our shores as possible by integrating intelligence, surveillance, observation and navigation systems into a common operating picture accessible throughout the [United States] Government' (President of the United States,

---

* Automatic identification systems (AIS) are designed to be capable of providing information about the ship to other ships and to coastal authorities automatically. Regulation 19 of SOLAS Chapter V – Carriage requirements for shipborne navigational systems and equipment – sets out navigational equipment to be carried on board ships, according to ship type. http://www .imo.org/en/OurWork/Safety/Navigation/Pages/AIS.aspx [accessed 26 March 2016]. See also Chapter 14 by Weintrit of the present volume.

† The long-range identification and tracking (LRIT) system provides for the global identification and tracking of ships. The obligations of ships to transmit LRIT information and the rights and obligations of SOLAS Contracting Governments and of search and rescue services to receive LRIT information are established in regulation V/19-1 of the 1974 SOLAS Convention. http:// www.imo.org/en/OurWork/Safety/Navigation/Pages/LRIT.aspx (accessed 26 March 2016).

‡ Link 22 is the next-generation NATO Tactical Data Link, and is also referred to as the NATO Improved Link Eleven (NILE). Link 22 is a multi-national development programme that will produce a 'J' series message standard in a time domain multiple access architecture over extended ranges. http://fas.org/irp/program/disseminate/tadil.htm (accessed 26 March 2016).

2004, p. 5). MDA is essential to ensure the safety of all maritime activity from a host of threats, and this is especially true in areas of high density activity (Shahir et al., 2015), such as the nearshore and offshore waters of Libya.

The compilation of the common operating picture, referred to as the recognized maritime picture (RMP), allows for the understanding in near real time of activity in the area of interest to that state, body or agency. Having been aware of the significance and volume of activity that occurs in the waters that Irish warships* operate in, whether in the North Atlantic Ocean, or further afield conducting operations such as the humanitarian operations in the Central Mediterranean, the establishment of a COP in the Naval Service has allowed for Irish Defence Forces Command to maintain this near real-time picture. Numerous sources of data, from multiple actors and agencies are fed into the RMP. This alone has been a significant advancement of the MDA in the Irish domain. However, the RMP brings significant gains with the technological advancements. With the volume and density of activity, there is an ongoing requirement to monitor vessels of interest (VOI). The RMP enables rapid sharing of such data, but critically it filters such data, thereby enhancing the surveillance and monitoring capability of the Irish state.

15.4 Sourcing and Managing Maritime Information

In the INS, the information provided to the RMP can range from open-source AIS information to highly sensitive intelligence information pertaining to illicit activity. The less-sensitive information can be shared openly with relevant agencies/authorities but, as the sensitivity of information increases, so too does the security and sharing difficulties applied to the information. The Irish Defence Forces classifies information based on sensitivity which creates information layers that can be best described by the diagram in Figure 15.4.

---

* 'A ship belonging to the armed forces of a State bearing the external marks distinguishing such ships of its nationality, under the command of an officer duly commissioned by the government of the State and whose name appears in the appropriate service list or its equivalent, and manned by a crew which is under regular armed forces discipline'. UN Convention on the Law of the Sea 1982, Part II, Article 29. http://www.un.org/depts/los /convention_agreements/texts/unclos/unclos_e.pdf (accessed 26 March 2016).

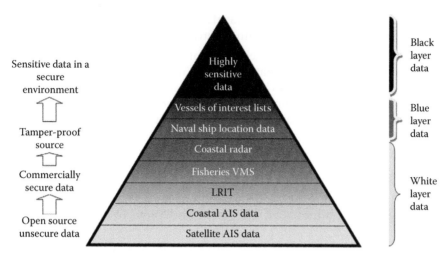

FIGURE 15.4
Data layers as distinguished by Defence Forces Ireland. (From the Irish Naval Service.)

15.4.1 Information Layers

White Layer Information: A white layer information set could be pre-scribed and promoted to all relevant maritime agencies. This white layer information would contain:

- Satellite AIS information provided by EMSA* or a commercial provider
- Terrestrial AIS from Coastal AIS receivers provided by the Irish Coastguard
- Fisheries Vessel Monitoring System (VMS) information provided by the Irish Naval Service
- Long-range identification and tracking (LRIT) provided by the Marine Surveyors Office
- Coastal radar not presently available in Ireland

The list above serves as an example of information sets available and is not exhaustive.

Blue Layer Information: Blue layer information is data of a restricted nature for security reasons or otherwise. An example of blue layer

---

* The European Maritime Safety Agency (EMSA) is one of the EU's decentralized agencies. Based in Lisbon, the Agency provides technical assistance and support to the European Commission and Member States in the development and implementation of EU legislation on maritime safety, pollution by ships and maritime security. It has also been given operational tasks in the field of oil pollution response, vessel monitoring and in long-range identification and tracking of vessels. http://www.emsa.europa.eu/about.html (accessed 01 May 2016).

information is the location of Naval Service ships on patrol, or vessel of interest lists suspected of illicit activities.

Black Layer Information: Black layer information can be considered as highly sensitive information such as intelligence relating to a specific operation which can be time-bound, where the 'leaking' of such information could compromise a particular operation. An example of black layer information could be reliable information pertaining to a vessel engaged in illegal operations.

15.4.1.1 Sharing Information Layers

Information availability and security requirements will determine the methodology of sharing information and the network protocols required:

White Information – This information can be shared openly using protocols such as secure virtual private networks (VPNs) over the Internet infrastructure.

Blue Information – Within the Irish Defence Forces, this information is shared over a restricted secure wide area network (WAN) known as the DF CIS network. A similar secure network environment would be required to share such information amongst various agencies. Enabling inter-agency access to this type of information can be obtained through a joint agency secure location such as a Maritime Information Centre where the information is physically located and controlled on available respective systems and thus information network integration is not required.

Black Information – This information needs a sufficiently secured network and user security clearance access.

15.4.2 Information Sharing between Naval Operations Centre and Ships

A secure reliable satellite communication connection that allows ships on patrol to access the RMP in real time is vital to deliver a holistic COP and live tactical chat capability. Ship sensor information and command and control layers are securely shared within the RMP to the Naval Operations Centre.

The satellite communications connection used by the INS is a commercial VSAT solution operating on the Ku-Band.* This connection is encrypted

---

* Ku-Band satellites operate in the 12–14 GHz band and provide spot beam, or regional coverage to certain areas on Earth. They require small, 1 m, very small aperture terminal (VSAT) antennas and provide more economical coverage to vessels that operate in a particular region for some period of time. When the vessel relocates, it will need to change to a different satellite and a different satellite beam. Only high traffic areas are covered with Ku-Band, with the first Transatlantic beam available since 2009. http://www.marinesatellitesystems .com/index.php?page_id=113 (accessed 01 May 2016).

FIGURE 15.5
Ship's satellite dome. (From the Irish Naval Service.)

sufficiently to permit the extension of the DF CIS restricted network to NS ships deployed operationally, with potentially worldwide coverage.

A ship is a moving platform that is pitching, rolling and moving on the sea surface. The satellite dish onboard must point at the satellite continuously to maintain the connection. This is achieved by gyroscopic stabilization of the satellite dish which adjusts with the ship's movement. Ship's officers must be mindful of preventing the ship's structure from physically blocking the line of sight between the satellite azimuth and the satellite dish where possible, known as a blind arc. These satellite dishes are housed in a dome for protection against the difficult environmental conditions at sea (Figure 15.5).

15.4.3 RMP High-Level Architecture

The INS RMP solution comprises of an RMP server on each ship and one at the Naval Base. These servers exist on the same secure military network. Information is managed in layers and distributed through layer contracts between servers.

15.4.3.1 RMP System Administration

The Naval Base server is the contract originator as the system is administered from that location. Contracts are established from the Naval Base server and information layers are either pushed or pulled to/from the ship's servers.

15.4.3.2 Information Layer Types

There are four different layer types:

- **Source Layers**: These layers are attributed to vessel track information. Onboard ships the source layers are the sensors sending information to the server for distribution throughout the RMP, namely Warship AIS, and ship's radar target information. At the Naval Base server, the source layers include all external source layer information such as coastal AIS, fisheries VMS, satellite AIS and LRIT, all of which is sent to the ships.
- **Globally Significant Layers**: These layers contain static information unlikely to change, required to support the tactical picture. An example is the territorial limits of the state, exclusive fishery limits or search and rescue regions.
- **Command and Control (C2) Layers**: These layers contain information to amplify the picture and aid decision making from a C2 perspective. This information can change regularly. In the INS, NATO maritime symbology is applied in accordance with the NATO standard MIL-STD-2525B (U.S. Department of Defense, 1999). Graphics are applied in accordance with NATO vector graphics (NVG).
- **Smart/Filter Layers**: Smart layers are composed of customizable rules based on track attributes or a geolocation. Smart layers permit filtration of source layers or the monitoring of a certain area.

15.4.3.3 Information Security

As this is a military system, access to, and control of information is carefully managed. Information security is managed in two ways:

- **Layer Security Permissions**: Only specified users are authorized to view or edit layers.
- **User Permissions**: User profiles assign the level of user access to the RMP and the functions they are permitted to perform, such as delete or modify information layers.

15.5 Case Study – The Use of Geoinformatics to Support the Irish Naval Service's Rescue Effort during the Mediterranean Sea Migrant Crisis in 2015

15.5.1 Overview

The ongoing crisis in the failed state of Libya has meant that there is no authority ashore in Libya to act as the coordinating organization for SAR. As

FIGURE 15.6
L.E. Eithne during rescue operations in the Central Mediterranean Sea. (From the Irish Naval Service.)

such the Italian Marine Rescue Coordination Centre in Rome has, by default, become the responsible Marine Rescue Coordination Centre (MRCC) for the Libyan Search and Rescue Region (SRR). With no stable government in place, all the structures in place to meet the SOLAS and GMDSS requirements for Libya, as a signatory to these agreements, have been non-functioning.

While awaiting the establishment of a European solution to the developing crisis in the Central Mediterranean, the Italian government established a national naval mission in the area. Its role is to protect Italian national interests and support any SAR requirements that may arise. This mission is known as 'Operation Mare Sicuro'.* In April 2015, the Irish government signed a *Note Verbale* with the Italian government that would allow for liaison and cooperation between the Italian Navy and Irish Navy. It would also allow for an Irish Naval ship to land any rescued persons recovered in the Central Mediterranean to Italian soil for processing.

In May 2015, the Irish government responded to the migrant crisis in the Mediterranean Sea when the Irish Naval vessel *L.E. Eithne* was the first ship deployed to assist in the humanitarian crisis (Figure 15.6).

The RMP has contributed extensively to the Irish Naval Service efforts in the Mediterranean Sea, where over 8500 people have been rescued at sea by Irish Naval ships deployed to the area in 2015. On location in the Central Mediterranean Sea, the RMP has improved operational efficiencies for Irish Naval ships and has enabled a coordinated approach to activity planning

---

* http://www.marina.difesa.it/cosa-facciamo/operazioni-in-corso/Pagine/MareSicuro.aspx

and response. In Ireland, the RMP enabled naval command to monitor activity in the Mediterranean area of operations in near real time and provide direct support to the deployed naval ship operating overseas as required.

15.5.2 Geoinformatics Enabling People Trafficking

The people smugglers orchestrating the voyages of migrants in unseaworthy craft also used geoinformatics technologies. They were familiar with GMDSS, and provided the means to the migrants on how to seek assistance using a standard commercial satellite phone and GPS. The satellite phone provided to the migrants was pre-programmed with the number for the MRCC in Rome allowing the migrants in distress at sea to alert authorities and enable a coordinated response.

15.5.3 Effective Distribution of Rescue Assets

The lead authority coordinating the rescue effort was the Italian Coast Guard. Rescue assets were assigned patrol areas to provide the greatest coverage across the likely areas where casualties may be encountered. This standard approach would normally be managed using a COP. Figure 15.7 shows the patrol areas as displayed on the INS RMP.

15.5.4 Use of RMP Onboard Ship

In a multinational operation such as this, with so many factors to consider, situation awareness is critical. The COP provides a single platform where all the information is displayed, enabling superior and more effective decision making. Such considerations for the INS in theatre include:

- Geolocation of the distress and jurisdiction concerns such as distance from territorial waters of the failed-state Libya.
- Rescue assets in the vicinity to provide further support if required.
- Distance from distress, speed available and expected time of arrival on scene.
- Keeping MRCC Rome and Naval Operations in Ireland aware of the situation in near real time.

15.5.5 Use of RMP in Naval Operations Centre, Ireland

Monitoring the activity of the deployed INS ship in the Mediterranean Sea in near real time was a principal operational requirement in order to provide necessary remote support and coordinate jurisdictional matters as necessary. The RMP provided this capability through the COP and the secure tactical chat facility. In Ireland, as the RMP exists as the naval

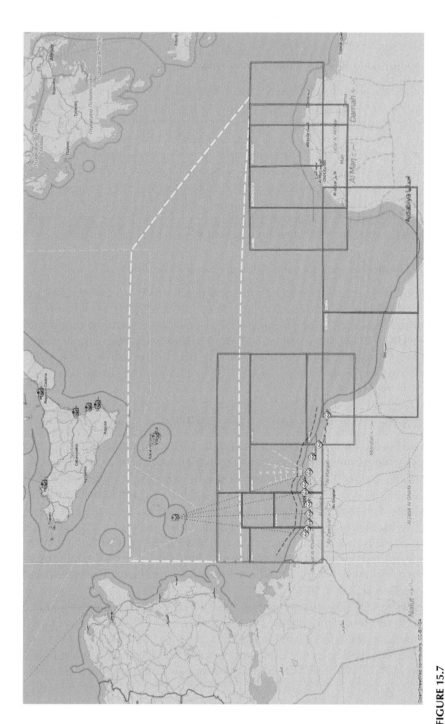

FIGURE 15.7
Patrol areas off the Libyan coast, as displayed on the Irish Naval Service RMP. (From the Irish Naval Service.)

element of the JCOP, dissemination of this information to required DFI offices is seamless.

15.5.6 Monitoring the Movement of Migrant Vessels

A typical Naval patrol vessel may detect a small fishing vessel or barge at 6 to 10 miles on radar, depending on a number of factors such as the height of the radar scanner, the weather conditions, target vessel hull composition, freeboard, superstructure and hull form. A small rubber craft is difficult to detect by radar and a patrol vessel would have to rely on optical cameras, infra-red cameras or visual lookouts to detect such vessels. With only a sensing radius of up to 10 miles for detecting migrant vessels in this operation, it is clear that a patrol vessel is not the most suitable platform for detection, and should be mainly considered with the response role. In contrast, remote sensing could provide much greater migrant vessel detection capability.

15.5.6.1 Remote Sensing through Airborne Sensor Platforms

An airborne vessel detection platform such as a maritime patrol aircraft or helicopter can significantly enhance the detection of migrant vessels, covering a large area quickly. Maritime patrol aircraft are important platforms for detection of migrant vessels during the humanitarian rescue operations in the Mediterranean Sea. Once a migrant vessel is detected by a maritime patrol aircraft, the location is passed to MRCC Rome and a surface vessel is deployed in response.

15.5.6.2 Remote Sensing through Satellite Capability

Copernicus is the European Earth Observation Programme, which provides Europe with continuous, independent and reliable access to data and information based on satellite observation. Through its security service supporting security missions, Copernicus delivers vital information to help monitor and address developing emergency situations, such as the ones that arise during the Mediterranean Sea refugee crisis. Copernicus provides satellite imagery which is used to monitor ports and beaches identified as departure points for migrant vessels. Under optimum conditions, it can also help track vessels in combination with surveillance planes and information from ships (European Commission, 2016).

In October 2015, 350 people were saved when Copernicus observation helped to detect four rubber boats leaving the coast of Libya. The exact positions of the boats were forwarded to the authorities and the people were saved by vessels participating in the Common Security and Defence Policy (CSDP) operation EUNAVFOR MED Op Sophia (European Commission, 2016).

15.6 Conclusion

The oceans are the last great global commons (Love, 2010) amounting to two and a half times the land surface of the planet (Murphy, 2010). Monitoring activity in these great expanses is a significant challenge. Furthermore, shipping must be considered a unique mode of transport given the number of associated potential variables: for instance, the destination ports and arrival times can change; a ship can change nationality at sea; ownership can change during a voyage; crews can rotate at sea and cargoes can be transhipped (such as fishing transhipments). 'Moreover given the sheer volume and vitality of seaborne trade ... the large number of crucial ports, transhipment nodes and ... trading links, the maritime environment offers an almost limitless range of tempting, high payoff targets' (Herbert-Burns, 2005). Technological advances as described in this chapter, and mandatory reporting systems, have assisted in solving the complexity of maritime domain awareness generation. Satellite-based technologies such as Earth observation have provided further remote surveillance capabilities to assist authorities faced with the challenge of trying to extrapolate illicit actors at sea from all the legitimate seafarers.

The application of geoinformatics to further support intelligence gathering of illicit activity is considered beyond the scope of this chapter; however, it does play a major part. A typical example includes establishing the routine movements of a vessel and generating a 'pattern of life' of this vessel through monitoring. Once routine behavioural analysis information is available, a non-routine activity of a vessel can generate alarms and be investigated. Such intelligence gathering is available to authorities by commercial providers, who gather geospatial data from as many sources as available. This activity fusion analysis demonstrates the recent advances in maritime situational awareness generation.

It is clear that geoinformatics exists at every level in the maritime domain from private leisure sailors navigating a safe route, to commercial shipping and government authorities policing, managing and monitoring their jurisdiction of the maritime domain. Indeed the application of geoinformatics could be considered scalable depending on the required utility for which it is to be employed.

Geoinformatics has revolutionized safety at sea through enabling the foundation of GMDSS as described. In terms of maritime emergency management support, the case study examining the Irish Naval Service's response to the Central Mediterranean Sea migrant crisis clearly portrays the critical role played by geoinformatics in detecting stricken vessels, coordinating rescue efforts and ultimately saving lives.

References

European Commission, 2016. *European Commission Growth News.* [Online] Available at: http://ec.europa.eu/growth/tools-databases/newsroom/cf/itemdetail.cfm?item_id=8706&lang=en (accessed 02 April 2016).

Government of Ireland, 2012. *Harnessing Our Ocean Wealth – An Integrated Marine Plan for Ireland.* Dublin: Government of Ireland.

Government of Ireland, 2015. *White Paper on Defence.* Newbridge: Department of Defence.

Herbert-Burns, R., 2005. Terrorism in the early 21st century maritime domain. In: J. Ho and C. Z. Raymond, Eds. *The Best of Times, the Worst of Times: Maritime Security in the Asia-Pacific.* Singapore: World Scientific Publishing, pp. 155–157.

International Maritime Organisation, 2016. *IMO Profile Overview.* [Online] Available at: https://business.un.org/en/entities/13 (accessed 24 March 2016).

International Maritime Organization, 2013. *IAMSAR Manual – Mission Coordination – Volume II.* London: IMO.

Irish Coastguard, 2010. Irish National Maritime Search and Rescue (SAR) Framework. 61pp. [Online]. Available at: http://www.dttas.ie/sites/default/files/publications/maritime/english/irish-national-maritime-search-and-rescue-sar-framework/sar-framework.pdf (accessed 19 August 2016).

Love, P., 2010. *Fisheries, While Stocks Last.* Paris: OECD.

Mahan, A., 1987. *The Influence of Sea Power upon History, 1660–1783.* New York: Dover.

MIP, n.d. *MIP Public Home.* [Online] Available at: https://mipsite.lsec.dnd.ca/pages/whatismip_3.aspx (accessed 26 March 2016).

Murphy, M., 2009. Lifeline or pipedream? Origins, purposes and benefits of AIS, LRIT and MDA. In: R. Herbert-Burns, S. Bateman and P. Lehr, Eds. *Lloyd's MIU Handbook of Maritime Security.* Boca Raton, FL: CRC Press, pp. 13–28.

Murphy, M., 2010. *Small Boats, Weak States, Dirty Money.* 2nd ed. London: Hurst & Co.

Oskin, B., 2014. *Livescience.* [Online] Available at: http://www.livescience.com/48788-ocean-shipping-big-increase-satellites.html (accessed 24 March 2016).

President of the United States, 2004. *National Security Presidential Directive NSPD-41,* Washington, DC: U.S. Government.

Raymond, C. Z. and Morrien, A., 2009. Security in the maritime domain and its evolution since 9/11. In: R. Herbert-Burns, S. Bateman and P. Lehr, Eds. *Lloyd's MIU Handbook of Maritime Security.* Boca Raton, FL: CRC Press, pp. 3–12.

Scott, R., 2016. Command view: Rethinking interactions in the ops room. *IHS Jane's Navy International,* 121(3), pp. 14–19.

Shahir, H. Y., Glasser, U., Shahir, A. Y. and Wehn, H., 2015. *Maritime Situation Analysis Framework.* Santa Clara, CA: IEEE International Conference on Big Data.

UK Hydrographic Office, 2015. *Admiralty List of Radio Signals.* Taunton: UK Hydrographic Office.

U.S. Department of Defense, 1999. *Common Warfighting Symbology.* Reston, VA: Defense Standardization Program Office.

Vice Admiral Mellett, M., 2015. *Irish Times.* [Online] Available at: http://www.irishtimes.com/news/ireland/irish-news/new-chief-of-staff-sets-out-national-security-priorities-1.2433671 (accessed 25 March 2016).

16

Spatial Analysis for Coastal Vulnerability Assessment

Jarbas Bonetti and Colin Woodroffe

CONTENTS

16.1 Coastal Hazards and Their Effects

Coastal systems are complex. They are variable across a wide range of spatial and temporal scales. Coastal processes cannot always be described by the basic laws of physics because the multiplicity of components, some hardly predictable, interact dynamically along a highly interconnected surface. Bak (1997) argued that complex behaviour in nature reflects the tendency of some systems to evolve into a 'poised, critical state, way out of balance', where minor disturbances may initiate a wide range of events. This seems to apply particularly well to coastal systems.

The complexity that characterizes coastal systems has been widely recognized (Carter and Woodroffe, 1994; Haslett, 2000; Masselink and Hughes, 2003; Masselink and Gehrels, 2014). Even with the recent advances in data acquisition instrumentation and analytical approaches many coastal processes are still poorly understood (Bonetti et al., 2013a).

Considering this, it may be expected that coastal areas will regularly experience disturbances in their equilibrium state; continuously adjusting their morphology and sediment distribution to newly imposed dynamic conditions (Woodroffe, 2002). As a result of extreme events, for example, after high-energy waves have pounded the coast, beaches, dunes and mangroves may appear threatened. However, in most cases, it can be anticipated that they will recover to a state close to the one prior to the change. Thus, the behavior of many natural systems may often be seen as cyclical, and generally oscillating across a state of greater or lesser dynamic equilibrium. The exposure of human assets to a coastal hazard, by contrast, tends to result in more permanent, linear changes and damage.

Settlements and human activities are concentrated in the coastal zone, a trend that seems inevitable to continue due to population growth and economic development (Neumann et al., 2015). In a steady climate condition, complexity would be enough to impede accurate long-term prediction and quantification of the effects of possible hazards in specific coastal locations. This difficulty is compounded if we take into account that the world's softer coastlines seem to be experiencing increased impacts as a result of climate change (Reguero et al., 2015).

Coastal communities are exposed to coastal hazards, and this exposure is expected to be amplified if extreme climate events become more frequent. According to the Intergovernmental Panel on Climate Change (IPCC, 2014), there is very high confidence that coastal systems will increasingly experience flooding and erosion during the course of the twenty-first century and beyond. Coastal hazards include processes that range from short-term events, such as severe storms, to slow trends, such as multi-century sea-level rise (Church and White, 2011; IPCC, 2014), and climate change is projected to amplify risks for people, assets, economies and ecosystems, particularly in urbanized areas. Both as individual events, and when considered as a series of occurrences, this is likely to lead to property damage, loss of life and environmental degradation, such as domestic and industrial wastes, salinization of groundwater, river inundation and so on.

In this chapter, sea-level rise and storm surges will be described. These hazards can be related directly to coastal retreat, episodic flooding and erosion, processes that are closely related but that operate on different time scales (Bonetti et al., 2013b). Moreover, due to the spatial representativeness of these climate-related drivers, they can be very well described and analysed by means of spatial analysis techniques using geographical information systems (GIS).

Sea-level rise manifests incrementally over relatively long periods of time, and can be considered a slow-onset hazard. A comprehensive review on this can be found in Houston (2013). There are natural exothermic processes, associated with the amount of heat transferred to Earth from the Sun, and many scientists believe that an observed increase in global air temperature, measured through the twentieth century, may be related to human activities (IPCC, 2014). Gradual erosion, longer-term coastal retreat and inundation can be considered consequences of this trend. The environmental, social and economic effects of sea-level rise on coasts have been discussed recently by Bosello and De Cian (2013), Brown et al. (2013), Allison and Basset (2015) and Neumann et al. (2015).

Many analyses indicate that sea levels have risen globally at increasing rates in recent decades (Church and Menéndez, 2010; White, 2011). IPCC (2014) claims that global mean sea level rose by 0.17–0.21 m over the period from 1901 to 2010. Long-term projections of mean sea level for the twenty-first century anticipate a further rise, in a range between 0.26 and 0.55 m (rigorous mitigation scenario) to 0.45 and 0.82 m (very high greenhouse gases emission scenario) by 2100 (Church et al., 2013). It is important to emphasize that this sea-level rise is not expected to be uniform across regions but, according to IPCC's AR5 Assessment Report (IPCC, 2014), about 70% of the world's coastlines are likely to experience a sea level change within ±20% of the global mean.

Storm surges, on the other hand, are rapid-onset hazards since they typically manifest over hours to a few days. They can be defined as the temporary increase in the sea level, at a particular locality, due to extreme meteorological conditions (most commonly low atmospheric pressure and/or strong winds [IPCC, 2014]). They can lead to rapid erosion of the coast and flooding (Bonetti et al., 2013b), primarily as a consequence of extreme sea levels during an event, aggregating tidal variation, barometric surge, wind setup, wave setup and wave runup (Dall'Osso and Dominey-Howes, 2013).

There is little evidence of changes in tropical or extratropical storm frequency or intensity, and a significant increase in the occurrence of these arising from climate change cannot necessarily be expected, other than those associated with sea-level rise (Rhein et al., 2013). However, in recent years, different sources have reported an increase in the number of emergency events associated with extreme high sea levels, mostly related to large storms. The EM-DAT International Disaster Database (Guha-Sapir, 2016), for example, reveals a significant increase in the number and effects of storms since the 1960s and a high-frequency oscillation around 100 events in the past 25 years (Figure 16.1). Since these data are based on reports from different kinds of institutions (UN agencies, non-governmental organizations, insurance companies, research institutes and press agencies), it can be expected that the increase in storm frequency is biased by the incentives to register and report damages, and might not necessarily be attributable to climate change.

In any scenario of sea-level rise rate or intensification of storm effects, it is strategic to develop predictive tools that could allow the assessment of coastal exposure. This is particularly critical in highly developed areas, where proximity of human assets to the shoreline exposes infrastructure and requires the formulation of effective policies for reducing the magnitude of adverse impacts and minimising economic losses. White et al. (2001) verified, in their research, that the loss of lives from natural disasters has declined worldwide, while property losses have been growing.

In describing approaches for managing risks of climate change through adaptation, the IPCC (2014) suggests a series of actions to reduce vulnerability and exposure, including vulnerability mapping. Vulnerability is a concept not always employed in the same way by different authors and a conceptual framework for it and related terminologies is proposed in the next section.

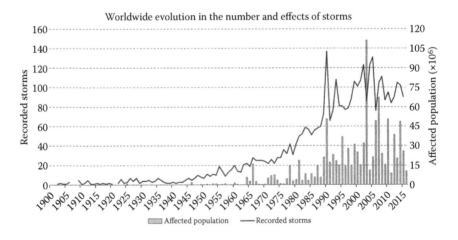

FIGURE 16.1
Total number of worldwide reported damage-related storms between 1900 and 2015. (From EM-DAT International Disaster Database, http://www.emdat.be/.)

16.2 Conceptual Framework for Vulnerability: Definitions and Interrelations

White et al. (2001) observe that *vulnerability* has emerged as a key concept in hazards and disasters literature. Vulnerability can be broadly considered a potential for loss, and its assessment can help individuals and societies be prepared, respond and adjust to environmental hazards. The IPCC (2014) defines it as the degree to which a system is susceptible to injury, damage or harm by climate stimuli. However, the interpretation of this concept varies widely among researchers, which reflects both the extensive use of the term, as well as the lack of consensual definition for it, including how vulnerability can be measured. For example, Cutter (1996) reviewed the definition of vulnerability, differentiating its use in social and biophysical contexts, which, when combined, would lead to recognition of 'place vulnerability'. Other comprehensive, more recent reviews on the matter were presented by Birkmann (2013) and Nguyen et al. (2016).

McFadden and Green (2007) maintain that vulnerability should be used as 'a flexible and adaptable concept', an idea shared by Nguyen et al. (2016), concluding that no single optimal definition of vulnerability would fit all assessment contexts. These authors note, moreover, that vulnerability has been related, not always in a clear way, to concepts such as exposure, susceptibility, sensitivity, coping capacity, resilience and risk, among others.

Exposure generally refers to the elements at risk from a given hazard, their quantity or extent in a certain spatial domain. This includes ecosystems, people, properties and so on located in hazard zones that are thereby subject to potential losses (UNISDR, 2009). Füssel (2005) indicates that exposure is determined by external biophysical factors, such as severe storms and sea level change, while sensitivity would be related to internal elements of the system, like topography or environmental characteristics and so on.

The concept of *sensitivity* (or susceptibility which, in practice, is often used to refer to the same phenomenon) emerged in studies aiming to characterize the natural likelihood of a system being affected by a certain threat. It can be considered equivalent to the 'biophysical vulnerability' proposed by Cutter (1996), and refers to the natural propensity of a system to be impacted by a hazard (Muler and Bonetti, 2014). It is largely independent of human influences, and is primarily derived from the environmental (physical) conditions (UNDP, 2004), denoting the degree to which a system is affected, either adversely or beneficially, by climate variability (Nguyen et al., 2016).

Resilience and *adaptive (or coping) capacity* are strongly related concepts. Resilience can be considered the capacity of a community to absorb and respond to an impact, and takes into account the resources of human settlements (Cardona, 2003). In a wider sense, it reflects the regenerative capabilities of a system, including 'the ability to learn and adapt to incremental changes and sudden shocks while maintaining its major functions' (Birkmann, 2013). The natural resilience of coastal systems, their morphological states and types of equilibrium have been discussed by Woodroffe (2006). Adaptive capacity, in turn, is 'the combination of all strengths and resources available within a community or organization that can reduce the level of risk, or the effects of a disaster' (UNISDR, 2009). According to Wamsley et al. (2015), adaptive capacity 'describes a system's ability to evolve, either naturally or through engineered maintenance activities, in such a way as to preserve or enhance the system's valued functions'.

Risk is another concept used with considerable liberty in literature, and sometimes interchanged with vulnerability (Kearney, 2013). Some authors (such as UNDRO, 1979; Cutter, 1996) present it as the likelihood of occurrence of a hazard, being thus directly related to its probability of incidence and to the quantification of expected losses or damages. IPCC (2014) defines risk as the interaction between hazard (triggered by an event or trend related to climate change), vulnerability (propensity to be harmed) and exposure (of people, assets or ecosystems at risk). In other words, according to IPCC (2014), risk would be represented as 'the probability of occurrence of hazardous events or trends multiplied by the magnitude of the consequences if these events occur'.

A scheme for integrating these concepts is presented in Figure 16.2. On it, the three key components for climate-change vulnerability assessment, exposure, sensitivity and adaptive capacity are illustrated. As described by Nguyen et al. (2016), exposure and sensitivity may determine a potential impact, which may be ameliorated by aspects of adaptive capacity to give overall vulnerability. Those components have very good potential to be spatially represented in geographic information systems, and their mapping and integrated analyses are frequently the basis of vulnerability mapping.

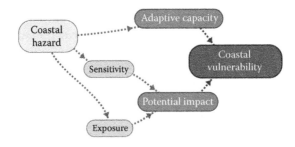

FIGURE 16.2
Conceptual framework for vulnerability and related terminology. (Modified from Wamsley, T. V., Collier, Z. A., Brodie, K., Dunkin, L. M., Raff, D. and Rosati, J. D. 2015. *Journal of Coastal Research*, 31(6): 1521–1530; Nguyen, T. T. X., Bonetti, J., Rogers, K. and Woodroffe, C. D. 2016. *Ocean and Coastal Management*, 123: 18–43.)

16.3 Spatial Analysis Techniques and Methodological Approaches for Coastal Vulnerability Assessment

In the early 1990s, Bartlett (1993) produced an annotated bibliography of research into GIS applications in the coastal zone. That review comprised 150 entries and, amongst them, at least 10 dealt with themes such as coastal erosion, storms, sea-level rise, exposure and vulnerability. Most of these aimed for assessment and development of tools to aid decision making and planning.

More than 20 years later, a list of worldwide vulnerability researches was provided by Rangel-Buitrago and Anfuso (2015), who described 45 studies specifically dedicated to this subject. This latter review showed that these original objectives remained, and that vulnerability assessments are extensively obtained by means of spatial analysis in GIS, sometimes integrated with computer-assisted multivariate statistics and numerical modeling. The development of studies based on GIS has thus been gaining a progressive acceptance, and at the present day the use of this technology for coastal-related research is customary (Bartlett, 2000; Bartlett and Smith, 2005), from the simple posting of sampling points to complex integrated analysis using artificial neural networks or genetic algorithms (for details on these two techniques see, e.g. Smith et al., 2007).

Resources have grown exponentially, accompanied by a reduction of hardware prices and consolidation of very good quality open-source software. The portal Geospatial Analysis Online (http://www.spatialanalysisonline .com/, visited on 15 February 2015), for example, lists and provides links to 177 examples of GIS and related software packages and tool sets, many of them being freeware and/or open source.

Taking advantage of these, several methodologies using different approaches have been proposed over the years to assess coastal vulnerabilities to hazards, in particular to the potential impacts of sea-level rise and storm surges on developed areas. Typically, they consider data representative of both biophysical and social attributes of coasts, as well as their mutual interaction, and promote their integration and transformation to adequately set up relevant policies for adaptation (Nguyen et al., 2016).

Many of the frameworks proposed for assessing coastal vulnerability to climate change effects were described in detail by Abuodha and Woodroffe (2010b), Ramieri et al. (2011) and Nguyen et al. (2016). A few selected methodologies will be illustrated here, presenting contrasting approaches, developed to operate at diverse scales and using different spatial data models. They have been grouped, according to the spatial scale in which they most efficiently perform analysis, into (1) *broad-scale*, usually based on raster or polygon (administrative units) and predefined datasets; (2) *multiscale*, using line segmentation procedures and (3) *local-scale*, for which basic input is point data obtained from community-level census data or field surveys.

16.3.1 Broad-Scale Assessment

Some frameworks were idealized to operate primarily at a global scale and use predefined datasets. Many were derived from the 'Common Methodology', originally proposed by IPCC CZMS (1992), which was accompanied by techniques to build an inventory and to delineate areas vulnerable to sea-level rise. This approach aimed to provide a framework for multi-component assessment of vulnerability by identifying priority sectors at global, national and regional scales. Although a number of studies have been developed using this approach and derivations, and large amounts of data generated, this methodology has been criticized due to its high degree of simplification and generalization (Abuodha and Woodroffe, 2010b). Romieu (2010) emphasized that such approaches are based on many assumptions that limit their relevance for larger scales and, therefore, inhibit their applicability for local or regional settings.

Other global exploratory computer models and tools conceived for supporting risk management decisions were described by Ramieri et al. (2011) and Zanuttigh et al. (2014). Some of them are flexible in their capabilities to deal with different data resolutions, but presented better results in scales from regional to global. Some examples are: Global Vulnerability Analysis (GVA), climate Framework of Uncertainty, Negotiation and Distribution (FUND), Dynamic and Interactive Vulnerability Assessment (DIVA), DESYCO (decision support system for coastal climate change), regional impact simulator (RegIS) and THESEUS DSS (innovative technologies for safer European coasts in a changing climate). One of those (DIVA), based on segmentation analysis, will be detailed below.

16.3.2 Multiscale Approaches

Methodologies based on line segmentation tend to be flexible in terms of scale representativeness since the statistical self-similarity of shorelines (i.e. fractal geometry; Kappraff, 1986) allows their accurate representation across different extents.

The shoreline is often the most fundamental geographical feature on maps, and a line can be used to conveniently represent the coastal zone (Monmonier, 2008). A segmentation procedure to linearly divide the shoreline was pioneered by Bartlett et al. (1997), and has been widely adopted since. However, the manner in which it has been used to portray potential vulnerability of the coast diverges between different researchers. Regarding this procedure, Sherin (2000) describes linear reference systems as well as dynamic segmentation techniques in GIS with application to coastal and marine studies.

Abuodha and Woodroffe (2006) summarized numerous approaches based on segmentation techniques within which sections of the shoreline are ranked with higher or lower propensity to be affected by coastal hazards based on spatial analysis of multivariate data.

A graphical methodology has been widely used in the United States, called the Coastal Vulnerability Index (CVI), based on ranking of fundamental parameters. Modifications of this approach have been applied to coasts elsewhere in the world. A global segmentation of the shoreline was proposed by the European DINAS-Coast project and the associated DIVA application in development of a coastal assessment support tool. By contrast, a segmentation of the shoreline that assigns attributes on the basis of form and fabric (Smartline) has been devised in Australia.

The rationale underlying the selection of a linear data model for referencing coastal information and three methods used for the implementation of such models with the use of GIS are briefly outlined below.

16.3.2.1 Coastal Vulnerability Indices

An early attempt to develop a CVI by Gornitz and Kanciruk (1989) and Gornitz (1991) considered susceptibility to erosion for sections of the United States coastline. A modified form of this was adopted by the USGS in 2001, and it has been used to conduct scientific assessments of coastal vulnerability to sea (and lake) level change (Pendleton et al., 2010).

The CVI provides a simple numerical basis for ranking sections of shoreline by estimating their potential for change and can be applied to identify specific locations where risks may be relatively high (Ramieri et al., 2011). Variables are selected taking into account their representativeness to synthesize physical characteristics on the coast. The following key descriptors, denoting significant driving processes that shape coastal evolution and vulnerability, were originally proposed: geomorphology, historical shoreline change rate, coastal slope, relative sea-level change rate, maximum wave height and tidal range (Figure 16.3a).

After spatialisation each factor receives a value for vulnerability, the classification being generally ranked and based on the definition of semi-quantitative scores according to a 1–5 scale (very low to very high vulnerability). Overall vulnerability is then obtained through analytical integration of these layers by performing map algebra to calculate a final single index value for each assigned coastal segment.

This approach has been widely used and modified by many authors, who have incorporated or supressed variables due to the need to adapt it to a local scale or as a consequence of data availability for specific sites. Many of these studies were extensively listed and synthetized by Abuodha and Woodroffe (2010b), Rangel-Buitrago and Anfuso (2015) and Nguyen et al. (2016).

It must be underlined that the CVI original formulation considers vulnerability exclusively related to physical factors, and does not address

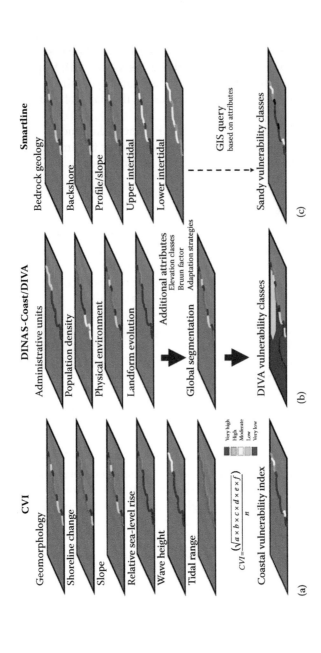

FIGURE 16.3
A schematic comparison of the input parameters and the principles on which (a) CVI, (b) DIVA and (c) Smartline/GSM methodologies are based.

socioeconomic aspects (such as economic loss, number of people affected, potentially damaged infrastructure, etc.).

For practical use it is essential to include human factors, taking into account that property damage occurs solely in the presence of assets. This has motivated the inclusion of human-related variables in the model. Some studies have therefore incorporated socioeconomic data (for example, McLaughlin and Cooper, 2010) and, sometimes, changed the acronym that refers to the model. For example, Boruff et al. (2005) preferred the abbreviation SoVI (social vulnerability index), to reflect the particular focus of their model. In contrast, Abuodha and Woodroffe (2010a) adopted Coastal Sensitivity Index (CSI) to state the biophysical nature of the model that they developed. In both studies, however, the general procedure of ranking and integration of variables was preserved.

Szlafsztein and Sterr (2007), Pereira and Coelho (2013) and Nguyen and Woodroffe (2015) also considered social factors in their development of indices. Moreover, these authors applied some weighting method, determined arithmetically or through the analytical hierarchy process (AHP, Saaty, 1980) to propose a hierarchy of the contributing variables to evaluate potential impacts. Nguyen and Woodroffe (2015), furthermore, incorporated a series of variables into the model to specifically express adaptive capacity.

16.3.2.2 DINAS-DIVA

Despite an apparent abundance of data, mainly originating from the numerous Earth observation systems orbiting the Earth (see Chapter 4 of the present volume), there is a characteristic lack of coastal databases at regional and global scales. At the global scale, the DINAS-Coast database ('Dynamic and Interactive Assessment of National, Regional and Global Vulnerability of Coastal Zones to Climate Change and Sea-Level Rise') and the DIVA tool ('Dynamic and Interactive Vulnerability Assessment') provide a comparative methodology that can be used to assess the apparent susceptibility of predefined segments of shoreline (Vafeidis et al., 2008). The tool encompasses a global database, an integrated model and a graphical interface that allows users to produce quantitative information about a range of coastal vulnerability indicators and thereby to select climatic and socioeconomic scenarios and adaptation strategies (Hinkel and Klein, 2006, 2009). The model enables the evaluation of costs and benefits related to the analysed impacts as well as predefined adaptation strategies. It was specifically designed and developed to support policy and decision makers in interpreting coastal vulnerability assessment, and in addressing related measures (Vafeidis et al., 2008). Moreover, it analyses a range of mitigation and adaptation options based on IPCC's Special Report on Emissions Scenarios (SRES).

Although linked to a global database, DIVA has been designed to provide output at a range of spatial scales focused on administrative units (Figure 16.3b). DIVA, for which the average coastal segment is approximately 70 km in length, is an efficient tool for regional scale analysis but may not be useful for site-specific local coastal planning. It has been used, for example, to compare the vulnerability of shorelines in the Coral Triangle region, incorporating Indonesia, New Guinea, the Philippines and the Solomon Islands (Mcleod et al., 2010). More recently, the wetland module has been refined and extended at a global scale (Spencer et al., 2016).

16.3.2.3 Smartline – Geomorphic Stability Mapping

The geomorphic stability mapping (GSM) approach was developed by Sharples (2006) who mapped the coast of Tasmania. The 'Smartline' is essentially a method of efficiently capturing detailed descriptions of coastal landform variability in a segmented GIS line format (http://www.ozcoasts.gov .au/coastal/introduction.jsp). These segments can be readily queried and used for many coastal research and management purposes, which to date have included identifying alongshore variability in sensitivity to oil spills and inherent susceptibility to erosion.

The mapping approach identifies a series of shoreline segments based on form and fabric, using a geomorphological classification of backshore, intertidal and bedrock, as well as slope (Sharples et al., 2009). For this method, the shoreline is segmented, bringing multiple pieces of information attached to the vector lines representing the coast. This then enables classification by GIS query into broad categories corresponding to different degrees and styles of susceptibility to sea-level rise impacts (Figure 16.3c). It proposes a pragmatic approach aiming at indicative mapping, intended as the first stage in a hierarchical sequence of assessments, which identifies shoreline landform types that are potentially vulnerable to sea-level rise.

The methodology has undergone several refinements and been adapted for application to the entire coast of the Australian mainland as part of a national coastal vulnerability assessment.

Both GSM and CVI approaches were applied at the Illawarra coast by Abuodha (2009; Figure 16.4).

16.3.3 Local-Scale Evaluation

One framework focused on the local level was adopted by NOAA's Coastal Services Center, which developed 'The Community Vulnerability Assessment Tool' – CVAT (Flax et al., 2002). CVAT deals with socioeconomic factors and allows the creation of combined hazard maps, using a score system which gives the greatest values to the most vulnerable areas. It consists of a map overlay procedure that enables a relative risk or vulnerability analysis of coastal communities to a series of existing threats. Similarly to the Common Methodology, CVAT proposes the completion of a series of incremental steps, which can be adapted to meet users' needs and which can be implemented using either a GIS or static maps as appropriate.

FIGURE 16.4

Representation of the Illawarra Coast sensitivity applying different methods: Smartline/GSM (a) and CVI (b). (Modified from Abuodha, P. A. O. 2009. Application and evaluation of shoreline segmentation mapping approaches to assessing response to climate change on the Illawarra Coast. Australia. Unpublished PhD thesis, School of Earth and Environmental Sciences, University of Wollongong, p. 286. http://ro.uow.edu.au/theses/852, accessed 14 May 2014.)

A different approach, based on the use of geoindicators, was proposed by Bush et al. (1999) as a management tool for rapid assessment of natural hazard risk potential. It is site-specific and relies on the description of features such as elevation, dune height, fetch exposure, presence of vegetation on the backshore, shoreline stability, inlet proximity and so on. The approach can adopt up to about 30 variables that can be selected on a case-by-case basis as primary indicators of property-threatening hazards.

In this methodology, selected variables are identified in the field, ranked in risk levels and tabulated, resulting in an index expressing local natural vulnerability. The basic idea behind geoindicators is that a selected set of variables can express short-term environmental dynamics and be used, therefore, as proxies for the multiplicity of factors acting over coasts.

A positive aspect of this strategy is that it is directly related to local conditions, since variables are observed, not retrieved from a generic dataset. Moreover, it facilitates permanent updating through long-term monitoring in contrast to the more static picture provided by other methodologies. On the other hand, some of the proposed geoindicators are very transitory and do not necessarily represent a trend or an average state of the system.

16.4 Examples of Application

As a general rule, vulnerability models are built through the integration of coastal descriptors aiming to recognize relationships and to generate new information. Spatial analysis using GIS is the preferred tool adopted for this task and, in recent years, a huge number of vulnerability assessments have been published using this technology (Rangel-Buitrago and Anfuso, 2015; Nguyen et al., 2016).

The two local (beach) scale examples below demonstrate different spatial analysis procedures in GIS. The first used the index-based methodology proposed by Gornitz (1991), with modifications, with input data being associated with coastal segments and classes of vulnerability expressed as linear features. The second, inspired by Bush et al. (1999), is an indicator-based approach which relies on the recognition of features indicative of coastal erosion and retreat using point data, with a fuzzy representation of the final classes. Both cases are located in Florianópolis (Santa Catarina, Brazil), and aimed to assess vulnerability and susceptibility, respectively, by quantifying the likelihood that physical changes may occur based on the analysis of selected variables. Florianópolis was chosen because it is one of the most threatened coastal municipalities in Santa Catarina state. There, storm surges regularly induce damages due to coastal flood and erosion which severely impact coastal communities (Rudorff et al., 2014).

16.4.1 Index-Based Approach – Armação and Barra da Lagoa-Moçambique Beach

Index-based approaches express coastal vulnerability by a one-dimensional, and generally unitless, value calculated through the quantitative or semi-quantitative evaluation and combination of different descriptors (Ramieri et al., 2011; Nguyen et al., 2016). Most studies that applied CVI and its variants adopted this procedure, as shown in the example presented below originally developed by Muler (2012) and Bonetti et al. (2013b).

The vulnerability of two beaches located at the eastern coast of Florianópolis, Armação Beach and Barra da Lagoa-Moçambique Beach, were assessed based on the CVI methodology. The original variables proposed by Gornitz (1991) were adapted for the local scale, also considering the availability of data for the study sites.

Five descriptors were used in the assessment: (1) Backshore landforms (land use, vegetation, dune features, etc.) obtained from visual classification of high-resolution satellite images; (2) backshore elevation, acquired from altimetry maps at the scale of 1:2000; (3) shoreline change rate over a five-decade time period, extracted from aerial photographs and processed with the ArcGIS extension digital shoreline analysis system (DSAS); Thieler et al., 2009; (4) shoreline exposure to waves, determined by the association of segmented mean shoreline orientation with a 50-year statistical analysis of extreme wave frequency and direction (Muler and Bonetti, 2014) and (5) assets, infrastructure and communities at risk, obtained by mapping infrastructure from high-resolution satellite images up to an inland distance determined by the position of the sea as predicted for the next 50 years. Census data allowed estimation of the number of residents in the area under risk.

The index was calculated using the formula:

$$CVI = \frac{\left(\sqrt{a} \times b \times c \times d \times e\right)}{n}$$

where a, b, c, d, e are the variables described above after their ranking in vulnerability intervals (from 1 to 5) and n is the number of considered descriptors.

Final vulnerability classification (from very low = 1 to very high = 5) was based on percentiles obtained from the CVI results (Figure 16.5).

16.4.2 Indicator-Based Approach – Armação Beach

Indicator-based approaches assess exposure, sensitivity and adaptive capacity of the coast by the spatial representation of a set of independent elements (i.e. the indicators) that characterize key coastal issues (Ramieri et al., 2011). In some cases, these indicators can be combined into a final summary indicator to express vulnerability. An example of this approach developed by Rudorff and Bonetti (2010) will be presented here.

The susceptibility of Armação Beach to coastal erosion and flooding induced by storms was obtained from the integration, using AHP (Saaty, 1980), of geoindicators evaluated *in situ*, a digital terrain model and maps of distance from the coastline. The geoindicators used in the analysis were selected from a list proposed by Bush et al. (1999), and included: (1) erosional state, obtained from a specific sub-model with indicators such as scarping of dunes, deficit of sediment at the beach, fallen vegetation in the backshore, exposure of peat layers and so on; (2) frontal dune height, dune state and dune type, observed *in situ*; (3) presence of vegetation, observed *in situ*; (4) wave energy, an average of one-year survey obtained from the literature;

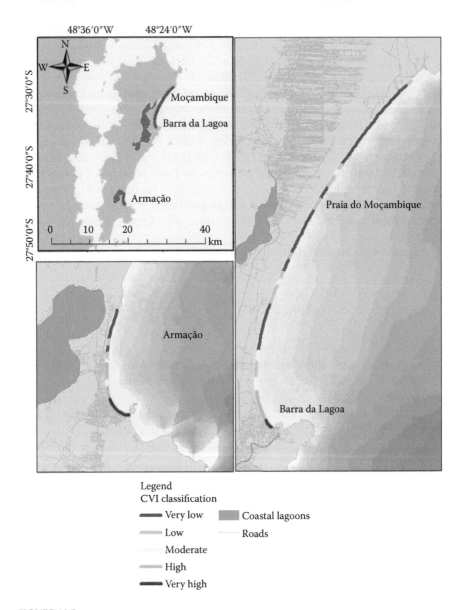

FIGURE 16.5
Vulnerability index maps obtained for Armação and Barra da Lagoa-Moçambique beaches using the CVI methodology. (Reprinted from Bonetti, J., Klein, A. H. F., Muler, M., de Luca, C. B., Silva, G. V., Toldo, E. E. Jr and González, M. 2013b. Spatial and numerical methodologies on coastal erosion and flooding risk assessment. In: Finkl, C. (Ed.). *Coastal Hazards*. Coastal Research Library Series. Springer, Dordrecht, the Netherlands, pp. 423–442.)

(5) beach width, observed *in situ* and complemented by a time-series of high-resolution satellite images and (6) presence of engineered shoreline protection structures, observed *in situ*.

In the field, the total length of the beach was surveyed at spaced intervals, or when any obvious change in its characteristics was noticed. Each interval was assigned a georeferenced point with attributing *in situ* measurements and observations. Each category of geoindicator (e.g. shoreline), the geoindicators themselves (e.g. frontal dune vegetation) and their attributes (e.g. presence or absence) were AHP weighted and ranked to express expected susceptibility. This resulted in a final computation for each surveyed point along the shore. The sum of all geoindicator scores allowed for a maximum value of 10 (maximum susceptibility), to facilitate interpretation of results.

The geoindicator scores were then organized in a GIS and their numerical values were interpolated to a grid composed of 1 m² cells using the inverse distance weighting (IDW) interpolator. They were combined with a digital terrain model obtained from an aerial survey at the scale of 1:2000 using the same spatial resolution as the geoindicator's surface. Using fuzzy logic (a technique outlined by Burrough and McDonnell, 1998), altimetry values were transformed into a scale from 0–1, where points with altitudes close to 0 m were related to a maximum susceptibility (fuzzy susceptibility potential of 1) while altitudes close to 4 m (susceptibility potential of 0.5) present lower susceptibility, a tendency that increased towards the transformed value of 1 (zero susceptibility).

A similar procedure was applied to the shoreline layer, considering that susceptibility decreases the more distant from the shoreline a point is located. Fuzzy logic was applied resulting in a probabilistic distance map. A distance of zero metres from the shoreline received a score of 0 (lower susceptibility potential) and at 33 m (a legal mark for public land extents in Brazil) received a fuzzy value of 0.5. This tendency, once more, increased towards the transformed value of 10 (zero susceptibility).

Map algebra was applied to combine geoindicators, elevation and distance, and each dataset was associated with a weighting factor, using the empirical expression:

$$\text{Susceptibility} = 0.8 \, (\text{Geoindicators} \times \text{Distance}) + 0.2 \, (\text{Elevation})$$

The resulting map is a fuzzy surface (Figure 16.6 illustrates the southernmost sector of the studied beach) where the distribution of lower to higher susceptibilities in a continuous colour ramp replaces the traditional deterministic representation.

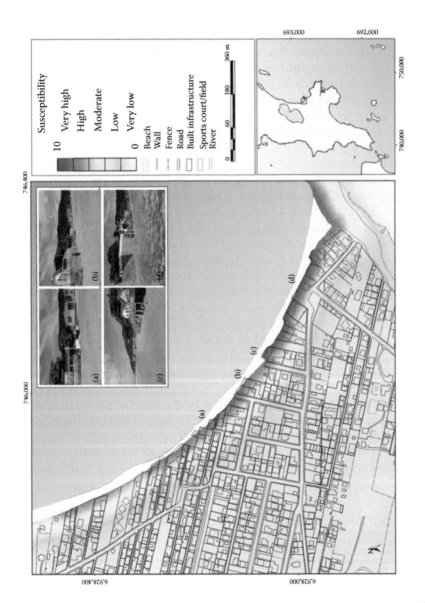

FIGURE 16.6
Fuzzy representation of the southern sector of Armação Beach susceptibility. (Modified from Rudorff, F. M. and Bonetti, J. 2010. *Brazilian Journal of Aquatic Science and Technology*, 14: 9–20.)

16.5 Existing Challenges and Critical Evaluation of Vulnerability Mapping

While there has been a recent proliferation of tools for vulnerability assessment and adaptation planning (Abuodha and Woodroffe, 2010b), there has been limited research on lessons learned from their application. Nguyen et al. (2016) examined more than 50 applications of vulnerability indices and found little methodological consistency between them, and also a lack of standardisation that also applied to the the concept of vulnerability itself.

The diversity of analytical strategies for assessing coastal vulnerability is evident. The main differences relate to scale of analysis, selection of variables, their ranking and methods used to combine indicators. Given that all those aspects directly influence the outcome of the analysis, comparability between different models and studies is poor. Hinkel (2011) demonstrated that inter-comparison of studies from geographically diverse areas is possible only when a consistent methodology has been used. It appears that no such methodology is yet available.

Existing concerns about the development of vulnerability assessments using GIS analysis can be expressed by a set of questions. Those questions can also be addressed as guidance during the design of an intended vulnerability study.

16.5.1 What Is the Best Methodological Approach to Adopt?

Until now, no consistent framework for developing studies to quantify and compare vulnerability to climate change at spatially variable scales has been proposed. Therefore, more effort needs to be directed towards identification and selection of models or strategies to be adopted in diverse contexts.

For example, although different geospatial data models have already been tested (points, segments, raster, fuzzy), it has been only very limited discussion of which one is better at cartographically representing coastal vulnerability for practical use in any particular coastal setting. Clearly no single model can claim to be the best option for any vulnerability study, and no one approach to vulnerability assessment fits every need (Abuodha and Woodroffe, 2010b). The choice of a framework must take into account the objectives of the assessment, the scales involved, availability of data, purpose of application and so on. In general, a correlation between the resolution of data and applicability for effective management policies can be recognized at the local scale. Figure 16.7 illustrates how some of the models presented in this chapter can be compared, taking into account their general characteristics.

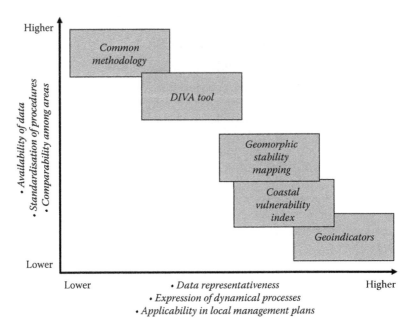

FIGURE 16.7
Main characteristics and applicability of vulnerability assessment models.

16.5.2 Specific Models Represent Coastal Vulnerability to What?

It is often unclear which coastal hazards (i.e. slow-onset vs. rapid-onset) or processes (coastal retreat, flooding, erosion) a proposed assessment is intended to represent. The driver, and the effect under analysis, is sometimes simply referred to as 'climatic change'. Since progressive sea-level rise and short-term storms operate over different spatial and temporal scales, they will not necessarily be efficiently represented by the same set of descriptors and analytical integration strategies.

It is important to clearly define the scope of the study, which ultimately will lead to the definition of variables and framework to be adopted. As seen before, the non-linearity of processes operating within the coastal system produces a complex biophysical environment. Although customarily represented that way, coasts are not single linear entities. On the contrary, they constitute a far more complex zone of interaction within which a wide range of processes and activities occur (Abuodha and Woodroffe, 2010b).

16.5.3 How Should Descriptors Be Selected?

The choice of descriptors to be used in vulnerability models seems poorly resolved, particularly when social variables are incorporated.

Many adjustments and adaptations have been proposed from the original six-variable CVI model proposed by Gornitz (1991), leading to a wide list of descriptors that could represent coastal vulnerability. Rangel-Buitrago and Anfuso (2015) described many of the factors adopted in different studies. For example, Dal Cin and Simeoni (1994), in a study performed in Italy, proposed the use of 15 variables including sediment characteristics and presence of defensive structures. Özyurt and Ergin (2009) carried out a study in Turkey, in which they considered five main types of variables, and a total of 22 sub-variables. Nguyen and Woodroffe (2015), in turn, determined vulnerability of the western Mekong River Delta using a hierarchical approach composed from models run at four different levels, incorporating three components with eight sub-components that used a total of 22 variables and 24 sub-variables, covering both natural and socioeconomic aspects.

A list of vulnerability descriptors and their possible metrics for ranking purposes was recently provided by Wamsley et al. (2015). The authors proposed an alternative approach for identifying and comprehensively defining meaningful metrics to enable assessment of vulnerability for a wide range of systems and hazards at multiple scales. The method includes five steps: identify the purpose, create a vulnerability profile, define system components and valued functions, link factors to functions and establish metrics.

This is an interesting approach to explore. However, some multivariate statistical analysis must be incorporated to the scheme in order to evaluate redundancy, as well as weighting methods such as the previously mentioned AHP applied to address the relative importance of variables. This was demonstrated by Pendleton et al. (2010), in their study performed in 22 national parks in the United States, where a principal component analysis (PCA) revealed that 99% of the index variability could be explained by only four variables: geomorphology, regional coastal slope, water-level change rate and mean significant wave-height.

It seems that site-specific analysis is still the best option for the recognition of a representative set of descriptors of vulnerability, even if this precludes inter-site comparison.

16.5.4 How Can Descriptor Scale Dependency in Space and Time Be Addressed?

In most cases, vulnerability frameworks rely on data availability and their suitability to be spatially represented, not necessarily by adoption of ideal criteria to express exposure and sensitivity. Availability and resolution of data, presented in distinct geometries, varies among different areas. As a result, it is not uncommon to find a combination of data originally produced in different scales and resolutions, which can lead to significant inconsistences.

Abuodha and Woodroffe (2010b) noted that one of the greatest challenges is for vulnerability analysis to be effective at different scales, both spatially and over short- to long-term perspectives. The authors argued that once an assessment is undertaken at relevant spatial and temporal scales, the results may only be appropriate at those scales.

The implications of spatial scale in representing coastal vulnerability by indices were investigated by McLaughlin and Cooper (2010). They concluded that, although the same factors can be used in different levels of spatial analysis (i.e. national, regional or local levels), more refined data are necessary at larger scales if management use of results is intended. Although this may seem to be common sense, the literature contains various examples where the resolution of variables is inadequate to accurately represent vulnerability, since simplifications at bigger scales can mask important local variations.

Another point to consider is the limitation in the representation of dynamical processes and temporal changes in existing frameworks. Variables are usually represented as if static, sometimes expressing an average or extreme (worst-case) condition. This is particularly common practice in relation to waves, for which the most frequent direction is often chosen as an input on vulnerability models (Muler and Bonetti, 2014), ignoring the fact that it may be waves from the less common direction that can cause the greatest damage or pose the greater threat.

Once more, actual methodologies seem to work better in case-specific studies, where the need for spatial and temporal representativeness can be more easily addressed. Multiscale analysis has been shown to demand different sets of data to be used (McLaughlin and Cooper, 2010) and, in practical terms, a similar framework can often not be applied.

16.5.5 How Can Variables Be Accurately Ranked and Models Compared?

Comparability between sites seems to be practicable only when methodologies developed for larger scale analysis (such as DIVA) are applied. In these cases, a common framework provides standardized results, enabling a general picture of vulnerabilities. Finer scale studies based on indicator or index-based analysis, however, tend to be more flexible in processing variables and integration strategies. Nguyen et al. (2016) summarized several physical and social variables, and the ranges that have been used in coastal vulnerability studies. It can be noted that even if the same variable is chosen to represent a component of vulnerability, its rank (i.e. the cutoff values that represent specific classes of vulnerability, usually: Very Low, Low, Moderate, High and Very High) differ significantly. Picking one of those descriptors, for example, shoreline displacement, this review shows that seven studies that were examined had between them adopted five different possibilities for cutoff value, given in metres per year, distributed over five proposed classes.

A commonly used approach for classification of numerical factors is the quantile, which is the determination of cut points for the classification of a set of observations into groups of equal population. This strategy, even though it guarantees the best distribution of a group of cases among all available classes, inhibits any quantitative comparison between sites, since thresholds are specifically determined for each dataset. In most cases, coastal segments of different sites assigned to the same vulnerability class ('Very High', for example) are not necessarily subject to the same degree of threat, since calculated scores are based solely on the input data at each site. This can lead to misinterpretation of their use for management purposes. It is questionable whether consistent threshold values could be determined and universally applied. Efforts, in this sense, should be addressed towards strategies for evaluation of internal consistency of models to allow qualitative comparisons between areas.

16.6 Concluding Remarks

Over the past few decades, assessments of vulnerability of the coast to climate change, particularly sea-level rise, have been undertaken at a range of different spatial scales, and a contradiction is evident between larger-scale studies that attempt to characterize coastlines at global or regional level, and more localized or site-specific studies. The former demand consistency in order that the vulnerability of one section of a coast can be compared to that of another, whereas the latter need to consider a wider range of hazards so that planning and management can proceed for that particular stretch of coastline.

As shown by the discussion in this chapter, several general points regarding vulnerability assessment in coastal zones can be made, reinforcing the conclusions by Abuodha and Woodroffe (2010b):

- The coastal zone does not behave homogeneously, and techniques need to focus on the geographical variability of shoreline response to the principal drivers such as sea-level rise or storm surges;
- Tools used to evaluate vulnerability need to integrate different kinds of information, preferably encompassing both physical and social variables;
- The definition and quantification of vulnerability should not be associated with subjective elements, but needs to be undertaken in a way that different observers working on different parts of the coast can derive similar values or indices and
- Results should provide meaningful data suitable for proper coastal zone planning and management.

Vulnerability assessments remain highly context-specific. Nguyen et al. (2016) argued that a key need is implementation of strategies to facilitate comparability between outputs for coasts from different areas, since there is no indication that a standard methodology will be widely adopted for vulnerability mapping in the near future. Several requirements will need to be fulfilled in order to accomplish this, as attempts at validation have been particularly lacking:

- Clearer identification and scale of the physical processes in the area under investigation, in relation to exposure to a particular hazard
- Specification of the assumptions adopted in ranking of variables, and an indication of how weightings have been assigned across relevant scales
- Adoption of some calibration procedures in order to test the effectiveness of the model

References

Abuodha, P. A. O. 2009. Application and evaluation of shoreline segmentation mapping approaches to assessing response to climate change on the Illawarra Coast. Australia. Unpublished PhD thesis, School of Earth and Environmental Sciences, University of Wollongong, 286 p. http://ro.uow.edu.au/theses/852, accessed 14 May 2014.

Abuodha, P. A. and Woodroffe, C. D. 2006. Assessing vulnerability of coasts to climate change: A review of approaches and their application to the Australian coast. In: Woodroffe, C. D., Bruce, E., Puotinen, M. and Furness, R. A. (Eds.). *GIS for the Coastal Zone: A Selection of Papers from CoastGIS 2006*, Australian National Centre for Ocean Resources and Security/University of Wollongong, Wollongong, Australia, 2007.

Abuodha, P. A. and Woodroffe, C. D. 2010a. Assessing vulnerability to sea-level rise using a coastal sensitivity index: A case study from southeast Australia. *Journal of Coastal Conservation Planning and Management*, 14: 189–205.

Abuodha, P. A. and Woodroffe, C. D. 2010b. Vulnerability assessment. In: Green, D. R. (Ed.). *Coastal Zone Management*. Thomas Telford Ltd., London, pp. 262–290.

Allison, E. H. and Basset, H. R. 2015. Climate change in the oceans: Human impacts and responses. *Science*, 350(6262): 778–782.

Bak, P. 1997. *How Nature Works: The Science of Self-Organized Criticality*. Oxford University Press, Oxford, UK.

Bartlett, D. J. 1993. *GIS and the Coastal Zone: An Annotated Bibliography*. Technical report, 93–9. National Center for Geographic Information and Analysis, Santa Barbara, CA.

Bartlett, D. J. 2000. Working on the frontiers of science: Applying GIS to the coastal zone. In: Wright, D. and Bartlett, D. (Eds.). *Marine and Coastal Geographic Information Systems*. Taylor & Francis, London.

Bartlett, D. and Smith, J. 2005. *GIS for Coastal Zone Management*. CRC Press, Boca Raton, FL.

Bartlett, D., Devoy, R., McCall, S. and O'Connor, I. 1997. A dynamically-segmented linear data model of the coast. *Marine Geodesy*, 20: 137–151.

Birkmann, J. 2013. Measuring vulnerability to promote disaster-resilient societies: Conceptual frameworks and definitions. In: Birkmann, J. (Ed.). *Measuring Vulnerability to Natural Hazards: Towards Disaster Resilient Societies.* 2nd ed. United Nations University Press, Tokyo, Japan, pp. 9–54.

Bonetti, J., Klein, A. H. F. and Sperb, R. M. 2013a. Preface – Geotechnologies applied to coastal studies. *Journal of Coastal Conservation: Planning and Management*, 17(2): 197–199.

Bonetti, J., Klein, A. H. F., Muler, M., de Luca, C. B., Silva, G. V., Toldo, E. E. Jr and González, M. 2013b. Spatial and numerical methodologies on coastal erosion and flooding risk assessment. In: Finkl, C. (Ed.). *Coastal Hazards*. Coastal Research Library Series. Springer, Dordrecht, the Netherlands, pp. 423–442.

Boruff, B. J., Emrich, C. and Cutter, S. L. 2005. Erosion hazard vulnerability of US coastal counties. *Journal of Coastal Research*, 21(5): 932–943.

Bosello, F. and De Cian, E. 2014. Climate change, sea level rise, and coastal disasters. A review of modeling practices, *Energy Economics*, 46: 593–605.

Brown, S., Nicholls, R. J., Woodroffe, C. D., Hanson, S., Hinkel, J., Kebede, A. S., Neumann, B. and Vafeidis, A. T. 2013. Sea-level rise impacts and responses: A global perspective. In: Finkl, C. (Ed.). *Coastal Hazards*. Coastal Research Library Series. Springer, Dordrecht, the Netherlands, pp. 117–140.

Burrough, P. A. and McDonnell, R. A. 1998. *Principles of Geographical Information Systems*. Oxford University Press, Oxford, UK.

Bush, D. M., Neal, W. J., Young, R. S. and Pilkey, O. H. 1999. Utilization of geoindicators for rapid assessment of coastal-hazard risk and mitigation. *Ocean and Coastal Management*, 42: 647–670.

Cardona, O. D. 2003. The need of rethinking the concepts of vulnerability and risk from a holistic perspective: A necessary review and criticism for effective risk management. In: Bankoff, G., Frerks, G. and Hilhorst, D. (Eds.). *Mapping Vulnerability: Disasters, Development and People*. Earthscan Publishers, London.

Carter, R. W. G. and Woodroffe, C. D. (Eds.). 1994. *Coastal Evolution. Late Quaternary Shoreline Morphodynamics*. Cambridge University Press, Cambridge, UK.

Church, J. A. and White, N. J. 2011. Sea-level rise from the late 19th to the early 21st century. *Surveys in Geophysics*, 32: 585–602.

Church, J. A., Clark, P. U., Cazenave, A., Gregory, J. M., Jevrejeva, S., Levermann, A., Merrifield, M. A. et al. 2013. Sea level change. In: Stocker, T. F., Qin, D., Plattner, G.-K., Tignor, M., Allen, S. K., Boschung, J., Nauels, A. et al. (Eds.). *Climate Change 2013: The Physical Science Basis. Contribution of Working Group I to the Fifth Assessment Report of the Intergovernmental Panel on Climate Change*. Cambridge University Press, Cambridge, UK, pp. 1137–1216.

Cutter, S. 1996. Vulnerability to environmental hazards. *Progress in Human Geography*, 20(4): 529–589.

Dal Cin, R. and Simeoni, U. 1994. A model for determining the classification, vulnerability and risk in the southern coastal zone of the Marche (Italy). *Journal of Coastal Research*, 10: 18–29.

Dall'Osso, F. and Dominey-Howes, D. 2013. Coastal vulnerability to multiple inundation sources – COVERMAR project – Literature Review (Second Edition). Report prepared for the Sydney Coastal Councils Group Inc. UNSW, Sydney, Australia.

Flax, L. K., Jackson, R. W. and Stein, D. N. 2002. Community vulnerability assessment tool methodology. *Natural Hazards Review*, 34(163): 1527–6988.

Füssel, H.-M. 2005. Vulnerability in climate change research: A comprehensive conceptual framework. In: *Breslauer Symposium*, University of California International and Area Studies. UC Berkeley, Berkeley, CA.

Gornitz, V. 1991. Global coastal hazards from future sea level rise. *Palaeogeography, Palaeoclimatology and Palaeoecology*, 89: 379–398.

Gornitz, V. and Kanciruk, P. 1989. Assessment of global coastal hazards from sea-level rise. *Proceedings of the 6th Symposium on Coastal and Ocean management*, ASCE, 11–14 July 1989, Charleston, SC, pp. 1345–1359.

Guha-Sapir, D., Below, R. and Hoyois, Ph. – EM-DAT. 2016. The OFDA/CRED International Disaster Database. Université Catholique de Louvain, Belgium. http://www.emdat.be/, accessed 20 January 2016.

Haslett, S. K. 2000. *Coastal Systems*. Routledge Introduction to Environmental Series. Routledge, London.

Hinkel, J. 2011. Indicators of vulnerability and adaptive capacity: Towards a clarification of the science-policy interface. *Global Environmental Change*, 21: 198–208.

Hinkel, J. and Klein, R. J. T. 2006. Integrating knowledge for assessing coastal vulnerability to climate change. In: McFadden, L., Nicholls, R. J. and Penning-Rowsell, E. C. (Eds.). *Managing Coastal Vulnerability*. Elsevier Science, Amsterdam, the Netherlands, pp. 61–78.

Hinkel J. and Klein, R. J. T. 2009. Integrating knowledge to assess coastal vulnerability to sea-level rise: The development of the DIVA tool. *Global Environmental Change*, 19: 384–395.

Houston, J. 2013. Sea level rise. In: Finkl, C. (Ed.). *Coastal Hazards*. Coastal Research Library Series. Springer, Dordrecht, the Netherlands, pp. 245–266.

IPCC. 2014. Climate Change 2014: Synthesis Report. Contribution of Working Groups I, II and III to the Fifth Assessment Report of the Intergovernmental Panel on Climate Change (Core Writing Team, R.K. Pachauri and L.A. Meyer [Eds.]). IPCC, Geneva, Switzerland.

IPCC CZMS. 1992. Global Climate Change and the Rising Challenge of the Sea. Report of the Coastal Zone Management Subgroup (CZMS), Response Strategies Working Group of the Intergovernmental Panel on Climate Change. Ministry of Transport, Public Works and Water Management, Directorate General Rijkswaterstaat, Tidal Waters Division, the Netherlands.

Kappraff, J. 1986. The geometry of coastlines: A study in fractals. *Computers and Mathematics with Applications*, 12B(3/4): 655–671.

Kearney, M. S. 2013. Coastal risk versus vulnerability in an uncertain sea level future. In: Finkl, C. (Ed.). *Coastal Hazards*. Coastal Research Library Series. Springer, Dordrecht, the Netherlands, pp. 101–115.

Masselink, G., and Hughes, M. G. 2003. *Introduction to Coastal Processes and Geomorphology*. Hodder Arnold, London.

Masselink, G. and Gehrels, R. 2014. *Coastal Environments and Global Change*. John Wiley & Sons, Chichester, UK.

McFadden, L. and Green, C. 2007. Defining 'vulnerability': Conflicts, complexities and implications for coastal zone management. *Journal of Coastal Research*, SI 50 (Proceedings of the 9th International Coastal Symposium): 120–124.

McLaughlin, S. and Cooper, J. A. G. 2010. A multi-scale vulnerability index: A tool for coastal managers? *Environmental Hazards*, 9: 233–248.

McLeod, E., Hinkel, J., Vafeidis, A. T., Nicholls, R. J. and Salm, R. 2010. Sea-level rise vulnerability in the countries of the Coral Triangle. *Sustainability Science*, 5: 207–222.

Monmonier, M. S. 2008. *Coast Lines: How Mapmakers Frame the World and Chart Environmental Change.* University of Chicago Press, Chicago.

Muler, M. 2012. Avaliação da vulnerabilidade de praias da Ilha de Santa Catarina a perigos costeiros através da aplicação de um índice multicritério. Unpublished MSc. thesis, Department of Geosciences, Federal University of Santa Catarina, Florianópolis, Brazil.

Muler, M. and Bonetti, J. 2014. An integrated approach to assess wave exposure in coastal areas for vulnerability analysis. *Marine Geodesy*, 37(2): 220–237.

Neumann, B., Vafeidis, A. T., Zimmermann, J. and Nicholls, R. J. 2015. Future coastal population growth and exposure to sea-level rise and coastal flooding – A global assessment. *PLoS ONE*, 10(3): e0118571.

Nguyen, T. T. X. and Woodroffe, C. D. 2015. Assessing relative vulnerability to sea-level rise in the western part of the Mekong River Delta in Vietnam. *Sustainability Science*, September 2015: 1–15.

Nguyen, T. T. X., Bonetti, J., Rogers, K. and Woodroffe, C. D. 2016. Indicator-based assessment of climate-change impacts on coasts: A review of concepts, approaches and vulnerability indices. *Ocean and Coastal Management*, 123: 18–43.

Özyurt, G. and Ergin, A. 2009. Application of sea level rise vulnerability assessment model to selected coastal areas of Turkey. *Journal of Coastal Research*, 51: 248–251.

Pendleton, E. A., Thieler, E. R. and Williams, S. J. 2010. Importance of coastal change variables in determining vulnerability to sea- and lake-level change. *Journal of Coastal Research*, 26: 176–183.

Pereira, C. and Coelho, C. 2013. Mapas de risco das zonas costeiras por efeito da ação energética do mar. *Journal of Integrated Coastal Zone Management*, 13(1): 27–43.

Ramieri, E., Hartley, A., Barbanti, A., Santos, F. D., Gomes, A., Hilden, M., Laihonen, P., Marinova, N. and Santini, M. 2011. Methods for assessing coastal vulnerability to climate change. *ETC CCA Technical Paper 1/2011.* European Topic Centre on Climate Change Impacts, Vulnerability and Adaptation, Bologna, Italy.

Rangel-Buitrago, N. and Anfuso, G. 2015. Review of the existing risk assessment methods. In: Rangel-Buitrago and Anfuso (Eds.). *Risk Assessment of Storms in Coastal Zones: Case Studies from Cartagena (Colombia) and Cadiz (Spain).* Springer, Dordrecht, the Netherlands, pp. 7–13.

Reguero, B. G., Losada, I. J., Díaz-Simal, P., Méndez, F. J. and Beck, M. W. 2015. Effects of climate change on exposure to coastal flooding in Latin America and the Caribbean. *PLoS ONE*, 10(7): e0133409.

Rhein, M., Rintoul, S. R., Aoki S., Campos, E., Chambers D., Feely, R. A., Gulev, S. et al. 2013. Observations: Ocean. In: Stocker, T. F., Qin, D., Plattner, G.-K., Tignor, M., Allen, S. K., Boschung, J., Nauels, A. et al. (Eds.). *Cambridge Climate Change 2013: The Physical Science Basis. Contribution of Working Group I to the Fifth Assessment Report of the Intergovernmental Panel on Climate Change.* University Press, Cambridge, United Kingdom, pp. 255–315.

Romieu, E., Welle, T., Schneiderbauer, S., Pelling, M. and Vinchon, C. 2010. Vulnerability assessment within climate change and natural hazard contexts: Revealing gaps and synergies through coastal applications. *Sustainability Science*, 5: 159–170.

Rudorff, F. M. and Bonetti, J. 2010. Avaliação da suscetibilidade à erosão costeira de praias da Ilha de Santa Catarina com base em geoindicadores e técnicas de análise espacial de dados. *Brazilian Journal of Aquatic Science and Technology*, 14: 9–20.

Rudorff, F. M., Bonetti Filho, J., Moreno, D. A., Oliveira, C. A. F. and Murara, P. G. 2014. Maré de tempestade. In: Herrmann, M. L. P. *Atlas de desastres naturais do Estado de Santa Catarina: período de 1980 a 2010*. 2. Ed. IHGSC/Cadernos Geográficos, Florianópolis, pp. 151–154.

Saaty, T. L. 1980. *The Analytic Hierarchy Process: Planning, Priority Setting, Resource Allocation*, McGraw-Hill, New York.

Sharples, C. 2006. Indicative mapping of Tasmanian coastal vulnerability to climate change and sea-level rise: Explanatory Report. 2nd Edition. Consultant Report to Department of Primary Industries and Water, Tasmania. Hobart, Australia.

Sharples, C., Mount, R. and Pedersen, T. 2009. *The Australian Coastal Smartline Geomorphic and Stability Map. Version 1: Manual and Data Dictionary. V. 1.0*. Report for Geoscience Australia and the Department of Climate Change, by School of Geography and Environmental Studies, University of Tasmania. Hobart, Australia.

Sherin, A. G. 2000. Linear reference data models and dynamic segmentation: Application to coastal and marine data. In: Wright, D. and Bartlett, D. (Eds.). *Marine and Coastal Geographic Information Systems*. Taylor & Francis, London, pp. 95–115.

Smith, M. J., Goodchild, M. F. and Longley, P. A. 2007. *Geospatial Analysis. A Comprehensive Guide to Principles, Techniques and Software Tools*. 2nd ed. Matador, Leicester, UK.

Spencer, T., Schuerch, M., Nicholls, R. J., Hinkel, J., Lincke, D., Vafeidis, A. T., Reef, R. et al. 2016. Global coastal wetland change under sea-level rise and related stresses: The DIVA Wetland Change Model. *Global and Planetary Change*, 139: 15–30.

Szlafsztein, F. and Sterr, A. 2007. A GIS-based vulnerability assessment of coastal natural hazards, state of Pará, Brazil. *Journal of Coastal Research*, 11(1): 53–66.

Thieler, E. R., Himmelstoss, E. A., Zichichi, J. L. and Ergul, A. 2009. Digital shoreline analysis system (DSAS) version 4.0 – An ArcGIS extension for calculating shoreline change. *U.S. Geological Survey Open-File Report 2008-1278*. USGS, Woods Hole. Available on: http://woodshole.er.usgs.gov/project-pages/DSAS/version4/index.html.

UNDP (United Nations Development Programme). 2004. *Reducing Disaster Risk: A Challenge for Development*. John S. Swift, New York.

UNDRO (Office of the United Nations Disaster Relief Co-ordinator). 1979. *Natural Disasters and Vulnerability Analysis*. Report of Expert Group Meeting, 9–12 July 1979, UNDRO Geneva, Switzerland.

UNISDR (United Nations International Strategy for Disaster Reduction). 2009. UNISDR *Terminology on Disaster Reduction*. United Nations, Geneva, Switzerland.

Vafeidis, A. T., Nicholls, R. J., McFadden, L., Tol, R. S. J., Hinkel, J., Spencer, T., Grashoff, P. S. et al. 2008. A new global coastal database for impact and vulnerability analysis to sea-level rise. *Journal of Coastal Research*, 24(4): 917–924.

Wamsley, T. V., Collier, Z. A., Brodie, K., Dunkin, L. M., Raff, D. and Rosati, J. D. 2015. Guidance for developing coastal vulnerability metrics. *Journal of Coastal Research*, 31(6): 1521–1530.

White, G. F., Kates, R. W. and Burton, I. 2001. Knowing better and losing even more: The use of knowledge in hazards management. *Environmental Hazards*, 3: 81–92.

Woodroffe, C. D. 2002. *Coasts: Form, Process, and Evolution.* Cambridge University Press, Cambridge, UK.

Woodroffe, C. D. 2006. The natural resilience of coastal systems: Primary concepts. In: McFadden, L., Nicholls, R. J. and Penning-Rowsell, E. C. (Eds.). *Managing Coastal Vulnerability.* Elsevier Science, Amsterdam, the Netherlands, pp. 45–60.

Zanuttigh, B., Simcic, D., Bagli, S., Bozzeda, F., Pietrantoni, L., Zagonari, F., Hoggart, S. and Nicholls, R. J. 2014. THESEUS decision support system for coastal risk management. *Coastal Engineering* (87): 218–239.

Index

Page numbers followed by f, t, and n indicate figures, tables, and notes, respectively.

Printed and bound by CPI Group (UK) Ltd, Croydon, CR0 4YY

01/11/2024

01782617-0016